VOLUME THREE HUNDRED AND SIXTY TWO

INTERNATIONAL REVIEW OF
CELL AND MOLECULAR BIOLOGY

Inter-Organellar Ca^{2+} Signaling in Health and Disease - Part A

INTERNATIONAL REVIEW OF CELL AND MOLECULAR BIOLOGY

Series Editors

VOLUME THREE HUNDRED AND SIXTY TWO

International Review of
CELL AND MOLECULAR BIOLOGY

Inter-Organellar Ca^{2+} Signaling in Health and Disease - Part A

Edited by

SAVERIO MARCHI
Marche Polytechnic University,
Ancona, Italy

LORENZO GALLUZZI
Weill Cornell Medical College,
New York, NY, United States

ACADEMIC PRESS
An imprint of Elsevier

ELSEVIER

Academic Press is an imprint of Elsevier
50 Hampshire Street, 5th Floor, Cambridge, MA 02139, United States
525 B Street, Suite 1650, San Diego, CA 92101, United States
The Boulevard, Langford Lane, Kidlington, Oxford OX5 1GB, United Kingdom
125 London Wall, London, EC2Y 5AS, United Kingdom

First edition 2021

ISBN: 978-0-12-824034-2
ISSN: 1937-6448

For information on all Academic Press publications
visit our website at https://www.elsevier.com/books-and-journals

Publisher: Zoe Kruze
Acquisitions Editor: Ashlie M. Jackman
Developmental Editor: Tara Nadera
Production Project Manager: Denny Mansingh
Cover Designer: Alan Studholme

Typeset by SPi Global, India

Working together
to grow libraries in
developing countries

www.elsevier.com • www.bookaid.org

Contents

4. Lysosomal calcium and autophagy   141
Diego L. Medina

5. Mitochondrial Ca^{2+} and cell cycle regulation   171
Haixin Zhao and Xin Pan

Contributors

Mayara Bertolini
Center for Tropical and Emerging Global Diseases and Department of Cellular Biology, University of Georgia, Athens, GA, United States

Massimo Bonora
Department of Medical Sciences, Section of Experimental Medicine, Laboratory for Technologies of Advanced Therapies (LTTA), University of Ferrara, Ferrara, Italy

Andreea Elena Burlacu
Department of Anatomy, Animal Physiology and Biophysics, Faculty of Biology, University of Bucharest, Bucharest, Romania

Miguel A. Chiurillo
Center for Tropical and Emerging Global Diseases and Department of Cellular Biology, University of Georgia, Athens, GA, United States

Agnese De Mario
Department of Biomedical Sciences, University of Padua, Padua, Italy

Roberto Docampo
Center for Tropical and Emerging Global Diseases and Department of Cellular Biology, University of Georgia, Athens, GA, United States

Armando Genazzani
Department of Pharmaceutical Sciences, Università del Piemonte Orientale, Novara, Italy

Gaia Gherardi
Department of Biomedical Sciences, University of Padua, Padua, Italy

Guozhong Huang
Center for Tropical and Emerging Global Diseases and Department of Cellular Biology, University of Georgia, Athens, GA, United States

Asrat Kahsay
Department of Medical Sciences, Section of Experimental Medicine, Laboratory for Technologies of Advanced Therapies (LTTA), University of Ferrara, Ferrara, Italy

Noelia Lander
Center for Tropical and Emerging Global Diseases and Department of Cellular Biology, University of Georgia, Athens, GA, United States

Dmitry Lim
Department of Pharmaceutical Sciences, Università del Piemonte Orientale, Novara, Italy

Cristina Mammucari
Department of Biomedical Sciences, University of Padua, Padua, Italy

Diego L. Medina
Telethon Institute of Genetics and Medicine (TIGEM), Pozzuoli; Medical Genetics Unit, Department of Medical and Translational Science, Federico II University, Naples, Italy

Xin Pan
State Key Laboratory of Proteomics, Institute of Basic Medical Sciences, National Center of Biomedical Analysis, Beijing, China

Paolo Pinton
Department of Medical Sciences, Section of Experimental Medicine, Laboratory for Technologies of Advanced Therapies (LTTA), University of Ferrara, Ferrara, Italy

Beatrice Mihaela Radu
Department of Anatomy, Animal Physiology and Biophysics, Faculty of Biology, University of Bucharest; Life, Environmental and Earth Sciences Division, Research Institute of the University of Bucharest, Bucharest, Romania

Mihai Radu
Department of Life and Environmental Physics, 'Horia Hulubei' National Institute for Physics and Nuclear Engineering, Măgurele, Romania

Călin Mircea Rusu
Department of Anatomy, Animal Physiology and Biophysics, Faculty of Biology, University of Bucharest, Bucharest; Department of Life and Environmental Physics, 'Horia Hulubei' National Institute for Physics and Nuclear Engineering, Măgurele, Romania

Alexey Semyanov
Shemyakin-Ovchinnikov Institute of Bioorganic Chemistry, Russian Academy of Sciences; Faculty of Biology, Moscow State University; Sechenov First Moscow State Medical University, Moscow, Russia

Cristina Elena Staicu
Department of Anatomy, Animal Physiology and Biophysics, Faculty of Biology, University of Bucharest, Bucharest; Center for Advanced Laser Technologies, National Institute for Laser, Plasma and Radiation Physics, Măgurele, Romania

Roberta Stoica
Department of Anatomy, Animal Physiology and Biophysics, Faculty of Biology, University of Bucharest, Bucharest; Department of Life and Environmental Physics, 'Horia Hulubei' National Institute for Physics and Nuclear Engineering, Măgurele, Romania

Anibal E. Vercesi
Departamento de Patologia Clinica, Universidade Estadual de Campinas, São Paulo, Brazil

Alexei Verkhratsky
Sechenov First Moscow State Medical University, Moscow, Russia; Faculty of Biology, Medicine and Health, The University of Manchester, Manchester, United Kingdom; Achucarro Centre for Neuroscience, IKERBASQUE, Basque Foundation for Science, Bilbao, Spain

Haixin Zhao
State Key Laboratory of Experimental Haematology, Institute of Hematology, Fifth Medical Center of Chinese PLA General Hospital, Beijing, China

Preface: Ca^{2+} in health and disease

Saverio Marchi[a,*] **and Lorenzo Galluzzi**[b,c,d,e,f,*]

[a]Department of Clinical and Molecular Sciences, Marche Polytechnic University, Ancona, Italy
[b]Department of Radiation Oncology, Weill Cornell Medical College, New York, NY, United States
[c]Sandra and Edward Meyer Cancer Center, New York, NY, United States
[d]Caryl and Israel Englander Institute for Precision Medicine, New York, NY, United States
[e]Department of Dermatology, Yale School of Medicine, New Haven, CT, United States
[f]Université de Paris, Paris, France
[*]Corresponding authors: e-mail address: s.marchi@univpm.it; deadoc80@gmail.com

Intracellular Ca^{2+} homeostasis relies on the coordinated activity of multiple Ca^{2+} channels, pumps, or exchangers, which altogether set a precise Ca^{2+} concentration ([Ca^{2+}]) not only in the cytoplasm but also within organelles (Berridge et al., 2003; Clapham, 2007). In resting conditions, the cytosolic [Ca^{2+}] of healthy cells is estimated around 100 nM (Bootman and Bultynck, 2020), but in response to hormones, growth factors, and other signaling molecules (Gaspers et al., 2014), free cytosolic Ca^{2+} rapidly increases (Giorgi et al., 2018a), regulating the activity of different enzymes and/or adaptor proteins. The major source for such cytosolic Ca^{2+} rise in excitable cells (i.e., neurons, cardiomyocytes) is the extracellular milieu (external [Ca^{2+}] is approximately 1 mM) (Hofer and Brown, 2003; Taylor, 2002), which enables a rapid influx of Ca^{2+} ions driven by electrochemical potential upon opening of transient receptor potential (TRP) superfamily members (Montell et al., 2002; Nilius and Szallasi, 2014). Ca^{2+} entry can also be mediated by voltage-dependent Ca^{2+} channels, a large group of proteins that comprises L-, R-, N-, P/Q-, and T-type plasma membrane (PM) transporters (Catterall, 2011). In non-excitable cells (e.g., epithelial cells), cytosolic Ca^{2+} elevations are most often initiated by Ca^{2+} release from intracellular stores, primarily the endoplasmic reticulum (ER), which exhibits a [Ca^{2+}] similar to the extracellular milieu and is provided with a tightly regulated machinery for Ca^{2+} trafficking (Krebs et al., 2015), including inositol 1,4,5-trisphosphate receptors (IP$_3$Rs) (Prole and Taylor, 2019; Rosa et al., 2020). Other intracellular sites of Ca^{2+} accumulation include the Golgi apparatus (GA) (Pizzo et al., 2011), which also possesses inositol 1,4,5-trisphosphate (IP$_3$)-dependent channels (Pinton et al., 1998), and lysosomes (Lloyd-Evans et al., 2008; Medina et al., 2015), which evoke Ca^{2+} signals mostly through transient receptor potential mucolipin 1 (TRPML1, also known as MCOLN1)

(Dong et al., 2008; Li et al., 2017). Thus, cytosolic Ca^{2+} waves can be triggered by the activity of PM channels or Ca^{2+} release from intracellular resources (Raffaello et al., 2016).

A sophisticated system of pumps and exchangers is designed to restore basal Ca^{2+} levels upon stimulation, including plasma membrane Ca^{2+} ATPases (PMCAs) (Stafford et al., 2017), which push Ca^{2+} outside the cell against its electrochemical gradient, sarcoendoplasmic reticular Ca^{2+} ATPases (SERCAs) (Primeau et al., 2018) and secretory pathway Ca^{2+} ATPases (SPCAs) (Vandecaetsbeek et al., 2011), which refill the ER and GA, respectively, as well as a still elusive Ca^{2+}/H^{+} exchanger, which is believed to promote lysosomal Ca^{2+} accumulation (Melchionda et al., 2016). However, since PMCAs are faster than ER and GA refilling systems, a Ca^{2+} flux from the extracellular milieu is required to replete intracellular Ca^{2+} stores (mainly the ER) after stimulation. This process is termed store-operated Ca^{2+} entry (SOCE) (Prakriya and Lewis, 2015) and involves members of the stromal interaction molecule (STIM) family, a class of reticular Ca^{2+} sensors (Soboloff et al., 2012; Stathopulos et al., 2008), and ORAI calcium release-activated calcium modulator (ORAI) proteins (Yoast et al., 2020), also known as Ca^{2+}-release activated Ca^{2+} (CRAC) PM channels. This inter-organellar Ca^{2+} connection is finely controlled by mitochondria (Rimessi et al., 2020), which harness their unique properties as key regulators of Ca^{2+} homeostasis (Giorgi et al., 2018b). In nonstimulated cells, mitochondrial Ca^{2+} levels are similar to cytosolic levels, but upon Ca^{2+} release from the ER, mitochondria accumulate Ca^{2+} up to 10-fold higher concentrations (Williams et al., 2013), largely due to the activity of the mitochondrial calcium uniporter (MCU) complex (Kamer and Mootha, 2015; Marchi and Pinton, 2014). However, mitochondria cannot store Ca^{2+} in the matrix for prolonged periods of time (which explains why they are generally not referred to as "Ca^{2+} deposits") (Rizzuto et al., 2012) because Ca^{2+} elevations are rapidly terminated by both Na^{+}-dependent and -independent Ca^{2+} efflux systems (Jiang et al., 2009; Palty et al., 2010). In summary, the complex machinery that controls Ca^{2+} fluxes at different subcellular sites is crucial to preserve the Ca^{2+} homeostasis and hence allow Ca^{2+} ions to regulate numerous biological reactions and cellular processes (Berridge, 2016; Parys and Bultynck, 2018). In line with this notion, defects in Ca^{2+} fluxes are involved in the etiology of a numerous human disorders (Berridge, 2012).

In neurodegenerative disorders, including Alzheimer's disease, Parkinson's disease, and amyotrophic lateral sclerosis (Dionísio et al., 2020; Schrank et al., 2020), alterations in Ca^{2+} signaling are particularly prominent at the ER-mitochondria interface (Prinz et al., 2020), a specific site commonly known as mitochondria-associated ER membranes (MAMs) (Missiroli et al., 2018; Perrone et al., 2020). In muscular dystrophies, such as dystrophinopathies, myotonic dystrophy, and Duchenne muscular dystrophy, a global impairment of Ca^{2+} homeostasis heavily affects mitochondrial Ca^{2+} signals and organelle functions (Bravo-Sagua et al., 2020; Vallejo-Illarramendi et al., 2014). In cancer, Ca^{2+} transporters have been shown to support malignant transformation and intersect with multiple oncogenic pathways (Marchi et al., 2020; Monteith et al., 2017), overall impacting on tumor heterogeneity (Vitale et al., 2021) and response to therapy (Cocetta et al., 2020; Maso et al., 2019; Petroni et al., 2021). Of note, several oncopromoting and oncosuppressive factors, including members of the Bcl-2 protein family (Chong et al., 2020; Glab et al., 2020; Rodriguez-Ruiz et al., 2019) and the PI3K-AKT1 signal transduction machinery (Borcoman et al., 2019; Ghigo et al., 2017; Marchi et al., 2019), influence cell fate also by manipulating Ca^{2+} homeostasis. In pancreatic β cells, dysregulated Ca^{2+} fluxes affect insulin secretion and hence promote the development of diabetes (Georgiadou and Rutter, 2020; Gil-Rivera et al., 2021). In ischemic conditions such as myocardial infarction and stroke, disrupted intracellular Ca^{2+} homeostasis alters the sensitivity of the mitochondrial permeability transition pore complex (PTPC) (Bonora et al., 2019; Dridi et al., 2020; Martins et al., 2011; Semyanov et al., 2020), a multi-protein channel that drives mitochondrial permeability transition (MPT)-associated variants of regulated necrosis (Bonora et al., 2017; Patel and Karch, 2020). Overall, the relevance of Ca^{2+} signaling in a plethora of pathological conditions reflects the capacity of Ca^{2+} ions to regulate numerous cellular processes, including survival (Miller et al., 2020), immunogenic cell death (ICD) (Galluzzi et al., 2020; Rodriguez-Ruiz et al., 2020; Trebak and Kinet, 2019; Yamazaki et al., 2020), bioenergetics (Porporato et al., 2018; Sica et al., 2019), redox homeostasis (Pervaiz et al., 2020), ER stress (Fucikova et al., 2020; Thangaraj et al., 2020; Wu Chuang et al., 2020), and autophagy (Arensman et al., 2020; Fairlie et al., 2020; Rimessi et al., 2013; Rybstein et al., 2018).

In *Inter-organellar Ca^{2+} signaling in health and disease* (a special issue of the *International Review of Cell and Molecular Biology* series) a panel of leading experts in the field gathered to discuss molecular and cellular aspects of intracellular Ca^{2+} fluxes as well as the implication of Ca^{2+} signaling in various human disorders. Each of the 13 chapters of *Inter-organellar Ca^{2+} signaling in health and disease* addresses a specific aspect of Ca^{2+} biology, ranging from the role of Ca^{2+} signaling in specific physiological processes to the mechanistic contribution of deranged Ca^{2+} fluxes to the etiology of neurodegenerative conditions, cancer, diabetes, and cardiovascular diseases. The book (which comes in two separate volumes, i.e., parts A and B) is equally suitable for experienced scientists who may want to obtain additional insights into a specific facet of Ca^{2+} biology, as well as to students or less experienced individual who are moving their first steps in this rapidly changing field of investigation.

Acknowledgments

S.M. is supported by the Italian Ministry of Health (GR-2016-02364602), Nanoblend (Rovereto, Italy), and local funds from Marche Polytechnic University (Ancona, Italy). The L.G. lab is supported by a Breakthrough Level 2 grant from the US DoD BRCP (#BC180476P1), by the 2019 Laura Ziskin Prize in Translational Research (#ZP-6177, PI: Formenti) from the Stand Up to Cancer (SU2C), by a Mantle Cell Lymphoma Research Initiative (MCL-RI, PI: Chen-Kiang) grant from the Leukemia and Lymphoma Society (LLS), by a startup grant from the Department of Radiation Oncology at Weill Cornell Medicine (New York, USA), by a Rapid Response Grant from the Functional Genomics Initiative (New York, USA), by industrial collaborations with Lytix Biopharma (Oslo, Norway) and Phosplatin (New York, USA), and by donations from Phosplatin (New York, USA), the Luke Heller TECPR2 Foundation (Boston, USA), Sotio a.s. (Prague, Czech Republic), and Onxeo (Paris, France).

Conflict of interest statement

L.G. has received research funding from Lytix Biopharma and Phosplatin, as well as consulting/advisory honoraria from Boehringer Ingelheim, AstraZeneca, OmniSEQ, Onxeo, The Longevity Labs, Inzen, and the Luke Heller TECPR2 Foundation.

References

Arensman, M.D., et al., 2020. Anti-tumor immunity influences cancer cell reliance upon ATG7. Oncoimmunology 9 (1), 1800162.

Berridge, M.J., 2012. Calcium signalling remodelling and disease. Biochem. Soc. Trans. 40 (2), 297–309.

Berridge, M.J., 2016. The inositol trisphosphate/calcium signaling pathway in health and disease. Physiol. Rev. 96 (4), 1261–1296.

Berridge, M.J., et al., 2003. Calcium signalling: dynamics, homeostasis and remodelling. Nat. Rev. Mol. Cell Biol. 4 (7), 517–529.

Bonora, M., et al., 2017. Mitochondrial permeability transition involves dissociation of F(1)F (O) ATP synthase dimers and C-ring conformation. EMBO Rep. 18 (7), 1077–1089.

Bonora, M., et al., 2019. Targeting mitochondria for cardiovascular disorders: therapeutic potential and obstacles. Nat. Rev. Cardiol. 16 (1), 33–55.

Bootman, M.D., Bultynck, G., 2020. Fundamentals of cellular calcium signaling: a primer. Cold Spring Harb. Perspect. Biol. 12 (1), a038802.

Borcoman, E., et al., 2019. Inhibition of PI3K pathway increases immune infiltrate in muscle-invasive bladder cancer. Oncoimmunology 8 (5), e1581556.

Bravo-Sagua, R., et al., 2020. Sarcoplasmic reticulum and calcium signaling in muscle cells: homeostasis and disease. Int. Rev. Cell Mol. Biol. 350, 197–264.

Catterall, W.A., 2011. Voltage-gated calcium channels. Cold Spring Harb. Perspect. Biol. 3 (8), a003947.

Chong, S.J.F., et al., 2020. Noncanonical cell fate regulation by Bcl-2 proteins. Trends Cell Biol. 30 (7), 537–555.

Clapham, D.E., 2007. Calcium signaling. Cell 131 (6), 1047–1058.

Cocetta, V., et al., 2020. Links between cancer metabolism and cisplatin resistance. Int. Rev. Cell Mol. Biol. 354, 107–164.

Dionísio, P.A., et al., 2020. Molecular mechanisms of necroptosis and relevance for neurodegenerative diseases. Int. Rev. Cell Mol. Biol. 353, 31–82.

Dong, X.P., et al., 2008. The type IV mucolipidosis-associated protein TRPML1 is an endolysosomal iron release channel. Nature 455 (7215), 992–996.

Dridi, H., et al., 2020. Intracellular calcium leak in heart failure and atrial fibrillation: a unifying mechanism and therapeutic target. Nat. Rev. Cardiol. 17 (11), 732–747.

Fairlie, W.D., et al., 2020. Crosstalk between apoptosis and autophagy signaling pathways. Int. Rev. Cell Mol. Biol. 352, 115–158.

Fucikova, J., et al., 2020. Calreticulin arms NK cells against leukemia. Oncoimmunology 9 (1), 1671763.

Galluzzi, L., et al., 2020. Consensus guidelines for the definition, detection and interpretation of immunogenic cell death. J. Immunother. Cancer 8 (1), e000337.

Gaspers, L.D., et al., 2014. Hormone-induced calcium oscillations depend on cross-coupling with inositol 1,4,5-trisphosphate oscillations. Cell Rep. 9 (4), 1209–1218.

Georgiadou, E., Rutter, G.A., 2020. Control by Ca(2+) of mitochondrial structure and function in pancreatic β-cells. Cell Calcium 91, 102282.

Ghigo, A., et al., 2017. PI3K and calcium signaling in cardiovascular disease. Circ. Res. 121 (3), 282–292.

Gil-Rivera, M., et al., 2021. Physiology of pancreatic β-cells: ion channels and molecular mechanisms implicated in stimulus-secretion coupling. Int. Rev. Cell Mol. Biol. 359, 287–323.

Giorgi, C., et al., 2018a. Calcium dynamics as a machine for decoding signals. Trends Cell Biol. 28 (4), 258–273.

Giorgi, C., et al., 2018b. The machineries, regulation and cellular functions of mitochondrial calcium. Nat. Rev. Mol. Cell Biol. 19 (11), 713–730.

Glab, J.A., et al., 2020. Bcl-2 family proteins, beyond the veil. Int. Rev. Cell Mol. Biol. 351, 1–22.

Hofer, A.M., Brown, E.M., 2003. Extracellular calcium sensing and signalling. Nat. Rev. Mol. Cell Biol. 4 (7), 530–538.

Jiang, D., et al., 2009. Genome-wide RNAi screen identifies Letm1 as a mitochondrial Ca2+/H+ antiporter. Science 326 (5949), 144–147.

Kamer, K.J., Mootha, V.K., 2015. The molecular era of the mitochondrial calcium uniporter. Nat. Rev. Mol. Cell Biol. 16 (9), 545–553.

Krebs, J., et al., 2015. Ca(2+) homeostasis and endoplasmic reticulum (ER) stress: an integrated view of calcium signaling. Biochem. Biophys. Res. Commun. 460 (1), 114–121.

Li, M., et al., 2017. Structural basis of dual Ca(2+)/pH regulation of the endolysosomal TRPML1 channel. Nat. Struct. Mol. Biol. 24 (3), 205–213.

Lloyd-Evans, E., et al., 2008. Niemann-Pick disease type C1 is a sphingosine storage disease that causes deregulation of lysosomal calcium. Nat. Med. 14 (11), 1247–1255.

Marchi, S., Pinton, P., 2014. The mitochondrial calcium uniporter complex: molecular components, structure and physiopathological implications. J. Physiol. 592 (5), 829–839.

Marchi, S., et al., 2019. Akt-mediated phosphorylation of MICU1 regulates mitochondrial Ca(2+) levels and tumor growth. EMBO J. 38 (2), e99435.

Marchi, S., et al., 2020. Ca(2+) fluxes and cancer. Mol. Cell 78 (6), 1055–1069.

Martins, I., et al., 2011. Hormesis, cell death and aging. Aging (Albany, NY) 3 (9), 821–828.

Maso, K., et al., 2019. Molecular platforms for targeted drug delivery. Int. Rev. Cell Mol. Biol. 346, 1–50.

Medina, D.L., et al., 2015. Lysosomal calcium signalling regulates autophagy through calcineurin and TFEB. Nat. Cell Biol. 17 (3), 288–299.

Melchionda, M., et al., 2016. Ca2+/H+ exchange by acidic organelles regulates cell migration in vivo. J. Cell Biol. 212 (7), 803–813.

Miller, D.R., et al., 2020. The interplay of autophagy and non-apoptotic cell death pathways. Int. Rev. Cell Mol. Biol. 352, 159–187.

Missiroli, S., et al., 2018. Mitochondria-associated membranes (MAMs) and inflammation. Cell Death Dis. 9 (3), 329.

Monteith, G.R., et al., 2017. The calcium-cancer signalling nexus. Nat. Rev. Cancer 17 (6), 367–380.

Montell, C., et al., 2002. A unified nomenclature for the superfamily of TRP cation channels. Mol. Cell 9 (2), 229–231.

Nilius, B., Szallasi, A., 2014. Transient receptor potential channels as drug targets: from the science of basic research to the art of medicine. Pharmacol. Rev. 66 (3), 676–814.

Palty, R., et al., 2010. NCLX is an essential component of mitochondrial Na+/Ca2+ exchange. Proc. Natl. Acad. Sci. U. S. A. 107 (1), 436–441.

Parys, J.B., Bultynck, G., 2018. Calcium signaling in health, disease and therapy. Biochim. Biophys. Acta, Mol. Cell Res. 1865 (11 Pt. B), 1657–1659.

Patel, P., Karch, J., 2020. Regulation of cell death in the cardiovascular system. Int. Rev. Cell Mol. Biol. 353, 153–209.

Perrone, M., et al., 2020. The role of mitochondria-associated membranes in cellular homeostasis and diseases. Int. Rev. Cell Mol. Biol. 350, 119–196.

Pervaiz, S., et al., 2020. Redox signaling in the pathogenesis of human disease and the regulatory role of autophagy. Int. Rev. Cell Mol. Biol. 352, 189–214.

Petroni, G., et al., 2021. Immunomodulation by targeted anticancer agents. Cancer Cell 39 (3), 310–345.

Pinton, P., et al., 1998. The Golgi apparatus is an inositol 1,4,5-trisphosphate-sensitive Ca2+ store, with functional properties distinct from those of the endoplasmic reticulum. EMBO J. 17 (18), 5298–5308.

Pizzo, P., et al., 2011. Ca(2+) signalling in the Golgi apparatus. Cell Calcium 50 (2), 184–192.

Porporato, P.E., et al., 2018. Mitochondrial metabolism and cancer. Cell Res. 28 (3), 265–280.

Prakriya, M., Lewis, R.S., 2015. Store-operated calcium channels. Physiol. Rev. 95 (4), 1383–1436.

Primeau, J.O., et al., 2018. The sarcoendoplasmic reticulum calcium ATPase. Subcell. Biochem. 87, 229–258.

Prinz, W.A., et al., 2020. The functional universe of membrane contact sites. Nat. Rev. Mol. Cell Biol. 21 (1), 7–24.

Prole, D.L., Taylor, C.W., 2019. Structure and function of IP(3) receptors. Cold Spring Harb. Perspect. Biol. 11 (4), a035063.

Raffaello, A., et al., 2016. Calcium at the center of cell signaling: interplay between endoplasmic reticulum, mitochondria, and lysosomes. Trends Biochem. Sci. 41 (12), 1035–1049.

Rimessi, A., et al., 2013. Perturbed mitochondrial Ca2+ signals as causes or consequences of mitophagy induction. Autophagy 9 (11), 1677–1686.

Rimessi, A., et al., 2020. Interorganellar calcium signaling in the regulation of cell metabolism: a cancer perspective. Semin. Cell Dev. Biol. 98, 167–180.

Rizzuto, R., et al., 2012. Mitochondria as sensors and regulators of calcium signalling. Nat. Rev. Mol. Cell Biol. 13 (9), 566–578.

Rodriguez-Ruiz, M.E., et al., 2019. Apoptotic caspases inhibit abscopal responses to radiation and identify a new prognostic biomarker for breast cancer patients. Oncoimmunology 8 (11), e1655964.

Rodriguez-Ruiz, M.E., et al., 2020. Immunological impact of cell death signaling driven by radiation on the tumor microenvironment. Nat. Immunol. 21 (2), 120–134.

Rosa, N., et al., 2020. Type 3 IP(3) receptors: the chameleon in cancer. Int. Rev. Cell Mol. Biol. 351, 101–148.

Rybstein, M.D., et al., 2018. The autophagic network and cancer. Nat. Cell Biol. 20 (3), 243–251.

Schrank, S., et al., 2020. Calcium-handling defects and neurodegenerative disease. Cold Spring Harb. Perspect. Biol. 12 (7), a035212.

Semyanov, A., et al., 2020. Making sense of astrocytic calcium signals—from acquisition to interpretation. Nat. Rev. Neurosci. 21 (10), 551–564.

Sica, V., et al., 2019. A strategy for poisoning cancer cell metabolism: inhibition of oxidative phosphorylation coupled to anaplerotic saturation. Int. Rev. Cell Mol. Biol. 347, 27–37.

Soboloff, J., et al., 2012. STIM proteins: dynamic calcium signal transducers. Nat. Rev. Mol. Cell Biol. 13 (9), 549–565.

Stafford, N., et al., 2017. The plasma membrane calcium ATPases and their role as major new players in human disease. Physiol. Rev. 97 (3), 1089–1125.

Stathopulos, P.B., et al., 2008. Structural and mechanistic insights into STIM1-mediated initiation of store-operated calcium entry. Cell 135 (1), 110–122.

Taylor, C.W., 2002. Controlling calcium entry. Cell 111 (6), 767–769.

Thangaraj, A., et al., 2020. Targeting endoplasmic reticulum stress and autophagy as therapeutic approaches for neurological diseases. Int. Rev. Cell Mol. Biol. 350, 285–325.

Trebak, M., Kinet, J.P., 2019. Calcium signalling in T cells. Nat. Rev. Immunol. 19 (3), 154–169.

Vallejo-Illarramendi, A., et al., 2014. Dysregulation of calcium homeostasis in muscular dystrophies. Expert Rev. Mol. Med. 16, e16.

Vandecaetsbeek, I., et al., 2011. The Ca2 + pumps of the endoplasmic reticulum and Golgi apparatus. Cold Spring Harb. Perspect. Biol. 3 (5), a004184.

Vitale, I., et al., 2021. Intratumoral heterogeneity in cancer progression and response to immunotherapy. Nat. Med. 27 (2), 212–224.

Williams, G.S., et al., 2013. Mitochondrial calcium uptake. Proc. Natl. Acad. Sci. U. S. A. 110 (26), 10479–10486.

Wu Chuang, A., et al., 2020. Endoplasmic reticulum stress in the cellular release of damage-associated molecular patterns. Int. Rev. Cell Mol. Biol. 350, 1–28.

Yamazaki, T., et al., 2020. Mitochondrial DNA drives abscopal responses to radiation that are inhibited by autophagy. Nat. Immunol. 21 (10), 1160–1171.

Yoast, R.E., et al., 2020. The native ORAI channel trio underlies the diversity of Ca(2 +) signaling events. Nat. Commun. 11 (1), 2444.

CHAPTER ONE

Calcium signaling in neuroglia

Dmitry Lim[a],*, Alexey Semyanov[b,c,d], Armando Genazzani[a], and Alexei Verkhratsky[d,e,f],*

[a]Department of Pharmaceutical Sciences, Università del Piemonte Orientale, Novara, Italy
[b]Shemyakin-Ovchinnikov Institute of Bioorganic Chemistry, Russian Academy of Sciences, Moscow, Russia
[c]Faculty of Biology, Moscow State University, Moscow, Russia
[d]Sechenov First Moscow State Medical University, Moscow, Russia
[e]Faculty of Biology, Medicine and Health, The University of Manchester, Manchester, United Kingdom
[f]Achucarro Centre for Neuroscience, IKERBASQUE, Basque Foundation for Science, Bilbao, Spain
*Corresponding authors: e-mail address: dmitry.lim@uniupo.it; alexej.verkhratsky@manchester.ac.uk

Contents

Abstract

Glial cells exploit calcium (Ca^{2+}) signals to perceive the information about the activity of the nervous tissue and the tissue environment to translate this information into an array of homeostatic, signaling and defensive reactions. Astrocytes, the best studied glial cells, use several Ca^{2+} signaling generation pathways that include Ca^{2+} entry through plasma membrane, release from endoplasmic reticulum (ER) and from mitochondria. Activation of metabotropic receptors on the plasma membrane of glial cells is coupled to an enzymatic cascade in which a second messenger, InsP3 is generated thus activating intracellular Ca^{2+} release channels in the ER endomembrane. Astrocytes also possess store-operated Ca^{2+} entry and express several ligand-gated Ca^{2+} channels. In vivo astrocytes generate heterogeneous Ca^{2+} signals, which are short and frequent in distal processes, but large and relatively rare in soma. In response to neuronal activity intracellular

International Review of Cell and Molecular Biology, Volume 362
ISSN 1937-6448
https://doi.org/10.1016/bs.ircmb.2021.01.003

and inter-cellular astrocytic Ca^{2+} waves can be produced. Astrocytic Ca^{2+} signals are involved in secretion, they regulate ion transport across cell membranes, and are contributing to cell morphological plasticity. Therefore, astrocytic Ca^{2+} signals are linked to fundamental functions of the central nervous system ranging from synaptic transmission to behavior. In oligodendrocytes, Ca^{2+} signals are generated by plasmalemmal Ca^{2+} influx, or by release from intracellular stores, or by combination of both. Microglial cells exploit Ca^{2+} permeable ionotropic purinergic receptors and transient receptor potential channels as well as ER Ca^{2+} release. In this contribution, basic morphology of glial cells, glial Ca^{2+} signaling toolkit, intracellular Ca^{2+} signals and Ca^{2+}-regulated functions are discussed with focus on astrocytes.

1. Introduction: Supportive and protective neuroglia

The complexity of human brain is remarkable: more than 200 billion of neural cells are packed within a limited volume of $\sim$1400 cm^3 to create the most sophisticated and most powerful computing device with several petabytes of operational memory and processing speed in an petaFLOP range at the expense of 10–20 W/h (Bartol et al., 2015). To a very big extent these remarkable capabilities of the brain are achieved because of an existence of specialized homeostatic system represented by neuroglia (Fig. 1, Verkhratsky and Butt, 2013). Neurons and neuroglia are, respectively, the executive and logistical arms of the nervous tissue; while neurons have been

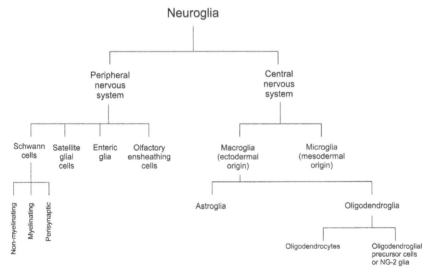

Fig. 1 Classification of neuroglia.

perfected by evolution to process information and govern the body, the neuroglial cells assumed full control over homoeostasis and defense of the nervous system (Verkhratsky and Nedergaard, 2016). The concept of supportive nervous tissue emerged in 19th century (see historic reviews, Chvatal and Verkhratsky, 2018; Kettenmann and Verkhratsky, 2008), and the neuroglia, as a connective tissue of the brain and the spinal cord, was conceptualized by Rudolf Virchow in 1850s (Virchow, 1856, 1858). From the very dawn of neuroscience the neuroglia was considered as a multifunctional supportive cell involved in homeostasis, nutritional support, regulation of blood flow, sleep and conscience (Golgi, 1870; Lugaro, 1907; Ramón y Cajal, 1895; Schleich, 1894). Neuroglia was also considered an important part of pathophysiology of neurological and psychiatric diseases (Achucarro, 1919; Alzheimer, 1910; Andriezen, 1893; Frommann, 1878). The first type of neuroglial cell, the astrocyte (the "star"-like cell; also referred to as astroglia or astroglial cells) has been defined by Lenhossék (1895). The oligodendroglia or oligodendrocytes (cells with few processes), the myelinating cells of the central nervous system, were characterized by Pio de Rio Hortega in 1920 (Del Río-Hortega, 1921); at the same time Hortega identified and characterized microglia, the innate immune cells of the CNS of mesenchymal origin (Del Rio-Hortega, 1919, 1932). The last type of neuroglia, the NG2 glia (cells which are also known as oligodendrocyte progenitor cells or poly-dendrocytes), were discovered by William Stallcup and his colleagues in 1980s (Stallcup, 1981). While this chapter considers glial cells of the central nervous system (CNS), glial cell in the peripheral nervous system (PNS) per-form similar homeostatic and signaling functions. There are four major types of glia in the PNS, all originating from precursors located in the neural crest. The peripheral glia are: myelinating, non-myelinating and perisynaptic Schwann cells, satellite glial cells, enteric glial cells (EGCs), and olfactory ensheathing cells.

Neuroglial cells of both neuroepithelial (astrocytes and oligodendro-cytes) and myeloid (microglia) origin are electrically non-excitable; these cells cannot generate action potentials, which remain a functional hallmark of neurons, while, as any other cell, they can generate slow (electrotonic or graded) potentials. Glial cells respond to external stimuli with a specific type of excitability associated with intracellular ions and second messengers. This type of excitability is defined as intracellular excitability of glia and, to a large extent, is mediated by fluctuations of free ions in the cytosol—the ionic excitability. The role for two major cations, Ca^{2+} and Na^+ in glial excitabil-ity is firmly established (Kirischuk et al., 2012; Rose and Verkhratsky, 2016;

Semyanov, 2019; Semyanov et al., 2020; Verkhratsky et al., 2012); possible contribution of other ions such as K^+, Cl^- or H^+ is under consideration (Verkhratsky et al., 2020b). In this chapter we shall focus on neuroglial Ca^{2+} signaling.

2. Glial cells

2.1 Astrocytes

Astrocytes, also known as astroglia, are primary homoeostatic cells of the CNS; astrocytes control homeostasis of the brain and the spinal cord at all levels of organization from molecular to the whole organ (Verkhratsky and Nedergaard, 2018). In addition, astrocytes contribute to the pathophysiology of the neurological disorders; astrogliopathology includes several complex processes from reactive astrogliosis to astrodegeneration, astroglial atrophy and loss of function (Pekny et al., 2016; Verkhratsky et al., 2017, 2019). Astrocytes are highly heterogeneous cells in morphology and function present throughout the CNS in both gray and white matter. The main types of astrocytes (Fig. 2) are discussed below.

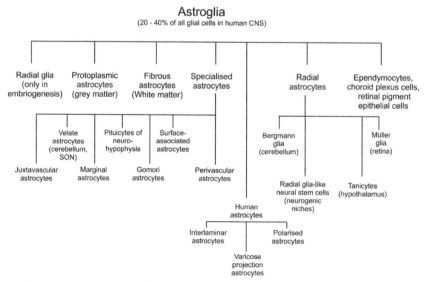

Fig. 2 Main types of astroglial cells.

2.1.1 Protoplasmic astrocytes

Protoplasmic astrocytes dominate the gray matter of the brain and of the spinal cord. In mice, protoplasmic astrocytes (Fig. 3) have small round somata ($\sim$10 µm in diameter) and 5–10 primary processes (up to 50 µm in length). The primary processes branch into secondary to tertiary processes, from which peripheral tiny processes emanate giving the protoplasmic astrocyte distinctive spongioform morphology (Bushong et al., 2002; Popov et al., 2021). Morphological compartments of protoplasmic astrocytes are classified into (i) soma; (ii) main processes also known as branches; (iii) secondary to tertiary processes designated as branchlets; (iv) peripheral processes known as leaflets, which form the perisynaptic astroglial cradle (Verkhratsky and Nedergaard, 2014) and (v) perivascular processes which terminate with endfeet plastering blood vessels and forming *glia limitans vascularis* (Gavrilov et al., 2018; Khakh and Sofroniew, 2015; Popov et al., 2021; Semyanov, 2019). Protoplasmic astrocytes parcellate (through the process known as tiling) the gray matter into neurovascular units (Iadecola, 2017). Protoplasmic astrocytes are integrated into syncytia through gap junctions formed by connexons (Giaume et al., 2010). The density of protoplasmic astrocytes in rodents ranges (depending on the brain region) between 10,000 and 30,000 per mm^3. The volume of a single rodent protoplasmic astrocyte is 50,000–80,000 µm^3. The surface area of an individual rodent protoplasmic astrocyte is $\sim$80,000 µm^2; with the major part of the surface area belonging to peripheral processes (Bushong et al., 2002; Ogata and Kosaka, 2002; Reichenbach et al., 2010).

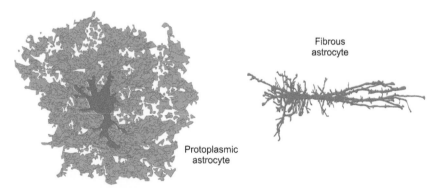

Fibrous
astrocyte

Protoplasmic
astrocyte

Fig. 3 Morphological profiles of protoplasmic and fibrous astrocytes. The profiles have been traced from original cel imaging provided by Prof. Milos Pekny (Gothenburg University) and Prof. Arthur Butt (University of Portsmouth).

2.1.2 Fibrous astrocytes

Fibrous astrocytes are localized in the white matter of the brain and of the spinal cord, in the optic nerve and in the nerve fiber layer of the retina. Fibrous astrocytes (Fig. 3) have several primary processes radially oriented in the direction of the axon bundles; these processes may reach up to 100 μm in length (Lundgaard et al., 2014). These processes form several perivascular or subpial endfeet and also send numerous extensions known as perinodal processes to the nodes of Ranvier. Morphological appearance of fibrous astrocytes is quite diverse; for example, in rodent optic nerve, fibrous astrocytes are subdivided into transverse, random, and longitudinal depending on the orientation of processes with respect to the long axis of the nerve (Butt et al., 1994).

2.1.3 Juxtavascular astrocytes

Juxtavascular astrocytes represent a subpopulation of protoplasmic astrocytes of the gray matter. Somata of juxtavascular astrocytes are in close apposition with blood vessels; these cells have some distinct physiological features and have significant proliferative potential in response to traumatic brain injury (Bardehle et al., 2013; Götz et al., 2020).

2.1.4 Surface-associated astrocytes

Surface-associated astrocytes are directly associated with the cortical surface in the posterior prefrontal and amygdaloid cortex. Somata of these cells are positioned at the cortical surface; these astrocytes send two types of processes: long parallel superficial process running beneath the pia matters vessels and shorter processes extending in all directions, with some of these processes descending towards cortical layer I (Feig and Haberly, 2011). The processes of surface-associated astrocytes demonstrate considerable overlap indicating absence of territorial domains. Cell bodies and superficial processes of surface-associated astrocytes form *glia limitans externa* in prefrontal cortex (Feig and Haberly, 2011).

2.1.5 Velate astrocytes

Velate astrocytes, named so by Ramon y Cajal in the early 20th century and by Chan-Palay and Palay (1972), are specialized parenchymal astrocytes which provide for coverage of densely packed neuronal populations in such regions as the olfactory bulb or in granular layer of cerebellar cortex. Velate astrocytes have a small soma with several primary processes from which a large leaflets (with high surface-to-volume ratio of $20-30\,\mu m^{-1}$)

are emanating to cover neuronal cell bodies. These perineuronal processes look like extended leaves, which defined their name derived from *vellum* (parchment in Latin) or *velatus* (which is Latin word for covered, wrapped, veiled). In cerebellum these veil-like processes ensheath somata of granule neurons (with a single astrocyte covering several neurons) as well as somata of Purkinje neurons (Chan-Palay and Palay, 1972). These processes also cover glomeruli composed of mossy fiber rosettes, boutons of Golgi neurons and granule cell dendrites, thus separating functionally distinct groups of neurons and synapses (Buffo and Rossi, 2013).

2.1.6 Pituicytes

Pituicytes are main type of astroglia in the neurohypophysis (Hatton, 1988). Pituicytes express astroglial markers GFAP and S100B; these cells are morphologicaly heterogeneous being sub-classified into major, dark, ependymal, oncocytic and granular pituicytes (Takey and Pearl, 1984). Pituicytes ensheath terminals of magnocellular neurons projecting into neurohypophysis and regulate neurosecretion through the release of taurine and by morphological plasticity of the coverage. Thus, in basal conditions ensheathing glial terminals reduce neuronal excitability restricting access of neuroactive substances to neuronal membranes. Upon physiological activation, morphological remodeling of glial processes facilitates neuron-neuron interaction and prolongs the action of neurotransmitters (Hatton, 1999; Rosso and Mienville, 2009). Of note, pituicytes receive synaptic contacts, and express many receptors for neurotransmitters and neurohormones (Hatton, 1999).

2.1.7 Perivascular and marginal astrocytes

Perivascular astrocytes are localized close to the *pia mater*, where they form numerous endfeet with blood vessels. The major function of these cells is to form *glia limitans* and accordingly they do not establish contacts with neurons (Liu et al., 2013).

2.1.8 Glia limitans

Glia limitans represent the parenchymal part of blood-brain and blood–cerebrospinal fluid barrier (BCSFB); it is further divided into *glia limitans externa* (bordering the pia matter) and *glia limitans perivascularis* (plastering the intra-parenchymal blood vessels). *Glia limitans* is formed mainly by astrocytic endfeet and also by surface-associated astrocytes in prefrontal cortex (see above) and by processes of juxtavascular microglial cells (Joost et al., 2019; Mathiisen et al., 2010; Quintana, 2017).

2.1.9 Gomori astrocytes

Specific set of astrocytes containing high concentrations of heme and hence stained with Gomori's chrome alum hematoxylin technique are present in hypothalamus and in hippocampus (Young et al., 1990, 1996).

2.1.10 Ependymocytes, choroid plexus cells and retinal pigment epithelial cells

These cells line up the ventricles and the subretinal space. The choroid plexus cells produce the cerebrospinal fluid, which fills the brain ventricles, spinal canal and the subarachnoid space (Reichenbach and Bringmann, 2017).

2.1.11 Radial astrocytes

Radial astrocytes, named so because of their radially extended processes, include (i) Retinal Müller cells; (ii) cerebellar Bergmann glia; (iii) radial astrocytes of the supra–optic nucleus; (iv) radial glia-like neural stem cells, present in neurogenic niches and (v) tanycytes present in the periventricular organs, in the hypothalamus, in the hypophysis and in the raphe part of the spinal cord (Reichenbach and Bringmann, 2017; Verkhratsky and Nedergaard, 2018).

2.1.12 Human-specific astroglia

In humans and in some higher primates protoplasmic and fibrous astrocytes are generally larger and more complex as compared to rodents (Oberheim et al., 2009; Verkhratsky and Nedergaard, 2018). In addition, the brain of humans and high primates possesses specific types of astroglia absent in other species. These include (i) *Interlaminar astrocytes*, which have small somata located in the upper cortical layer and sending long processes towards deeper layers (Colombo, 2017); (ii) *Polarized astrocytes*, cell bodies of which are positioned in the deep cortical layers, and several long processes penetrate into superficial cortical layers (Oberheim et al., 2009) and (iii) *Varicose projection astrocytes* characterized by several very long (up to 1 mm) unbranched processes bearing varicosities and extending in all directions through the deep cortical layers (Oberheim et al., 2009; Sosunov et al., 2014).

2.2 Oligodendrocytes

Oligodendrocytes are evolutionary specialized to form the insulating myelin sheaths around CNS axons (Verkhratsky and Butt, 2013). The myelin sheath is a fatty insulating layer that provides for the rapid conduction of nerve impulses. Oligodendrocytes are broadly classified into (i) myelinating oligodendrocytes; (ii) NG2 glia or adult oligodendrocyte progenitor cells and (iii) perineuronal or satellite oligodendrocytes, which are restricted to gray matter areas where they are directly opposed to neuronal cell bodies. NG2 glia is identified by the expression of nerve/glia antigen 2 (NG2) chondroitin sulfate proteoglycan (CSPG4). The ontogenesis and heterogeneity of NG2 glia has been described elsewhere (Butt et al., 2019; Dimou and Gallo, 2015). Myelinating oligodendrocytes are subdivided into types I–IV according to the shape and size of their soma, the number of their cellular processes, and the size of the axons they myelinate. In general oligodendrocytes type I/II support a large number of short myelin sheaths for axons with diameters $\leq 2\,\mu m$; type III/IV oligodendrocytes support a small number of long myelin sheaths for larger axons with diameters $\geq 4\,\mu m$ (see Verkhratsky and Butt, 2013 for details).

2.3 Microglia

Microglial cells are CNS tissue macrophages responsible for the innate immunity of the brain and the spinal cord. Microglia can be defined as a type of neuroglia of mesodermal origin, which forms the defensive system of the CNS (Kettenmann et al., 2011). Microglial cells are quite heterogeneous and their phenotypes are rather kaleidoscopic depending on the brain region, age and history of stresses and pathologies (i.e., exposome) (Dubbelaar et al., 2018). General classification of microglia distinguishes several phenotypes including: (i) ramified or surveilling microglia; (ii) satellite microglia, which contact neuronal somata and axons; (iii) rod microglia; (iv) Gitter cell-like microglia; (v) hypertrophic microglia; (vi) amoeboid microglia; (vii) dystrophic microglia populating aging brain; and (viii) dark microglia, associated with various types of neuropathology (Stratoulias et al., 2019).

3. Astrocytic Ca^{2+} signaling
3.1 Ca^{2+} as a second messenger

Ionized calcium (Ca^{2+}) is arguably the most versatile evolutionary conserved and ubiquitous intracellular and intercellular messenger in living organisms (Berridge et al., 2003; Cheng et al., 2006; Plattner and Verkhratsky, 2018). From an evolutionary prospective, the role of Ca^{2+} as an information carrying messenger is determined by its ability to reversibly form complexes with the proteins, thus accommodating irregularly shaped ion binding sites of numerous proteins (Carafoli and Krebs, 2016). Therefore, not only many proteins can complex with Ca^{2+}, but the binding is reversible. In the cytosol, Ca^{2+}-binding proteins contribute to set the resting Ca^{2+} ion concentration ($[Ca^{2+}]_i$) and shape amplitude, kinetics (rise and decay times) and propagation of Ca^{2+} transients (Semyanov et al., 2020). Notably, $[Ca^{2+}]_i$ is as low as 50–150 nM, which is five orders of magnitude lower than in sea water for marine organisms and four orders of magnitude lower than in extracellular fluids of vertebrate animals. Keeping low $[Ca^{2+}]i$ allows development of phosphate-based energetic metabolism which otherwise would be impossible due to low solubility of Ca^{2+} phosphate salts (Plattner and Verkhratsky, 2016). Complex molecular mechanism has been developed by cells to (i) keep low $[Ca^{2+}]_i$, (ii) allow controlled elevations of $[Ca^{2+}]_i$ and (iii) propagate Ca^{2+} signals through and between the cell(s). Steep transmembrane $[Ca^{2+}]$ gradient, exceptional ligand properties and mechanisms allowing fast temporally and space-organized increase in $[Ca^{2+}]_i$ as well as fast removal of Ca^{2+} provide the mechanistic background for the signaling role of Ca^{2+}.

3.2 Astroglial Ca^{2+} homeostasis

3.2.1 Intracellular Ca^{2+} stores and $[Ca^{2+}]_i$ in astrocytes

Resting $[Ca^{2+}]_i$ in the cytosol of astrocytes is similar to that of other cells being in the range of $\sim$50–100 nM of free Ca^{2+} (Shigetomi et al., 2010; Verkhratsky and Kettenmann, 1996; Zheng et al., 2015). Fluorescence-lifetime imaging microscopy of cortical and hippocampal astrocytes revealed $[Ca^{2+}]_i$ gradient between peripheral processes and soma. Resting $[Ca^{2+}]_i$ in the peripheral astrocyte processes was $\sim$125 nM to 200 nM, whereas $[Ca^{2+}]_i$ in the soma was $\sim$100 nM, which, incidentally, was about twice as high as in neighboring neurons (Zheng et al., 2015). Such intracellular $[Ca^{2+}]_i$ gradient may be explained by higher surface-to-volume ratio of peripheral processes which allows more Ca^{2+} to enter the cell through plasma membrane per volume of cytoplasm (Wu et al., 2019). Initiation of neuronal activity with theta-burst stimulation produced a decrease in the steady-state resting $[Ca^{2+}]_i$ in astrocytes (Mehina et al., 2017). This may have profound physiological significance: first, $[Ca^{2+}]_i$ changes are associated with vascular response (lowering the resting $[Ca^{2+}]_i$ induces vasodilation (Rosenegger et al., 2015)); second, it defines spontaneous Ca^{2+} activity and Ca^{2+} responsiveness of astrocytes to stimulation and neuronal activity in vivo (King et al., 2020). Finally, increase in resting $[Ca^{2+}]_i$ occurs in vivo and in vitro is association with neuropathology, e.g., AD (Kuchibhotla et al., 2009; Lim et al., 2013). Two sources provide for Ca^{2+} entry into the cytosol: (i) extracellular milieu represented by the interstitial fluid and (ii) intracellular Ca^{2+} stores.

Extracellular Ca^{2+} concentration ($[Ca^{2+}]_o$) is kept at a range of $\sim$1.5–2 mM while at a level of an organism, Ca^{2+} homeostasis is tightly controlled (Berridge et al., 2003). Because of the extremely compartmentalized structure of the brain tissue, different states of the brain activity may distinctly affect $[Ca^{2+}]_o$ in different compartments/extracellular regions (e.g., in the synaptic cleft) (Lopes and Cunha, 2019; Rusakov and Fine, 2003). The main intracellular Ca^{2+} store is associated with the endoplasmic reticulum (ER) present in soma and in astrocytic processes but not in the perisynaptic leaflets (Bernardinelli et al., 2014; Patrushev et al., 2013). Experiments with photoprotein aequorin (AEQ) targeted to the lumen of the ER showed that Ca^{2+} concentration in ER ($[Ca^{2+}]_{ER}$) in different cell types varies between 0.1 and $\sim$1 mM (Alonso et al., 1999; Solovyova et al., 2002). In cultured astrocytes $[Ca^{2+}]_{ER}$ lies at a lower range of this scale, being around 100–200 μM, as measured by low affinity Ca^{2+} probes (Golovina and Blaustein, 1997; Verkhratsky et al., 2002), and $\sim$110 μM as measured by ER-targeted aequorin (Dematteis et al., 2020b). The Ca^{2+} content of the ER is constantly monitored by STIM1 proteins localized in the endomembrane; depletion of the ER from the releasable Ca^{2+} instigates polymerization of STIM1 proteins with subsequent activation of plasmalemmal TRPC or ORAI channels, which both mediate store-operated Ca^{2+} entry (SOCE) operational in all glial cells including astrocytes (Verkhratsky and Parpura, 2014).

Special role in cellular Ca^{2+} homeostasis is played by mitochondria, which are not a *stricto sensu* Ca^{2+} store, they rather act as a capacious Ca^{2+} sink, buffering large Ca^{2+} loads and preventing cell from Ca^{2+} overload (a mitochondrial Ca^{2+} firewall), although they also may release Ca^{2+} through flickering of permeability transition pore (mPTP) or by mitochondrial Na^+/Ca^{2+} exchanger (NCX) (Agarwal et al., 2017; Walsh et al., 2009). Ca^{2+} is moved across mitochondrial membrane with Ca^{2+} uniporter (mCU) according to electrochemical gradient. In resting conditions, $[Ca^{2+}]$ in the mitochondrial matrix ($[Ca^{2+}]_m$) is significantly higher than that of $[Ca^{2+}]_i$ being around 7 μM (Golovina and Blaustein, 1997). Therefore, Ca^{2+} movement by mCU strongly depends on the potential of mitochondria inner membrane. Stimulation of astrocytes with ATP produces $[Ca^{2+}]_m$ transients with peak amplitude $\sim$20 μM (Dematteis et al., 2020b). Elevation of $[Ca^{2+}]_m$ accelerates tricarboxylic acid cycle (TCA cycle, or the Krebs cycle) (Traaseth et al., 2004). This provides additional ATP synthesis to fulfill metabolic demand triggered by Ca^{2+} activity in astrocyte (e.g., process outgrowth, Molotkov et al., 2013; Tanaka et al., 2013).

3.2.2 Astrocytes possess a special form of excitability

Neurons and neuroglia are endowed with distinct form of excitability, which reflects their functional specialization. Neurons are specialized in generation of regenerative action potentials (APs) which originate from a propagating wave of openings and closures of voltage-gated ion channels. The complement of these channels capable of triggering regenerative APs underlies "electrical excitability." In contrast, neuroglial cells, including astrocytes, have a very different repertoire of plasmalemmal channels, and hence are incapable of generating AP, being thus "electrically non-excitable" cells. Nonetheless, glial cells express multiple channels, receptors and transporters, which, upon physiological stimulation generate transmembrane ionic fluxes. These fluxes translate in spatiotemporally organized changes in the intracellular ion concentration and local non-regenerative astrocyte depolarization. These changes in turn regulate activity of multiple targets such as enzymes controlling gene expression, metabolism or secretion as well as multiple transporters associated with astroglial homeostatic responses. In principle every major ion including Na^+, Ca^{2+}, K^+, Cl^- or H^+ may act as an intracellular signaler; the role for Ca^{2+} and Na^+ signaling as substrates of astroglial excitability is firmly established (Rose and Verkhratsky, 2016; Verkhratsky et al., 2012, 2020b), whereas the role of other ions is considered (Verkhratsky et al., 2020b). In addition, astroglial excitability is assisted by coordinated changes in the intracellular second messengers (Verkhratsky et al., 2020a).

3.2.3 Astroglial Ca^{2+} signaling toolkit: Molecular basis for Ca^{2+} excitability

In neurons, conduction of electrical impulses occurs through nerve fibers (axons), while transmission of a signal from one neuron to another mainly utilizes chemical synapses. Therefore, signaling toolkit in neurons, including that for Ca^{2+} signaling, is highly compartmentalized between neuronal cell bodies, axons and synaptic compartments, the latter, in chemical synapses, is being composed of pre-synaptic terminal (bouton) and post-synaptic terminal. Presynaptic terminals exploit voltage-gated Ca^{2+} channels for generation of $[Ca^{2+}]_i$ microdomains regulating vesicular exocytosis of neurotransmitters, while in the postsynaptic membrane of glutamatergic excitatory synapses NMDA receptors are the principal Ca^{2+}-conducting channel. Postsynaptic neurons also express voltage-gated Ca^{2+} channels that can be activated by excitatory postsynaptic potential (EPSP) or back-propagating APs. Metabotropic $G_{\alpha q/11}$-coupled receptors, that trigger Ca^{2+} release from Ca^{2+} stores, are activated by an array of neurotransmitters including ATP, glutamate, GABA, acetylcholine and others; metabotropic receptors are expressed in a region- and neuronal type-specific fashion, and they are present at both, synaptic and extrasynaptic compartments to modulate neuronal excitability, release of neurotransmitters and post-synaptic signaling events (for reviews, see Augustine et al., 2003; Fernández de Sevilla et al., 2020; Ferraguti et al., 2008; Kabbani and Nichols, 2018; Niswender and Conn, 2010; North and Verkhratsky, 2006).

Very similarly to neurons, astrocytes display a remarkable degree of compartmentalization associated with complex morphology of these cells. Peripheral processes of protoplasmic astrocytes (see Section 2.1.1), the leaflets, are exceptionally thin ($\sim$100 nM in thickness) and are devoid of organelles; in contrast, soma, branches and branchlets contain ER and mito-chondria. These morphological differences determine distinct Ca^{2+} signaling pathways operating in different parts of the astrocyte. Conceptually, Ca^{2+} signaling in the soma, branches and branchlets mainly relies upon ER Ca^{2+} release and store-operated Ca^{2+} entry, whereas $[Ca^{2+}]_i$ increases in the leaflets are regulated by plasmalemmal Ca^{2+} influx through channels or Na^+/Ca^{2+} exchanger operating the reverse mode (see Khakh and Sofroniew, 2015; Lalo et al., 2011; Rose et al., 2020; Semyanov, 2019, 2020; Shigetomi et al., 2013a; Verkhratsky et al., 2020a). In a different type of astrocyte, in the cerebellar Bergmann glial cell, $[Ca^{2+}]_i$ dynamics in the perisynaptic cradle mostly results from Ca^{2+} release from the ER which populates the appendages providing for glial synaptic coverage (Grosche et al., 1999; Kirischuk et al., 1999).

3.2.3.1 Metabotropic receptors, ER Ca^{2+} release and store-operated Ca^{2+} entry

Metabotropic signaling associated with Ca^{2+} release from the ER stores is mediated by (mostly) $InsP_3$ receptors and (in some astrocytic subpopulations) by ryanodine receptors (Verkhratsky and Nedergaard, 2018). Both types of these Ca^{2+} release channels are sensitive to $[Ca^{2+}]_i$ and hence can be engaged in Ca^{2+}-induced Ca^{2+} release (CICR) which recruits neighboring channels thus creating a propagating wave of their activation. Thus the membrane of the ER in astrocytes acts as an excitable media mediating intracellular propagating Ca^{2+} wave. Astrocytes express numerous metabotropic G-protein-coupled receptors including those coupled to $G_{\alpha q/11}$-containing trimeric proteins (Kofuji and Araque, 2020; Verkhratsky and Kettenmann, 1996; Verkhratsky and Nedergaard, 2018). The repertoire of receptors varies between types of astrocytes, brain regions and is also subjected to developmental modifications; the combinations include $GABA_B$, α an β-adrenoceptors (which seem to be a predominant type in adult astrocytes), M cholinoceptors, serotonin, dopamine and histamine receptors, cannabinoid, vasopressin, VIP and bradykinin receptors (Araque et al., 2002; Ding et al., 2013; Dong et al., 2012; Navarrete and Araque, 2008; Verkhratsky and Kettenmann, 1996; Verkhratsky and Nedergaard, 2018).

In cultured astrocytes, the mGluR5, a member of group I metabotropic glutamate receptors is highly expressed (Aronica et al., 2003; Grolla et al., 2013a; Lim et al., 2013). In the in vivo settings, mGluR5 mediates whisker stimulation-induced and odor-evoked Ca^{2+} transients in cortical astrocytes (Petzold et al., 2008; Wang et al., 2006). Glutamatergic signaling in the adult cortical and hippocampal astrocytes also occurs through mGluR3 receptors (belonging to group II glutamate receptors) coupled to inhibitory $G_{\alpha i/o}$-containing trimeric G protein (Sun et al., 2013). Upregulation of mGluR5 in adult animal and human brain astrocytes may occur in pathological conditions, e.g., astrocytic tumors (Aronica et al., 2003), amyotrophic lateral sclerosis (ALS) (Martorana et al., 2012) or in astrocytes surrounding amyloid plaques in Alzheimer's disease patients and in cultured astrocytes exposed to β-amyloid (Lim et al., 2013).

Purinergic signaling, mediated by purines and pyrimidines, mainly represented by ATP and adenosine, is widely expressed throughout the CNS (Burnstock and Verkhratsky, 2012; North and Verkhratsky, 2006). Astrocytes express both metabotropic P2Y and ionotropic P2X receptors. Astrocytes in culture express full panel of eight subtypes of P2Y purinergic receptors (Abbracchio and Ceruti, 2006; Butt, 2011; Weisman et al., 2012), of which $P2Y_1$, $P2Y_2$, $P2Y_4$, $P2Y_6$ and $P2Y_{11}$ receptors are coupled to $G_{\alpha q/11}$, PLC and ER Ca^{2+} release (Barańska et al., 2017; Erb and Weisman, 2012). At a protein and functional levels $P2Y_1$ receptors have been detected in astrocytes in white matter tracts and optic nerve (Hamilton et al., 2008; Morán-Jiménez and Matute, 2000). In cultured astrocytes ATP-evoked Ca^{2+} transients are mediated mainly by $P2Y_1$ and $P2Y_2$ receptors (Fam et al., 2003; Muller and Taylor, 2017). Over-expression of $P2Y_1$ receptors specifically in astrocytes in mice, did not contribute to spontaneous astrocytic Ca^{2+} signals in distal processes (Shigetomi et al., 2018). Inhibition of $P2Y_1$ receptors in an AD mouse model exerted beneficial effect on neuronal network function, learning and memory (Reichenbach et al., 2018).

Intracellular Ca^{2+} release channels residing in the endomembrane include $InsP_3$ receptors and ryanodine receptors (RyRs) (Berridge et al., 2003). There are some indications for RyR-mediated Ca^{2+}-induced

Ca^{2+} release in astrocytes in culture and in brain slices (Matyash et al., 2002; Pankratov and Lalo, 2015; Simpson et al., 1998a, 1998b). The RyR3 seems to be a predominant astrocytic subtype (Chai et al., 2017; Matyash et al., 2002). Immunoreactivity for $InsP_3$ receptors was widely detected in astrocytes (Hertle and Yeckel, 2007; Sharp et al., 1999) with $InsP_3$ receptor type 2 being suggested as a principal subtype in the forebrain (Holtzclaw et al., 2002; Sheppard et al., 1997) and in the cerebellar Bergmann glia (Tamamushi et al., 2012). In cultured astrocytes from rat and mouse hippocampus and mouse entorhinal cortex the expression of $InsP_3$ receptor type 1 has been identified at an mRNA and protein levels (Grolla et al., 2013a, 2013b). Presence of $InsP_3R$ type 1 has been confirmed in vivo by astrocyte-specific actively translated transcriptome RNA-sequencing (Chai et al., 2017). In the context of Alzheimer's disease, the Ca^{2+}-calcineurin (CaN)-dependent overexpression of $InsP_3$ receptor type 1 has been linked to a global remodeling of astrocytic Ca^{2+} signaling toolkit (Lim et al., 2014, 2016b).

Depletion of the ER Ca^{2+} store in astrocytes triggers SOCE, which was functionally detected in cultured astrocytes and in astrocytes in brain slices (Verkhratsky and Parpura, 2014). Canonical SOCE requires two molecular components, stromal interacting molecule (STIM) and Orai channels, although, transient receptor potential (TRP) channels are also subject of STIM regulation (Cheng et al., 2013; Salido et al., 2009). All components of the SOCE molecular machinery, including STIM1,2, Orai1,2,3 and TRP channels have been detected in astrocytes (Golovina, 2005; Grimaldi et al., 2003; Malarkey et al., 2008; Pizzo et al., 2001; Ronco et al., 2014). Experiments with anti-TRPC1 antibodies suggest pivotal role of TRP channels (Malarkey et al., 2008; Reyes et al., 2013; Verkhratsky et al., 2014; Verkhratsky and Nedergaard, 2018), however evidence indicating substantial role for ORAI channels in astroglial SOCE are on the rise (Kwon et al., 2017; Toth et al., 2019).

3.2.3.2 Ca^{2+} entry

Transcriptome of astrocytes of rodents and humans contains all subunits of ionotropic NMDA receptors (Cahoy et al., 2008; Lee et al., 2010; Orre et al., 2014; Zhang et al., 2014), while NMDA receptor proteins have been detected in astrocytes in vitro (Lee et al., 2010). The GluN1 and GluN2A/B have been localized in astroglial peripheral processes by immunogold electron microscopy (Aoki et al., 1994; Conti et al., 1999); whereas NMDA receptors immunoreactivity was detected in astrocytes from basolateral amygdala, from nucleus of the *stria terminalis* and from the *locus coeruleus* (Farb et al., 1995; Gracy and Pickel, 1995; Van Bockstaele and Colago, 1996). Stimulation of astrocytes with NMDA triggers inward currents sensitive to NMDA receptor blocker and positively potentiated by glycine (Lalo et al., 2006). Astroglial NMDA receptors demonstrate idiosyncratic biophysical properties: they have relatively low Ca^{2+} permeability ($P_{Ca}/P_{monovalent} \sim 3$), and very weak Mg^{2+} block at $-80\,mV$ (this Mg^{2+} block fully develops at hyperpolarised membrane potentials $\sim -120\,mV$). These properties and sensitivity to memantine and to GluN2C/D subunit-selective antagonist UBP141 indicate the heterotetrameric assembly from GluN1, GluN2 C or D, and GluN3 subunits (Lalo et al., 2011; Palygin et al., 2010; Verkhratsky and Chvátal, 2020). Physiological stimulation of astrocytes by activation of synaptic inputs triggers $[Ca^{2+}]_i$ elevations mediated by Ca^{2+} influx through NMDA receptors (Palygin et al., 2010).

Ionotropic P2X receptors are expressed in astrocytes, albeit at different levels and in a brain region-specific manner (Abbracchio and Ceruti, 2006; Butt, 2011; Erb and Weisman, 2012; Fumagalli et al., 2003; Kukley et al., 2001). Numerous reports suggest that $P2X_7$ receptors are widely expressed in astrocytes and contribute to astrocytic ionotropic Ca^{2+} response to ATP (Carrasquero et al., 2009; Illes et al., 2012; Nobile et al., 2003; Salas et al., 2013). In addition, an mRNA and functional expression of heteromeric $P2X_{1/5}$ receptors with Ca^{2+} permeability has been demonstrated in cortical astrocytes (Lalo et al., 2008; Palygin et al., 2010). Astroglial purinergic signaling is prominently involved in the pathophysiology of brain disorders from ischemia and trauma to neurodegeneration, major depression and bipolar disorder (Franke et al., 2012; Illes et al., 2016, 2019). In AD models, the ATP-induced Ca^{2+} signaling has been shown to be enhanced in astrocytes both in culture (Grolla et al., 2013a, 2013b) and in vivo (Delekate et al., 2014).

The brain α7-containing Ca^{2+}-permeable nicotinic acetylcholine receptors (nAChR) have been shown to mediate Ca^{2+} signals in astrocytes (Oikawa et al., 2005; Sharma and Vijayaraghavan, 2001; Talantova et al., 2013). Although functional expression of α7-nAChRs has been documented in mouse, rat and human astrocytes, subunits composition and Ca^{2+} permeability of these receptors/channels in vivo remains somewhat unclear as other nAChRs subunits, α3, α4 and β2, are also present in astrocytes (Duffy et al., 2011; Graham et al., 2002, 2003; Teaktong et al., 2003; Wang et al., 2012b).

Dopamine has been shown to have a biphasic effect on astroglial Ca^{2+} in hippocampal CA1 where initial $[Ca^{2+}]_i$ rise followed by a decrease in $[Ca^{2+}]_i$ below the baseline (Jennings et al., 2017). The D_1/D_2 dopamine receptors are responsible for initial $[Ca^{2+}]_i$ elevation, while $[Ca^{2+}]_i$ decrease depends solely on D_2 receptors. In olfactory bulb astrocytes, however, dopamine triggered monophasic $[Ca^{2+}]_i$ elevation mediated by both D_1 and D_2 receptors (Fischer et al., 2020). Synaptically released dopamine also triggers Ca^{2+} response in astrocytes in the nucleus accumbens (a reward center in the brain) of freely moving mice (Corkrum et al., 2020). These astrocytic Ca^{2+} responses are enhanced by amphetamine.

Astroglial $[Ca^{2+}]_i$ dynamics are tightly coupled with changes in cytoplasmic Na^+ concentration (Verkhratsky et al., 2018). Physiological stimulation triggers substantial transient elevations in $[Na^+]_i$, which represent a substrate for Na^+ signaling regulating numerous astroglial homoeostatic transporters (Kirischuk et al., 2012; Rose and Verkhratsky, 2016; Verkhratsky et al., 2020b). Particularly important is astroglial NCX, which, when operating in the reverse mode substantially contributes to Ca^{2+} influx especially in terminal leaflets (Kirischuk et al., 1997; Reyes et al., 2012; Rose et al., 2020; Takuma et al., 1994). The reversal potential for NCX is close to astroglial membrane potential and even small depolarization favor the reverse operational mode. In addition, the reversal of NCX is promoted by increase in $[Na^+]_i$ associated with glutamate uptake by Na^+-dependent astroglial transporters (Kirischuk et al., 2007) (Fig. 4).

3.3 Astrocytic Ca^{2+} signals display a remarkable diversity

Mechanical stimulation or application of metabotropic receptors agonists (glutamate, DHPG, ATP) to cultured astrocytes typically evokes a biphasic $[Ca^{2+}]_i$ transient composed of an initial $[Ca^{2+}]_i$ peak followed by either sustained $[Ca^{2+}]_i$ plateau or $[Ca^{2+}]_i$ oscillations which may last for minutes after stimulations (Charles et al., 1991; Verkhratsky and Kettenmann, 1996). Initial Ca^{2+} transient represents Ca^{2+} release from intracellular Ca^{2+} stores. The second phase of $[Ca^{2+}]_i$ response, represented by a plateau and/or oscillations, is also linked to Ca^{2+} entry from external milieu as they are suppressed in Ca^{2+}-free media. These events may have a more complex nature and require simultaneous or simultaneous activation of metabotropic and ionotropic receptors and SOCE channels (Hashioka et al., 2014; Papanikolaou et al., 2017; Ronco et al., 2014; Taheri et al., 2017).

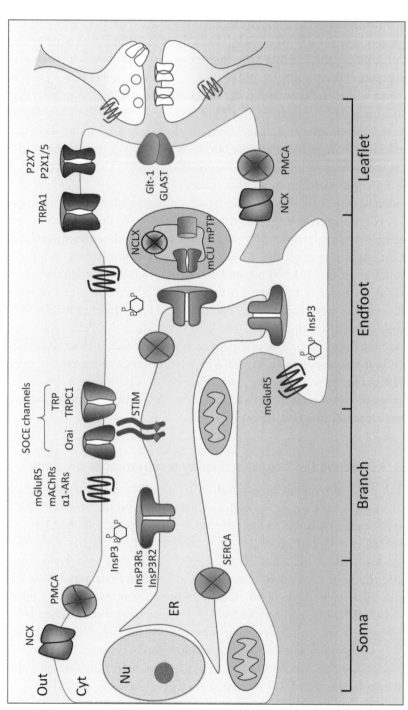

Fig. 4 See figure legend on opposite page.

Initiation of Ca^{2+} signal in an individual cultured astrocyte may trigger propagation of Ca^{2+} signal to adjacent cells in a form of a Ca^{2+} wave (Cornell-Bell and Finkbeiner, 1991; Enkvist and McCarthy, 1992; Finkbeiner, 1992; Fujii et al., 2017), which travels through astrocytic syncytium to a distance of over 100 µm with a speed of ∼10–20 µm/s (Bowser and Khakh, 2007; Dani et al., 1992; Verkhratsky and Kettenmann, 1996). Astrocytes communicate through gap junctions formed by connexons, formed by connexins Cx26, Cx30 and Cx43 expressed in astrocytes (Charvériat et al., 2017; De Bock et al., 2014). Propagation of the Ca^{2+} wave requires both functional $InsP_3$ receptors and Ca^{2+} entry from extracellular space and occurs via the diffusion of $InsP_3$ through gap junction and release of ATP (reviewed in Abbracchio and Ceruti, 2006; Butt, 2011; Scemes and Giaume, 2006). Intercellular astrocytic Ca^{2+} waves have been shown to be induced also by neuronal activity (Dani et al., 1992; Rizzoli et al., 2002; Rouach et al., 2000). However, Ca^{2+} waves propagating in astrocytic syncytium have not been always detected in brain slices (Porter and McCarthy, 1996). In freely moving animals, instead of propagating Ca^{2+} waves, synchronized Ca^{2+} activity of many astrocytes has been readily observed during locomotion (Bojarskaite et al., 2020; Dombeck et al., 2007), whisker (Sonoda et al., 2018) or visual stimulation (Stobart et al., 2018a). This activity relies on neuronal activation, hence, opening intriguing possibility that astrocytes can employ neuronal wiring for Ca^{2+} activity synchronization and propagation within astrocytic networks which are naturally devoid of fast signal propagation mechanisms.

Fig. 4 Astroglial Ca^{2+} signaling toolkit. Resting $[Ca^{2+}]_i$ is maintained by the activity of Ca^{2+} pumps plasma membrane Ca^{2+} ATPase (PMCA) and sarco-endoplasmic reticulum Ca^{2+} ATPase (SERCA) and Na^+/Ca^{2+} exchanger (NCX). NCX in astrocytes can work in a reverse mode, contributing to generation of intracellular Ca^{2+} signals. Activation of metabotropic receptors (mGluR5, mAChRs, α1-ARs) leads to generation of a diffusible second messenger inositol-1,4,5-trisphosphate ($InsP_3$) which, through activation to $InsP_3$ receptors ($InsP_3Rs$) located on the endomembrane, induces Ca^{2+} release from the endoplasmic reticulum (ER). Emptying of the ER activates entry of Ca^{2+} from extracellular milieu through the plasma membrane *via* store-operated Ca^{2+} entry mechanism (SOCE). $InsP_3R2$ is thought to be the principal $InsP_3R$ in astrocytes. Metabotropic Ca^{2+} release mechanism and SOCE operate mainly in soma and processes as branches, branchlets and endfeet endowed with the ER. In leaflets forming perisynaptic cradle and devoid of ER, Ca^{2+} signals are generated through Ca^{2+} entry through TRPA1 channels and ionotorpic glutamate (AMPA and NMDA), purinergic ($P2X_{1/5}$, $P2X_7$) and nicotinic cholinergic receptors and through reversed NCX. Mitochondria contribute to cytosolic Ca^{2+} signals by Ca^{2+} release through mitochondrial permeability transition pore (mPMT).

Spontaneous astrocytic $[Ca^{2+}]_i$ activity has been detected in brain slices from different brain regions (Di Castro et al., 2011; Haustein et al., 2014a; Nett et al., 2002; Panatier et al., 2011; Shigetomi et al., 2013a; Volterra et al., 2014) as well as in vivo in anesthetized (Kanemaru et al., 2014; Kuchibhotla et al., 2009; Mathiesen et al., 2013; Otsu et al., 2015; Stobart et al., 2018b; Wang et al., 2006) and awake animals (Ding et al., 2013; King et al., 2020; Nimmerjahn et al., 2009; Paukert et al., 2014; Srinivasan et al., 2015) including rats (Takata and Hirase, 2008) and ferrets (Schummers et al., 2008). Most of spontaneous Ca^{2+} activity occurs as localized transient elevations of $[Ca^{2+}]_i$ (Kanemaru et al., 2014). They are detected in branches, branchlets and endfeet and occur independently in the soma; somatic signals have lower frequency but higher amplitude and duration. Peripheral localized Ca^{2+} signals occur independently from electrical neuronal activity, but may be linked to neurotransmitter release (Di Castro et al., 2011; Panatier et al., 2011; Takata and Hirase, 2008). The spontaneous $[Ca^{2+}]_i$ events arise from plasmalemmal Ca^{2+} influx because: (i) they are present in branchlets and endfeet of InsP$_3$ receptor type 2 knockout mice; (ii) they are sensitive to inhibition of ionotropic receptors (Lalo et al., 2011; Palygin et al., 2010), TRP channels (Reyes et al., 2013; Shigetomi et al., 2013b) or NCX (Kirischuk et al., 1997; Ziemens et al., 2019); and (iii) they can be enhanced by elevation of extracellular Ca^{2+} (Wu et al., 2019). Additionally, Ca^{2+} release from flickering of the mitochondrial permeability transition pore has been suggested to contribute to spontaneous and evoked Ca^{2+} signals in astrocytic processes (Agarwal et al., 2017). Somatic signals are often observed upon stimulation and are generated by Ca^{2+} release form the ER through InsP$_3$ receptor type 2 (Ding et al., 2013; Srinivasan et al., 2015; Stobart et al., 2018b) with subsequent Ca^{2+} entry through store-operated pathway (Reyes et al., 2013; Toth et al., 2019).

Ca^{2+} signaling in astrocytes in vivo (in anesthetized and awake animals) may be induced by various stimuli, for example, startle response by whisker or paw stimulation or voluntary locomotion, frequently used in experiments on awake animals (Bindocci et al., 2017; Di Castro et al., 2011; Dombeck et al., 2007; Lind et al., 2013; Otsu et al., 2015; Petzold et al., 2008; Wang et al., 2006; Winship et al., 2007). Generally, evoked $[Ca^{2+}]_i$ responses occur with higher frequency, have higher amplitude and duration and are more synchronized. They may involve many astrocytes (all or most of cells in the microscopic field) and generate intracellular waves (which are not observed without stimulation) (Dombeck et al., 2007). Elevations of $[Ca^{2+}]_i$ starts in branchlets and branches and propagates centripetally

(Kanemaru et al., 2014). Evoked Ca^{2+} signals depend on frequency of neuronal firing, and, in anesthetized animals, are mediated by GABA and glutamate (Haustein et al., 2014a). Deletion of InsP$_3$ receptor type 2 eliminates only somatic ER-dependent Ca^{2+} signals (Kanemaru et al., 2014; Petravicz et al., 2014), while leaving signals in branchlets and endfeet nearly intact (Haustein et al., 2014b; Kanemaru et al., 2014; Sherwood et al., 2017; Srinivasan et al., 2015; Stobart et al., 2018a, 2018b), reflecting the prevalence of Ca^{2+} influx (Dunn et al., 2013; Rungta et al., 2016; Srinivasan et al., 2015). Stimulation of hippocampal mossy fiber pathway evokes Ca^{2+} signals in astrocytes linked to an uptake of glutamate by glial glutamate transporters, which most likely favors reversal of the NCX (Haustein et al., 2014b; Rothstein et al., 1996).

Coordination of region-wide and synchronized astrocytic $[Ca^{2+}]_I$ elevations in sensory cortex in response to sensory stimulation occurs through noradrenergic innervation by activation of α_1-adrenoceptors (Ding et al., 2013; Oe et al., 2020; Paukert et al., 2014) and cholinergic innervation by activation of muscarinic AChRs; both mechanisms are linked to the ER Ca^{2+} release through InsP$_3$ receptor type 2 (Chen et al., 2012; Takata et al., 2011). Coordination by adrenergic innervation leads to sensitization of astrocytes to local network activity (Paukert et al., 2014) through the release of ATP (Pankratov and Lalo, 2015) and/or phosphorylation/redistribution of Cx43 (Nuriya et al., 2018) during arousal and prolonged vigilance (Oe et al., 2020), which may mediate awake brain state transitions (Kjaerby et al., 2017). Noradrenergic input-induced Ca^{2+} signaling in zebrafish radial glia has been implicated in suppression of futile swim attempt (Mu et al., 2019), while in chicks astrocytic Ca^{2+} signals, induced by noradrenergic stimulation, have been implicated in memory consolidation (Gibbs and Bowser, 2010).

In cerebellar Bergmann glial cells local Ca^{2+} events occur in appendages—lateral emanations from primary radial glial processes, which extend leaflets to form perisynaptic processes contacting 50–70 synapses (formed by terminals of granule neurons on the Purkinje neuron); the appendages contain mitochondria (Grosche et al., 1999; Hoogland et al., 2009). In Bergmann glia Ca^{2+} signals arise through activation of an array of metabotropic receptors including mGluR5, P2YRs, endothelin$_B$ receptors, α_1-adrenoreceptors and H$_1$ histamine receptors (Kirischuk et al., 1995a, 1996, 1999; Paukert et al., 2014; Tuschick et al., 1997). Spontaneous intercellular Ca^{2+} waves have been observed in vivo in anesthetized animals (Hoogland et al., 2009; Mathiesen et al., 2013). In awake animals during locomotion Ca^{2+} waves invade hundreds of Bergmann glial cells (Nimmerjahn et al., 2009). Injection of ATP triggered Ca^{2+} wave in Bergmann glia syncytium suggesting that propagation of the wave is mediated by paracrine release of ATP and activation of P2Y receptors (Hoogland et al., 2009; Mathiesen et al., 2013).

In Müller cells, the radial astrocytes of the retina, purinergic signaling through both metabotropic and ionotropic receptors are responsible for generation of Ca^{2+} signals (Reichenbach and Bringmann, 2016). Müller cells generate spontaneous and light-evoked Ca^{2+} transients as well as intercellular Ca^{2+} waves (Metea and Newman, 2006; Newman, 2001; Newman and Zahs, 1997). Ca^{2+} entry through non-selective Ca^{2+}/Na^{+}-permeable TRPV4 channels have been implicated in volume regulation (Jo et al., 2015). Müller cell endfeet contain densely packed stack-organized ER cisternae expressing high amounts of STIM1. Activation of metabotropic pathway, with consequent Ca^{2+} release from the ER, potently activates SOCE, which is mediated by synergistic activity of Orai and TRPC1/3 channels (Molnár et al., 2016). Ablation of TRPC-mediated SOCE enhanced reactive gliosis in a mouse glaucoma model (Molnár et al., 2016).

In summary, protoplasmic and radial astrocytes share general design of intracellular Ca^{2+} events: multiple, frequent and low intensity spontaneous localized events in distal processes, which are mediated by Ca^{2+} fluxes across the plasma membrane (protoplasmic astrocytes) or ER Ca^{2+} release and store-operated Ca^{2+} entry (radial astrocytes) (Wu et al., 2019). These events are often triggered by neurotransmitters. In soma and primary branches Ca^{2+} events are larger, less frequent and are initiated by the release of Ca^{2+} from the ER through InsP$_3$ receptor type 2 in response to stimulation of metabotropic receptors, with consequent SOCE-mediated Ca^{2+} entry (Fig. 5). In addition, Ca^{2+} signals in astrocytes display significant intra- and inter-brain region specificity, reflecting their morphological diversity and neural circuit specialization (Khakh and Deneen, 2019; Verkhratsky and Nedergaard, 2018).

3.4 Functional significance of Ca^{2+} signals in astrocytes

Diversity and heterogeneity of Ca^{2+} signals in astrocytes reflect multiple regulatory roles of $[Ca^{2+}]_i$ in both generic cellular activities and in astrocyte-specific functions. Generic functions (which are similar in all cells),

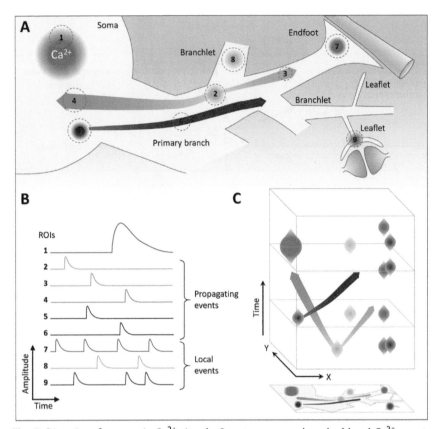

Fig. 5 Diversity of astrocytic Ca^{2+} signals. Spontaneous and evoked local Ca^{2+} events are frequent in peripheral districts of astrocytic processes like branches, branchlets, endfeet and leaflets. Peripheral Ca^{2+} events are independent of somatic Ca^{2+} signals which are less frequent and have higher amplitude and duration. Ca^{2+} waves can propagate both centrifugally and centripetally. (A) Color code: pink, blue and yellow spots represent local Ca^{2+} events; big red spot represents large somatic Ca^{2+} event; bidirectional (centrifugal and centripetal) propagation of Ca^{2+} wave originated in the branch; violet spot and arrow represent centrifugally propagating Ca^{2+} wave originated in soma. (B) The time course of fluorescent Ca^{2+} signal recorded in circled regions-of-interest (ROI) on panel A. Ca^{2+} transients detected in different ROIs represent both local Ca^{2+} events and invasion of propagating Ca^{2+} events. (C) Schematic X-Y-Time representation of Ca^{2+} events shown on panel A and traced on panel B.

include the cellular Ca^{2+} homeostasis (see Section 3.2), Ca^{2+} regulation of metabolism and bioenergetics (Rossi et al., 2019), autophagy and proteostasis (Bootman et al., 2018), gene transcription (Naranjo and Mellström, 2012; Oh-hora and Rao, 2008), motility and excitation-secretion coupling

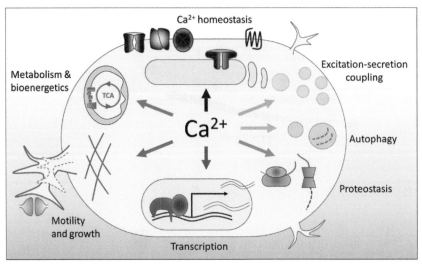

Fig. 6 Generic functions of intracellular Ca^{2+} signals in astroglial cells. Generic Ca^{2+}-regulated processes in astrocytes include transcriptional and functional regulation of cellular Ca^{2+} signaling toolkit, excitation-secretion coupling, autophagy, proteostasis, gene transcription, motility and growth, metabolism and cellular bioenergetics.

(Vardjan et al., 2016) and cellular morphological remodeling (Molotkov et al., 2013; Tanaka et al., 2013) (Fig. 6). Different aspects of these functions are tailored by cell- and context-specific gene and protein expression (Chai et al., 2017).

In the context of CNS homeostasis and activity, the primary role of astrocytic Ca^{2+} signals is the perception (see Section 3.2.2) and decoding of the information about the activity outside these cells and transmission of this information to adapt cellular biochemistry and physiology to satisfy the nervous tissue homeostatic, energetic and signaling demands. Downstream effects of Ca^{2+} signals include (in order of complexity) Ca^{2+}-regulated secretion, modulation of synaptic transmission, regulation of cerebral hemodynamic, systemic blood pressure (Marina et al., 2020) and contribution to higher brain functions including respiratory control, sleep, locomotion, memory and cognition (Fig. 7). Evidence for the role of astrocytes and astrocytic Ca^{2+} in these processes has been extensively discussed in several reviews (Augusto-Oliveira et al., 2020; Khakh and Sofroniew, 2015; Oliveira et al., 2015; Santello et al., 2019). A concept of spatiotemporal Ca^{2+} activity pattern in astrocytic network serving as guiding template for neuronal network activity has been proposed (Semyanov, 2008, 2019; Semyanov et al., 2020).

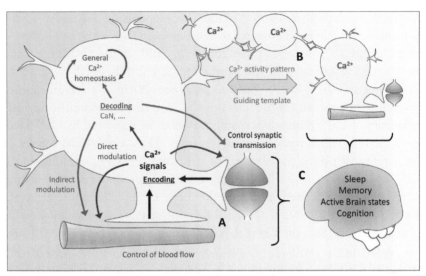

Fig. 7 Cell-specific functions of astrocytic Ca²⁺ signals. Functions regulated by astrocytic Ca²⁺ signals can be conditionally classified by (i) the complexity of the regulated phenomena and number of cells/circuits involved and (ii) direct/indirect regulation by Ca²⁺ ions. The first classification includes, in order of complexity, such functions as Ca²⁺-regulated secretion, modulation of synaptic transmission, regulation of local hemodynamic (A); the second order of complexity may include Ca²⁺ activity pattern in astrocytic syncytium and generations of guiding templates for neuronal networks (B); the third order of complexity include control of systemic blood pressure and contribution to higher brain functions including respiratory control, sleep, locomotion, memory and cognition (C). The second classification considers processes directly regulated by Ca²⁺ ions (violet arrows and text) like vesicular exocytosis (gliocrine function) and regulation of the activity of ion channels and transporters; while indirect regulation implicates decodification of the spatiotemporal pattern of [Ca2 +]ᵢ (blue arrows and text) by calmodulin and other Ca²⁺-binding enzymes (red arrows and text). To date, only calcineurin (CaN) has been identified in astrocytes as a Ca²⁺-sensitive molecular switch signals.

However, the mechanistic details of the Ca^{2+} regulation remain largely unexplored and the uncertainty of specific mechanism(s), and the "degree of freedom" in the mechanistic interpretation, increases with the complexity of the studied phenomena.

The best studied primary processes, in which Ca^{2+} ions exert direct modulation are exemplified by the Ca^{2+}-mediated secretion and the regulation of ion transporters and channels. Secretory function of astrocytes (gliosecretion or gliocrine function) implies a SNARE-23 (soluble *N*-ethylmaleimide-sensitive-factor attachment protein receptor 23)-mediated exocytosis.

Although mechanism and timing of the Ca^{2+}-dependent vesicular exocytosis in astrocytes differs as compared to neurons, being slower by two orders of magnitude, Ca^{2+} ions are required for the assembly of the SNARE zip and for the vesicle fusion with the plasma membrane (discussed in details in Vardjan et al., 2016, 2019; Vardjan and Zorec, 2015). Ca^{2+} regulation of ion transport, primarily of the fluxes of Ca^{2+} ion itself, exploits the intrinsic properties of Ca^{2+} channels and transporters to be regulated by Ca^{2+} at all levels (activity, transcription and localization). Thus, Ca^{2+}/CaM control of the activity of plasma membrane Ca^{2+} ATPase (PMCA), and differential Ca^{2+}-dependence of the $InsP_3$ receptors and ryanodine receptors gating have been characterized in details (Di Leva et al., 2008; Joseph et al., 2005; Yamaguchi, 2020). Ca^{2+} signaling is intimately linked to homeostasis of other ions (see Section 3.2) with functional interaction between Ca^{2+} and Na^+ signaling representing, perhaps, the best studied example (Verkhratsky et al., 2018, 2020b). In addition, Ca^{2+} regulates synaptic K^+ clearance by astrocytes (Tapella et al., 2020a; Wang et al., 2012a), and K^+ homeostasis regulates synaptic transmission (Lebedeva et al., 2018; Shih et al., 2013) and neuronal network function.

Several signaling cascades, such as, for example, Ca^{2+}-regulated gene transcription and context-dependent regulation of phosphorylation cascades, require participation of Ca^{2+}-activated molecular switches like Ca^{2+}/calmodulin (CaM)-activated kinase II (CaMKII) or calcineurin (CaN) phosphatase, with well-established neuronal function (Baumgärtel and Mansuy, 2012; Bayer and Schulman, 2019; Naranjo and Mellström, 2012). In astrocytes, little is known about processes which require decoding of Ca^{2+} signals by Ca^{2+}-binding proteins as well as about how Ca^{2+} signals in astrocytes are decoded and processed to be translated into a plethora of general and astrocyte-specific homeostatic and signaling responses. Expression of the universal Ca^{2+} detector, CaM, has been documented in astrocytes (Chai et al., 2017), however the information about its downstream effectors and links to astroglial function remains limited. Hypothermia-induced expression of AQP4 (Salman et al., 2017), activation of volume-regulated anion channel (VRAC) conductance and VRAC-mediated release of taurine has been shown to be regulated by Ca^{2+}-CaM (Li et al., 2002; Mongin et al., 1999; Olson et al., 2004). Two most studied effectors of Ca^{2+}-CaM signaling are CaMKII and phosphatase CaN. Expression and activity of CaMKII has been detected in astrocytes (Babcock-Atkinson et al., 1989; Bronstein et al., 1988;

Neary et al., 1986) and CaMKII activity in cultured astrocytes has been associated with release of inflammatory mediators (Watterson et al., 2001), InsP$_3$-kinase cascade (Communi et al., 1999) and functional expression of glutamate transporters (Ashpole et al., 2013; Chawla et al., 2017; Underhill et al., 2015). Activation of CaMKII in astrocytes has been also shown to occur following stimulation of endothelin (Kubes et al., 1998), thrombin (Lin et al., 2013) and P2X$_7$ receptors (Oliveira et al., 2019). Whether CaMKII exerts these functions in astrocytes in vivo remains an open question.

Activation of CaN in astrocytes has been demonstrated in several pathological conditions including Alzheimer's disease, experimental dementia and brain trauma (Dos Santos et al., 2020; Furman and Norris, 2014; Sompol and Norris, 2018). In pathology, CaN modulation involves transcriptional remodeling through NFAT and NF-κB. The CaN-NFAT axis has been shown to drive reactive astrogliosis and neuroinflammation (Furman and Norris, 2014; Sompol and Norris, 2018). Interaction of CaN with NF-κB pathway in Alzheimer's disease is complex and context-dependent and may be both pro- and anti-inflammatory (Fernandez et al., 2007, 2012; Lim et al., 2016a). Finally, NF-κB mediates oligomeric Aβ-induced astrocytic remodeling and secretion (Lim et al., 2013; Tapella et al., 2018).

Information about physiological role of astrocytic CaN remains limited. Conditional deletion of CaN in GFAP-expressing astrocytes results in the impairment of neuronal excitability in both hippocampal pyramidal and cerebellar granule neurons and inhibition of astrocytic Na$^+$/K$^+$ ATPase (NKA) (Tapella et al., 2020b). Deletion of CaN in astrocytes in culture results in an increase of [Na$^+$]$_i$ and reduction of secretory response (Dematteis et al., 2020a; Tapella et al., 2020b). In human and mouse cultured astrocytes, pharmacological or genetic inhibition of CaN dynamically regulates expression of GLAST/EAAT1 at a protein, but not at mRNA level (Dematteis et al., 2020a). Neuronal activity-dependent activation of astrocytic CaN has been also demonstrated in hippocampal co-cultures (Lim et al., 2018).

In summary, in spite of a substantial progress in our understanding of the generation and spatiotemporal properties of Ca^{2+} signals in astrocytes, their functional significance, particularly in context of complex and/or systemic brain activities, is studied largely at a phenomenological level, while mechanisms acting downstream of astrocytic Ca^{2+} signals remain poorly understood (Fig. 7).

4. Ca^{2+} signaling in oligodendrocytes

Ca^{2+} signaling plays important role in differentiation of oligodend-roglial precursor cells; in particular Ca^{2+} signals allow the OPCs to locate active axons and initiate myelination. Many oligodendroglial precursor cells receive glutamatergic synaptic inputs (Bergles et al., 2010), or are stimulated by glutamate or ATP released from firing axons (Haberlandt et al., 2011; Hamilton et al., 2010). The OPC/NG2 glial cells express AMPA receptors with some (although minor) Ca^{2+} permeability; activation of these receptors produces depolarization, which opens voltage-gated Ca^{2+} channels expressed in OPCs (Blankenfeld et al., 1992; Kirischuk et al., 1995b; Verkhratsky et al., 1990). Oligodendroglial precursors express low and high-threshold voltage-gated Ca^{2+} channels (Verkhratsky et al., 1990); the low-threshold channels are almost entirely concentrated in the tips of cellular processes and can be activated even by a moderate increase in extracellular [K$^+$], which are compatible with K$^+$ release from firing axons (Kirischuk et al., 1995b). Activation of these low-threshold Ca^{2+} channels triggers local Ca^{2+} signals which may represent another mechanism for processes guidance towards functionally active axons (Fig. 8). In addition, oligodendroglial precursors express NMDA receptors (Wang et al., 1996), functional role of which remains unclear, although excessive activation of these receptors may trigger pathological Ca^{2+} influx responsible for high vulnerability of oligodendroglial precursors in ischemic conditions (Matute et al., 2007). Oligodendroglial precursors also express purinoceptors which initiate [Ca^{2+}]$_i$ increases and induce myelination (Ishibashi et al., 2006).

A wide range of neurotransmitters, neurohormones and growth factors trigger Ca^{2+} signaling in oligodendrocytes (Butt et al., 2014). Similarly to astrocytes, oligodendrocytic Ca^{2+} signaling in generated by plasmalemmal Ca^{2+} influx or by release from intracellular stores, or by combination of both. Ca^{2+} influx in oligodendrocytes is mediated by an array of Ca^{2+} permeable ionotropic receptors, including NMDA receptors (Salter and Fern, 2005) and P2X receptors with prominent role for P2X$_7$ receptor (Hamilton et al., 2008). Mature oligodendrocytes are also endowed with metabotropic receptors to glutamate and to ATP, with high expression of P2Y$_2$ receptors, with intracellular Ca^{2+} channels represented by InsP$_3$ and ryanodine receptors (Verkhratsky and Butt, 2013) and with store-operated Ca^{2+} entry mediated by ORAI and TRPM channels (Papanikolaou et al., 2017).

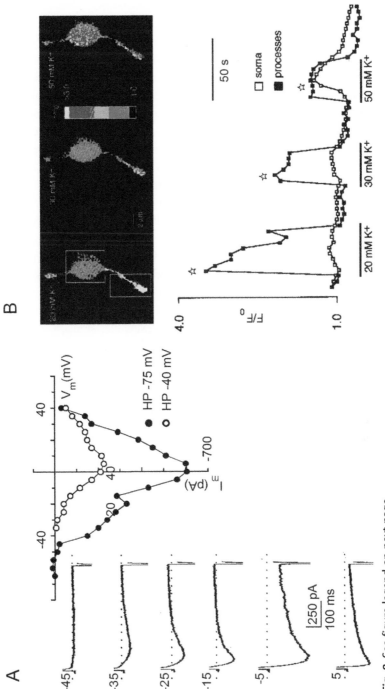

Fig. 8 See figure legend on next page.

5. Ca²⁺ signaling in microglia

The fetal macrophages, which are microglial precursors, after entering the CNS undergo a remarkable metamorphosis. These macrophages completely change their morphology and acquire "neuron-like" appearance with long ramified and constantly moving processes (Kettenmann et al., 2011). Furthermore, microglial cells express an extended repertoire of receptors, which include multiple receptors to neurotransmitters and neuromodulators as well as "immunological" receptors characteristic for myeloid cells; this expended receptor pattern arguably makes microglial cells the most "receptive" cells of the CNS (see Fig. 9 and Kettenmann et al., 2011; Pocock and Kettenmann, 2007; Ransohoff and Perry, 2009; Verkhratsky et al., 2015).

Many of these receptors are linked to Ca^{2+} signaling pathways; while Ca^{2+} signaling in microglia to a large extend controls microgliosis and pathological remodeling of these cells, which contribute to most, if not to all, neurological diseases (Brawek and Garaschuk, 2013; Ransohoff and Perry, 2009). Microglial cells are electrically non-excitable, similarly to all other glial cells, and their basic mechanism of Ca^{2+} signal generation relies on plasmalemmal Ca^{2+} influx and intercellular Ca^{2+} release from the ER. This

Fig. 8 Voltage-gated Ca^{2+} currents and related Ca^{2+} signals in cells of oligodendroglial lineage. (A) Current traces recorded from single cultured oligodendrocyte in response to depolarisation voltage steps from holding potential of −75 mV in a 90 mM Ba^{2+}-containing solution. Right panel shows current-voltage relations obtained from current recording from a single cultured oligodendrocyte at holding potentials (HP) -75 mV and −40 mV revealing activation of low- and high-voltage activated (T- and L-) Ca^{2+} channels. (B) Stimulation of cultured oligodendrocite precursor cell with different extarcellular K+ concentrations results in spatially distinct Ca^{2+} signals. Top panel shows pseudocolour images obtained from a precursor cell. The cell was superfused with three different [K^+] as indicated. Challenge with 20 mM [K^+] resulted in [Ca^{2+}]$_i$ increase only in the tips of processes, while stimulation with 50 mM [K^+] induced [Ca^{2+}]$_i$ increase across the entire cell. Lower panel shows time lapse [Ca^{2+}]$_i$ recordings from the processes and the soma (the regions of interest are delineated by squared at the top panel) in response to different [K^+]. Stars correspond to the images shown in the top panel. *Modified and reproduced, with permission, from panel A: Blankenfeld GV, Verkhratsky, A.N., Kettenmann, H., 1992. Ca2+ channel expression in the oligodendrocyte lineage. Eur. J. Neurosci. 4, 1035–1048. https://doi.org/10.1111/j.1460-9568.1992.tb00130.x and panel B: Kirischuk, S., Scherer, J., Möller, T., Verkhratsky, A., Kettenmann, H., 1995b. Subcellular heterogeneity of voltage-gated Ca2+ channels in cells of the oligodendrocyte lineage. Glia 13, 1–12. https://doi.org/10.1002/glia.440130102.*

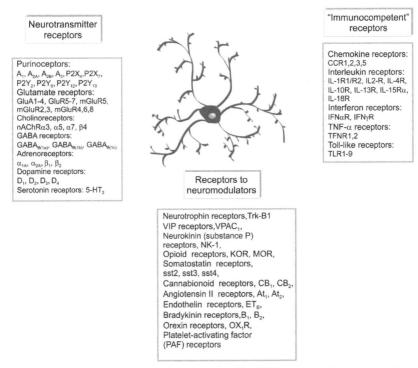

Fig. 9 Multiple receptors expressed in microglial cell (see text for further details).

latter mechanism is predominant and stimulation of numerous metabotropic receptors expressed in microglial cells triggers opening of ER Ca^{2+} channels. Microglial cells express both $InsP_3$ receptors and RyRs, although $InsP_3$ receptors seem to play the leading role (Kettenmann et al., 2011; Klegeris et al., 2007; Shideman et al., 2006).

Microglia express voltage-gated Ca^{2+} channels of $Ca_v1.2$ and $Ca_v2.2$ subtypes (Wang et al., 2019) although their exact contribution of $[Ca^{2+}]_i$ dynamics remains to be elucidated. Prominent Ca^{2+} entry pathway in microglial cells is associated with $P2X_7$ receptors, which are primary players in microglial pathophysiology (He et al., 2017). Microglial cells express TRPV, TRPM, and TRPC channels, which contribute to cytokine production, proliferation, osmotic regulation, and oxidative stress responses and microgliosis (Echeverry et al., 2016). The TRPV1 channels are localized in microglial mitochondrial membrane and are related to regulation of cellular migration (Miyake et al., 2015). Microglial cells have rather prominent SOCE, which is mediated by Stim1, Stim2 and Orai1 (Michaelis et al., 2015). Finally, microglial cells express NCX, which as a rule is operating in the reverse mode, because microglial membrane potential of -20 to $-50\,mV$ is quite far from NCX reversal potential; the NCX-mediated Ca^{2+} influx contributes to regulation of cell migration (Noda et al., 2013).

6. Conclusions

Progress, made during last decades in the field of gliobiology, revealed an unprecedented morpho-functional complexity of glial cells. Specifically, this holds true for glial Ca^{2+} signaling. Rich repertoire of membrane receptors and channels, linked to the generation of intracellular Ca^{2+} signals, warrants in depth monitoring of all aspects of physiological activity in the CNS, and the encoding of the information which is then translated into context-dependent developmental, homeostatic, signaling and defensive responses. Astrocytes are the most intensively studied neuroglial cell type, partly due to their relative abundance in the CNS and the ease to be studied in culture, but mostly due to their central role in the CNS health and disease. However, in spite of significant technological advance, reliable detection of astrocytic Ca^{2+} signals in intact brain tissue remains challenging (Semyanov et al., 2020), and even more so the interpretation of their spatiotemporal properties in terms of significance for synaptic transmission and other CNS activities, which unleashed intensive debates (see for example, Fiacco and McCarthy, 2018; Savtchouk and Volterra, 2018). In this complex and rapidly developing scenario, much effort is being dedicated to the detection of Ca^{2+} signals themselves, while very little is known on how spatiotemporal patterns of Ca^{2+} signals are decoded and translated into cellular activities. Without investigation of this latter aspect, our understanding of physiology and pathology of Ca^{2+} signaling in glial cells remains incomplete.

Acknowledgments

This work had the following financial supports: D.L. was supported by grant 2014-1094 from the Fondazione Cariplo and grants DSF-FAR-2016 and DSF-FAR-2019 from The Università del Piemonte Orientale. A.S. was supported by Russian Science Foundation grant 20-14-00241.

References

Abbracchio, M.P., Ceruti, S., 2006. Roles of P2 receptors in glial cells: focus on astrocytes. Purinergic Signal 2, 595–604. https://doi.org/10.1007/s11302-006-9016-0.
Achucarro, N., 1919. Some pathological findings in the neuroglia and in the ganglion cells of the cortex in senile conditions. Bull. Gov. Hosp. Insane 2, 81–90.

Agarwal, A., Wu, P.-H., Hughes, E.G., Fukaya, M., Tischfield, M.A., Langseth, A.J., Wirtz, D., Bergles, D.E., 2017. Transient opening of the mitochondrial permeability transition pore induces microdomain calcium transients in astrocyte processes. Neuron 93, 587–605.e7. https://doi.org/10.1016/j.neuron.2016.12.034.

Alonso, M.T., Barrero, M.J., Michelena, P., Carnicero, E., Cuchillo, I., García, A.G., García-Sancho, J., Montero, M., Alvarez, J., 1999. Ca^{2+}-induced Ca^{2+} release in chromaffin cells seen from inside the ER with targeted aequorin. J. Cell Biol. 144, 241–254. https://doi.org/10.1083/jcb.144.2.241.

Alzheimer, A., 1910. Beiträge zur Kenntnis der pathologischen Neuroglia und ihrer Beziehungen zu den Abbauvorgängen im Nervengewebe. In: Nissl, F., Alzheimer, A. (Eds.), Histologische und histopathologische Arbeiten über die Grosshirnrinde mit besonderer Berücksichtigung der pathologischen Anatomie der Geisteskrankheiten. Gustav Fischer, Jena, pp. 401–562.

Andriezen, W.L., 1893. The neuroglia elements in the human brain. Br. Med. J. 2, 227–230. https://doi.org/10.1136/bmj.2.1700.227.

Aoki, C., Venkatesan, C., Go, C.G., Mong, J.A., Dawson, T.M., 1994. Cellular and subcellular localization of NMDA-R1 subunit immunoreactivity in the visual cortex of adult and neonatal rats. J. Neurosci. 14, 5202–5222.

Araque, A., Martín, E.D., Perea, G., Arellano, J.I., Buño, W., 2002. Synaptically released acetylcholine evokes Ca^{2+} elevations in astrocytes in hippocampal slices. J. Neurosci. 22, 2443–2450. https://doi.org/20026212.

Aronica, E., Gorter, J.A., Ijlst-Keizers, H., Rozemuller, A.J., Yankaya, B., Leenstra, S., Troost, D., 2003. Expression and functional role of mGluR3 and mGluR5 in human astrocytes and glioma cells: opposite regulation of glutamate transporter proteins. Eur. J. Neurosci. 17, 2106–2118. https://doi.org/10.1046/j.1460-9568.2003.02657.x.

Ashpole, N.M., Chawla, A.R., Martin, M.P., Brustovetsky, T., Brustovetsky, N., Hudmon, A., 2013. Loss of calcium/calmodulin-dependent protein kinase II activity in cortical astrocytes decreases glutamate uptake and induces neurotoxic release of ATP. J. Biol. Chem. 288, 14599–14611. https://doi.org/10.1074/jbc.M113.466235.

Augustine, G.J., Santamaria, F., Tanaka, K., 2003. Local calcium signaling in neurons. Neuron 40, 331–346. https://doi.org/10.1016/s0896-6273(03)00639-1.

Augusto-Oliveira, M., Arrifano, G.P., Takeda, P.Y., Lopes-Araújo, A., Santos-Sacramento, L., Anthony, D.C., Verkhratsky, A., Crespo-Lopez, M.E., 2020. Astroglia-specific contributions to the regulation of synapses, cognition and behaviour. Neurosci. Biobehav. Rev. 118, 331–357. https://doi.org/10.1016/j.neubiorev.2020.07.039.

Babcock-Atkinson, E., Norenberg, M.D., Norenberg, L.O., Neary, J.T., 1989. Calcium/calmodulin-dependent protein kinase activity in primary astrocyte cultures. Glia 2, 112–118. https://doi.org/10.1002/glia.440020207.

Barańska, J., Czajkowski, R., Pomorski, P., 2017. P2Y1 receptors–properties and functional activities. Adv. Exp. Med. Biol. 1051, 71–89. https://doi.org/10.1007/5584_2017_57.

Bardehle, S., Krüger, M., Buggenthin, F., Schwausch, J., Ninkovic, J., Clevers, H., Snippert, H.J., Theis, F.J., Meyer-Luehmann, M., Bechmann, I., Dimou, L., Götz, M., 2013. Live imaging of astrocyte responses to acute injury reveals selective juxtavascular proliferation. Nat. Neurosci. 16, 580–586. https://doi.org/10.1038/nn.3371.

Bartol, T.M., Bromer, C., Kinney, J., Chirillo, M.A., Bourne, J.N., Harris, K.M., Sejnowski, T.J., 2015. Nanoconnectomic upper bound on the variability of synaptic plasticity. Elife 4, e10778. https://doi.org/10.7554/eLife.10778.

Baumgärtel, K., Mansuy, I.M., 2012. Neural functions of calcineurin in synaptic plasticity and memory. Learn. Mem. 19, 375–384. https://doi.org/10.1101/lm.027201.112.

Bayer, K.U., Schulman, H., 2019. CaM Kinase: still Inspiring at 40. Neuron 103, 380–394. https://doi.org/10.1016/j.neuron.2019.05.033.

Bergles, D.E., Jabs, R., Steinhäuser, C., 2010. Neuron-glia synapses in the brain. Brain Res. Rev. 63, 130–137. https://doi.org/10.1016/j.brainresrev.2009.12.003.

Bernardinelli, Y., Muller, D., Nikonenko, I., 2014. Astrocyte-synapse structural plasticity. Neural Plast. 2014, 232105. https://doi.org/10.1155/2014/232105.

Berridge, M.J., Bootman, M.D., Roderick, H.L., 2003. Calcium signalling: dynamics, homeostasis and remodelling. Nat. Rev. Mol. Cell Biol. 4, 517–529. https://doi.org/10.1038/nrm1155.

Bindocci, E., Savtchouk, I., Liaudet, N., Becker, D., Carriero, G., Volterra, A., 2017. Three-dimensional Ca^{2+} imaging advances understanding of astrocyte biology. Science 356, eaai8185. https://doi.org/10.1126/science.aai8185.

Blankenfeld, G.V., Verkhratsky, A.N., Kettenmann, H., 1992. Ca^{2+} channel expression in the oligodendrocyte lineage. Eur. J. Neurosci. 4, 1035–1048. https://doi.org/10.1111/j.1460-9568.1992.tb00130.x.

Bojarskaite, L., Bjørnstad, D.M., Pettersen, K.H., Cunen, C., Hermansen, G.H., Åbjørsbråten, K.S., Chambers, A.R., Sprengel, R., Vervaeke, K., Tang, W., Enger, R., Nagelhus, E.A., 2020. Astrocytic Ca2+ signaling is reduced during sleep and is involved in the regulation of slow wave sleep. Nat. Commun. 11, 3240. https://doi.org/10.1038/s41467-020-17062-2.

Bootman, M.D., Chehab, T., Bultynck, G., Parys, J.B., Rietdorf, K., 2018. The regulation of autophagy by calcium signals: do we have a consensus? Cell Calcium 70, 32–46. https://doi.org/10.1016/j.ceca.2017.08.005.

Bowser, D.N., Khakh, B.S., 2007. Vesicular ATP is the predominant cause of intercellular calcium waves in astrocytes. J. Gen. Physiol. 129, 485–491. https://doi.org/10.1085/jgp.200709780.

Brawek, B., Garaschuk, O., 2013. Microglial calcium signaling in the adult, aged and diseased brain. Cell Calcium 53, 159–169. https://doi.org/10.1016/j.ceca.2012.12.003.

Bronstein, J., Nishimura, R., Lasher, R., Cole, R., de Vellis, J., Farber, D., Wasterlain, C., 1988. Calmodulin kinase II in pure cultured astrocytes. J. Neurochem. 50, 45–49. https://doi.org/10.1111/j.1471-4159.1988.tb13227.x.

Buffo, A., Rossi, F., 2013. Origin, lineage and function of cerebellar glia. Prog. Neurobiol. 109, 42–63. https://doi.org/10.1016/j.pneurobio.2013.08.001.

Burnstock, G., Verkhratsky, A., 2012. Purinergic Signalling and the Nervous System. Springer-Verlag, Berlin Heidelberg. https://doi.org/10.1007/978-3-642-28863-0.

Bushong, E.A., Martone, M.E., Jones, Y.Z., Ellisman, M.H., 2002. Protoplasmic astrocytes in CA1 stratum radiatum occupy separate anatomical domains. J. Neurosci. 22, 183–192.

Butt, A.M., 2011. ATP: a ubiquitous gliotransmitter integrating neuron-glial networks. Semin. Cell Dev. Biol. 22, 205–213. https://doi.org/10.1016/j.semcdb.2011.02.023.

Butt, A.M., Colquhoun, K., Tutton, M., Berry, M., 1994. Three-dimensional morphology of astrocytes and oligodendrocytes in the intact mouse optic nerve. J. Neurocytol. 23, 469–485. https://doi.org/10.1007/BF01184071.

Butt, A.M., Fern, R.F., Matute, C., 2014. Neurotransmitter signaling in white matter. Glia 62, 1762–1779. https://doi.org/10.1002/glia.22674.

Butt, A.M., Papanikolaou, M., Rivera, A., 2019. Physiology of oligodendroglia. Adv. Exp. Med. Biol. 1175, 117–128. https://doi.org/10.1007/978-981-13-9913-8_5.

Cahoy, J.D., Emery, B., Kaushal, A., Foo, L.C., Zamanian, J.L., Christopherson, K.S., Xing, Y., Lubischer, J.L., Krieg, P.A., Krupenko, S.A., Thompson, W.J., Barres, B.A., 2008. A transcriptome database for astrocytes, neurons, and oligodendrocytes: a new resource for understanding brain development and function. J. Neurosci. 28, 264–278. https://doi.org/10.1523/JNEUROSCI.4178-07.2008.

Carafoli, E., Krebs, J., 2016. Why calcium? How calcium became the best communicator. J. Biol. Chem. 291, 20849–20857. https://doi.org/10.1074/jbc.R116.735894.

Carrasquero, L.M.G., Delicado, E.G., Bustillo, D., Gutiérrez-Martín, Y., Artalejo, A.R., Miras-Portugal, M.T., 2009. P2X7 and P2Y13 purinergic receptors mediate intracellular calcium responses to BzATP in rat cerebellar astrocytes. J. Neurochem. 110, 879–889. https://doi.org/10.1111/j.1471-4159.2009.06179.x.

Chai, H., Diaz-Castro, B., Shigetomi, E., Monte, E., Octeau, J.C., Yu, X., Cohn, W., Rajendran, P.S., Vondriska, T.M., Whitelegge, J.P., Coppola, G., Khakh, B.S., 2017. Neural circuit-specialized astrocytes: transcriptomic, proteomic, morphological, and functional evidence. Neuron 95, 531–549.e9. https://doi.org/10.1016/j.neuron.2017.06.029.

Chan-Palay, V., Palay, S.L., 1972. The form of velate astrocytes in the cerebellar cortex of monkey and rat: high voltage electron microscopy of rapid Golgi preparations. Z. Anat. Entwicklungsgesch. 138, 1–19. https://doi.org/10.1007/BF00519921.

Charles, A.C., Merrill, J.E., Dirksen, E.R., Sanderson, M.J., 1991. Intercellular signaling in glial cells: calcium waves and oscillations in response to mechanical stimulation and glutamate. Neuron 6, 983–992. https://doi.org/10.1016/0896-6273(91)90238-u.

Charvériat, M., Naus, C.C., Leybaert, L., Sáez, J.C., Giaume, C., 2017. Connexin-dependent Neuroglial networking as a new therapeutic target. Front. Cell. Neurosci. 11, 174. https://doi.org/10.3389/fncel.2017.00174.

Chawla, A.R., Johnson, D.E., Zybura, A.S., Leeds, B.P., Nelson, R.M., Hudmon, A., 2017. Constitutive regulation of the glutamate/aspartate transporter EAAT1 by calcium-calmodulin-dependent protein kinase II. J. Neurochem. 140, 421–434. https://doi.org/10.1111/jnc.13913.

Chen, N., Sugihara, H., Sharma, J., Perea, G., Petravicz, J., Le, C., Sur, M., 2012. Nucleus basalis-enabled stimulus-specific plasticity in the visual cortex is mediated by astrocytes. Proc. Natl. Acad. Sci. U. S. A. 109, E2832–E2841. https://doi.org/10.1073/pnas.1206557109.

Cheng, H., Wei, S., Wei, L., Verkhratsky, A., 2006. Calcium signaling in physiology and pathophysiology. Acta Pharmacol. Sin. 27, 767–772. https://doi.org/10.1111/j.1745-7254.2006.00399.x.

Cheng, K.T., Ong, H.L., Liu, X., Ambudkar, I.S., 2013. Contribution and regulation of TRPC channels in store-operated Ca2+ entry. Curr. Top Membr. 71, 149–179. https://doi.org/10.1016/B978-0-12-407870-3.00007-X.

Chvatal, A., Verkhratsky, A., 2018. Early History of Neuroglial Research. Personalities Neuroglia in Press.

Colombo, J.A., 2017. The interlaminar glia: from serendipity to hypothesis. Brain Struct. Funct. 222, 1109–1129. https://doi.org/10.1007/s00429-016-1332-8.

Communi, D., Dewaste, V., Erneux, C., 1999. Calcium-calmodulin-dependent protein kinase II and protein kinase C-mediated phosphorylation and activation of D-myo-inositol 1,4, 5-trisphosphate 3-kinase B in astrocytes. J. Biol. Chem. 274, 14734–14742. https://doi.org/10.1074/jbc.274.21.14734.

Conti, F., Barbaresi, P., Melone, M., Ducati, A., 1999. Neuronal and glial localization of NR1 and NR2A/B subunits of the NMDA receptor in the human cerebral cortex. Cereb. Cortex 9, 110–120. https://doi.org/10.1093/cercor/9.2.110.

Corkrum, M., Covelo, A., Lines, J., Bellocchio, L., Pisansky, M., Loke, K., Quintana, R., Rothwell, P.E., Lujan, R., Marsicano, G., Martin, E.D., Thomas, M.J., Kofuji, P., Araque, A., 2020. Dopamine-evoked synaptic regulation in the nucleus Accumbens requires astrocyte activity. Neuron 105, 1036–1047.e5. https://doi.org/10.1016/j.neuron.2019.12.026.

Cornell-Bell, A.H., Finkbeiner, S.M., 1991. Ca2+ waves in astrocytes. Cell Calcium 12, 185–204. https://doi.org/10.1016/0143-4160(91)90020-f.

Dani, J.W., Chernjavsky, A., Smith, S.J., 1992. Neuronal activity triggers calcium waves in hippocampal astrocyte networks. Neuron 8, 429–440. https://doi.org/10.1016/0896-6273(92)90271-e.

De Bock, M., Decrock, E., Wang, N., Bol, M., Vinken, M., Bultynck, G., Leybaert, L., 2014. The dual face of connexin-based astroglial ca(2+) communication: a key player in brain physiology and a prime target in pathology. Biochim. Biophys. Acta 1843, 2211–2232. https://doi.org/10.1016/j.bbamcr.2014.04.016.

Del Rio-Hortega, P., 1919. El tercer elemento de los centros nerviosos. I. La microglia en estado normal. II. Intervención de la microglia en los procesos patológicos. III. Naturaleza probable de la microglia. Bol de la Soc esp de biol 9, 69–120.

Del Río-Hortega, P., 1921. Estudios sobre la neuroglia. La glia de escasas radiaciones oligo-dendroglia. Biol. Soc. Esp. Biol. 21, 64–92.

Del Rio-Hortega, P., 1932. Microglia. In: Penfield, W. (Ed.), Cytology and Cellular Pathology of the Nervous System. Hoeber, New York, pp. 482–534.

Delekate, A., Füchtemeier, M., Schumacher, T., Ulbrich, C., Foddis, M., Petzold, G.C., 2014. Metabotropic P2Y1 receptor signalling mediates astrocytic hyperactivity in vivo in an Alzheimer's disease mouse model. Nat. Commun. 5, 5422. https://doi.org/10.1038/ncomms6422.

Dematteis, G., Restelli, E., Chiesa, R., Aronica, E., Genazzani, A.A., Lim, D., Tapella, L., 2020a. Calcineurin controls expression of EAAT1/GLAST in mouse and human cultured astrocytes through dynamic regulation of protein synthesis and degradation. Int. J. Mol. Sci. 21, 2213. https://doi.org/10.3390/ijms21062213.

Dematteis, G., Vydmantaitė, G., Ruffinatti, F.A., Chahin, M., Farruggio, S., Barberis, E., Ferrari, E., Marengo, E., Distasi, C., Morkūnienė, R., Genazzani, A.A., Grilli, M., Grossini, E., Corazzari, M., Manfredi, M., Lim, D., Jekabsone, A., Tapella, L., 2020b. Proteomic analysis links alterations of bioenergetics, mitochondria-ER interactions and proteostasis in hippocampal astrocytes from 3xTg-AD mice. Cell Death Dis. 11, 645. https://doi.org/10.1038/s41419-020-02911-1.

Di Castro, M.A., Chuquet, J., Liaudet, N., Bhaukaurally, K., Santello, M., Bouvier, D., Tiret, P., Volterra, A., 2011. Local Ca2+ detection and modulation of synaptic release by astrocytes. Nat. Neurosci. 14, 1276–1284. https://doi.org/10.1038/nn.2929.

Di Leva, F., Domi, T., Fedrizzi, L., Lim, D., Carafoli, E., 2008. The plasma membrane Ca2+ ATPase of animal cells: structure, function and regulation. Arch. Biochem. Biophys. 476, 65–74. https://doi.org/10.1016/j.abb.2008.02.026.

Dimou, L., Gallo, V., 2015. NG2-glia and their functions in the central nervous system. Glia 63, 1429–1451. https://doi.org/10.1002/glia.22859.

Ding, F., O'Donnell, J., Thrane, A.S., Zeppenfeld, D., Kang, H., Xie, L., Wang, F., Nedergaard, M., 2013. α1-adrenergic receptors mediate coordinated Ca2+ signaling of cortical astrocytes in awake, behaving mice. Cell Calcium 54, 387–394. https://doi.org/10.1016/j.ceca.2013.09.001.

Dombeck, D.A., Khabbaz, A.N., Collman, F., Adelman, T.L., Tank, D.W., 2007. Imaging large-scale neural activity with cellular resolution in awake, mobile mice. Neuron 56, 43–57. https://doi.org/10.1016/j.neuron.2007.08.003.

Dong, J., Chen, X., Cui, M., Yu, X., Pang, Q., Sun, J., 2012. β2-adrenergic receptor and astrocyte glucose metabolism. J. Mol. Neurosci. 48, 456–463. https://doi.org/10.1007/s12031-012-9742-4.

Dos Santos, J.P.A., Vizuete, A.F., Gonçalves, C.-A., 2020. Calcineurin-mediated hippocampal inflammatory alterations in Streptozotocin-induced model of dementia. Mol. Neurobiol. 57, 502–512. https://doi.org/10.1007/s12035-019-01718-2.

Dubbelaar, M.L., Kracht, L., Eggen, B.J.L., Boddeke, E.W.G.M., 2018. The kaleidoscope of microglial phenotypes. Front. Immunol. 9, 1753. https://doi.org/10.3389/fimmu.2018.01753.

Duffy, A.M., Fitzgerald, M.L., Chan, J., Robinson, D.C., Milner, T.A., Mackie, K., Pickel, V.M., 2011. Acetylcholine α7 nicotinic and dopamine D2 receptors are targeted to many of the same postsynaptic dendrites and astrocytes in the rodent prefrontal cortex. Synapse 65, 1350–1367. https://doi.org/10.1002/syn.20977.

Dunn, K.M., Hill-Eubanks, D.C., Liedtke, W.B., Nelson, M.T., 2013. TRPV4 channels stimulate Ca2+−induced Ca2+ release in astrocytic endfeet and amplify neurovascular coupling responses. Proc. Natl. Acad. Sci. U. S. A. 110, 6157–6162. https://doi.org/10.1073/pnas.1216514110.

Echeverry, S., Rodriguez, M.J., Torres, Y.P., 2016. Transient receptor potential channels in microglia: roles in physiology and disease. Neurotox. Res. 30, 467–478. https://doi.org/10.1007/s12640-016-9632-6.

Enkvist, M.O., McCarthy, K.D., 1992. Activation of protein kinase C blocks astroglial gap junction communication and inhibits the spread of calcium waves. J. Neurochem. 59, 519–526. https://doi.org/10.1111/j.1471-4159.1992.tb09401.x.

Erb, L., Weisman, G.A., 2012. Coupling of P2Y receptors to G proteins and other signaling pathways. Wiley Interdiscip. Rev. Membr. Transp. Signal 1, 789–803. https://doi.org/10.1002/wmts.62.

Fam, S.R., Gallagher, C.J., Kalia, L.V., Salter, M.W., 2003. Differential frequency dependence of P2Y1- and P2Y2- mediated Ca 2+ signaling in astrocytes. J. Neurosci. 23, 4437–4444.

Farb, C.R., Aoki, C., Ledoux, J.E., 1995. Differential localization of NMDA and AMPA receptor subunits in the lateral and basal nuclei of the amygdala: a light and electron microscopic study. J. Comp. Neurol. 362, 86–108. https://doi.org/10.1002/cne.903620106.

Feig, S.L., Haberly, L.B., 2011. Surface-associated astrocytes, not endfeet, form the glia limitans in posterior piriform cortex and have a spatially distributed, not a domain, organization. J. Comp. Neurol. 519, 1952–1969. https://doi.org/10.1002/cne.22615.

Fernandez, A.M., Fernandez, S., Carrero, P., Garcia-Garcia, M., Torres-Aleman, I., 2007. Calcineurin in reactive astrocytes plays a key role in the interplay between proinflammatory and anti-inflammatory signals. J. Neurosci. 27, 8745–8756. https://doi.org/10.1523/JNEUROSCI.1002-07.2007.

Fernandez, A.M., Jimenez, S., Mecha, M., Dávila, D., Guaza, C., Vitorica, J., Torres-Aleman, I., 2012. Regulation of the phosphatase calcineurin by insulin-like growth factor I unveils a key role of astrocytes in Alzheimer's pathology. Mol. Psychiatry 17, 705–718. https://doi.org/10.1038/mp.2011.128.

Fernández de Sevilla, D., Núñez, A., Buño, W., 2020. Muscarinic receptors, from synaptic plasticity to its role in network activity. Neuroscience 456, 60–70. https://doi.org/10.1016/j.neuroscience.2020.04.005.

Ferraguti, F., Crepaldi, L., Nicoletti, F., 2008. Metabotropic glutamate 1 receptor: current concepts and perspectives. Pharmacol. Rev. 60, 536–581. https://doi.org/10.1124/pr.108.000166.

Fiacco, T.A., McCarthy, K.D., 2018. Multiple Lines of evidence indicate that Gliotransmission does not occur under physiological conditions. J. Neurosci. 38, 3–13. https://doi.org/10.1523/JNEUROSCI.0016-17.2017.

Finkbeiner, S., 1992. Calcium waves in astrocytes-filling in the gaps. Neuron 8, 1101–1108. https://doi.org/10.1016/0896-6273(92)90131-v.

Fischer, T., Scheffler, P., Lohr, C., 2020. Dopamine-induced calcium signaling in olfactory bulb astrocytes. Sci. Rep. 10, 631. https://doi.org/10.1038/s41598-020-57462-4.

Franke, H., Verkhratsky, A., Burnstock, G., Illes, P., 2012. Pathophysiology of astroglial purinergic signalling. Purinergic Signal 8, 629–657. https://doi.org/10.1007/s11302-012-9300-0.

Frommann, C., 1878. Untersuchungen über die Gewebsveränderungen bei der Multiplen Sklerose des Gehirns und Rückenmarks. Verlag von Gustav Fischer, Jena.

Fujii, Y., Maekawa, S., Morita, M., 2017. Astrocyte calcium waves propagate proximally by gap junction and distally by extracellular diffusion of ATP released from volume-regulated anion channels. Sci. Rep. 7, 13115. https://doi.org/10.1038/s41598-017-13243-0.

Fumagalli, M., Brambilla, R., D'Ambrosi, N., Volonté, C., Matteoli, M., Verderio, C., Abbracchio, M.P., 2003. Nucleotide-mediated calcium signaling in rat cortical astrocytes: role of P2X and P2Y receptors. Glia 43, 218–303. https://doi.org/10.1002/glia.10248.

Furman, J.L., Norris, C.M., 2014. Calcineurin and glial signaling: neuroinflammation and beyond. J. Neuroinflammation 11, 158. https://doi.org/10.1186/s12974-014-0158-7.

Gavrilov, N., Golyagina, I., Brazhe, A., Scimemi, A., Turlapov, V., Semyanov, A., 2018. Astrocytic coverage of dendritic spines, dendritic shafts, and axonal Boutons in hippocampal neuropil. Front. Cell. Neurosci. 12, 248. https://doi.org/10.3389/fncel.2018.00248.

Giaume, C., Koulakoff, A., Roux, L., Holcman, D., Rouach, N., 2010. Astroglial networks: a step further in neuroglial and gliovascular interactions. Nat. Rev. Neurosci. 11, 87–99. https://doi.org/10.1038/nrn2757.

Gibbs, M.E., Bowser, D.N., 2010. Astrocytic adrenoceptors and learning: alpha1-adrenoceptors. Neurochem. Int. 57, 404–410. https://doi.org/10.1016/j.neuint.2010.03.020.

Golgi, C., 1870. Sulla sostanza connettiva del cervello (nevroglia). In: Rendiconti del R Instituto Lombardo di Scienze e Lettere serie 2. vol. 3, pp. 275–277.

Golovina, V.A., 2005. Visualization of localized store-operated calcium entry in mouse astrocytes. Close proximity to the endoplasmic reticulum. J. Physiol. (Lond.) 564, 737–749. https://doi.org/10.1113/jphysiol.2005.085035.

Golovina, V.A., Blaustein, M.P., 1997. Spatially and functionally distinct Ca2 + stores in sarcoplasmic and endoplasmic reticulum. Science 275, 1643–1648. https://doi.org/10.1126/science.275.5306.1643.

Götz, S., Bribian, A., López-Mascaraque, L., Götz, M., Grothe, B., Kunz, L., 2020. Heterogeneity of astrocytes: electrophysiological properties of juxtavascular astrocytes before and after brain injury. Glia 69, 346–361. https://doi.org/10.1002/glia.23900.

Gracy, K.N., Pickel, V.M., 1995. Comparative ultrastructural localization of the NMDAR1 glutamate receptor in the rat basolateral amygdala and bed nucleus of the stria terminalis. J. Comp. Neurol. 362, 71–85. https://doi.org/10.1002/cne.903620105.

Graham, A., Court, J.A., Martin-Ruiz, C.M., Jaros, E., Perry, R., Volsen, S.G., Bose, S., Evans, N., Ince, P., Kuryatov, A., Lindstrom, J., Gotti, C., Perry, E.K., 2002. Immunohistochemical localisation of nicotinic acetylcholine receptor subunits in human cerebellum. Neuroscience 113, 493–507. https://doi.org/10.1016/s0306-4522(02)00223-3.

Graham, A.J., Ray, M.A., Perry, E.K., Jaros, E., Perry, R.H., Volsen, S.G., Bose, S., Evans, N., Lindstrom, J., Court, J.A., 2003. Differential nicotinic acetylcholine receptor subunit expression in the human hippocampus. J. Chem. Neuroanat. 25, 97–113. https://doi.org/10.1016/s0891-0618(02)00100-x.

Grimaldi, M., Maratos, M., Verma, A., 2003. Transient receptor potential channel activation causes a novel form of [ca 2 +]I oscillations and is not involved in capacitative ca 2 + entry in glial cells. J. Neurosci. 23, 4737–4745.

Grolla, A.A., Fakhfouri, G., Balzaretti, G., Marcello, E., Gardoni, F., Canonico, P.L., DiLuca, M., Genazzani, A.A., Lim, D., 2013a. Aβ leads to Ca^{2+} signaling alterations and transcriptional changes in glial cells. Neurobiol. Aging 34, 511–522. https://doi.org/10.1016/j.neurobiolaging.2012.05.005.

Grolla, A.A., Sim, J.A., Lim, D., Rodriguez, J.J., Genazzani, A.A., Verkhratsky, A., 2013b. Amyloid-β and Alzheimer's disease type pathology differentially affects the calcium signalling toolkit in astrocytes from different brain regions. Cell Death Dis. 4, e623. https://doi.org/10.1038/cddis.2013.145.

Grosche, J., Matyash, V., Möller, T., Verkhratsky, A., Reichenbach, A., Kettenmann, H., 1999. Microdomains for neuron-glia interaction: parallel fiber signaling to Bergmann glial cells. Nat. Neurosci. 2, 139–143. https://doi.org/10.1038/5692.

Haberlandt, C., Derouiche, A., Wyczynski, A., Haseleu, J., Pohle, J., Karram, K., Trotter, J., Seifert, G., Frotscher, M., Steinhäuser, C., Jabs, R., 2011. Gray matter NG2 cells display multiple Ca2+−signaling pathways and highly motile processes. PLoS One 6, e17575. https://doi.org/10.1371/journal.pone.0017575.

Hamilton, N., Vayro, S., Kirchhoff, F., Verkhratsky, A., Robbins, J., Gorecki, D.C., Butt, A.M., 2008. Mechanisms of ATP- and glutamate-mediated calcium signaling in white matter astrocytes. Glia 56, 734–749. https://doi.org/10.1002/glia.20649.

Hamilton, N., Vayro, S., Wigley, R., Butt, A.M., 2010. Axons and astrocytes release ATP and glutamate to evoke calcium signals in NG2-glia. Glia 58, 66–79. https://doi.org/10.1002/glia.20902.

Hashioka, S., Wang, Y.F., Little, J.P., Choi, H.B., Klegeris, A., McGeer, P.L., McLarnon, J.G., 2014. Purinergic responses of calcium-dependent signaling pathways in cultured adult human astrocytes. BMC Neurosci. 15, 18. https://doi.org/10.1186/1471-2202-15-18.

Hatton, G.I., 1988. Pituicytes, glia and control of terminal secretion. J. Exp. Biol. 139, 67–79.

Hatton, G.I., 1999. Astroglial modulation of neurotransmitter/peptide release from the neurohypophysis: present status. J. Chem. Neuroanat. 16, 203–221. https://doi.org/10.1016/s0891-0618(98)00067-2.

Haustein, M.D., Kracun, S., Lu, X.-H., Shih, T., Jackson-Weaver, O., Tong, X., Xu, J., Yang, X.W., O'Dell, T.J., Marvin, J.S., Ellisman, M.H., Bushong, E.A., Looger, L.L., Khakh, B.S., 2014a. Conditions and constraints for astrocyte calcium signaling in the hippocampal mossy fiber pathway. Neuron 82, 413–429. https://doi.org/10.1016/j.neuron.2014.02.041.

Haustein, M.D., Kracun, S., Lu, X.-H., Shih, T., Jackson-Weaver, O., Tong, X., Xu, J., Yang, X.W., O'Dell, T.J., Marvin, J.S., Ellisman, M.H., Bushong, E.A., Looger, L.L., Khakh, B.S., 2014b. Conditions and constraints for astrocyte calcium signaling in the hippocampal mossy fiber pathway. Neuron 82, 413–429. https://doi.org/10.1016/j.neuron.2014.02.041.

He, Y., Taylor, N., Fourgeaud, L., Bhattacharya, A., 2017. The role of microglial P2X7: modulation of cell death and cytokine release. J. Neuroinflammation 14, 135. https://doi.org/10.1186/s12974-017-0904-8.

Hertle, D.N., Yeckel, M.F., 2007. Distribution of inositol-1,4,5-trisphosphate receptor isotypes and ryanodine receptor isotypes during maturation of the rat hippocampus. Neuroscience 150, 625–638. https://doi.org/10.1016/j.neuroscience.2007.09.058.

Holtzclaw, L.A., Pandhit, S., Bare, D.J., Mignery, G.A., Russell, J.T., 2002. Astrocytes in adult rat brain express type 2 inositol 1,4,5-trisphosphate receptors. Glia 39, 69–84. https://doi.org/10.1002/glia.10085.

Hoogland, T.M., Kuhn, B., Göbel, W., Huang, W., Nakai, J., Helmchen, F., Flint, J., Wang, S.S.-H., 2009. Radially expanding transglial calcium waves in the intact cerebellum. Proc. Natl. Acad. Sci. U. S. A. 106, 3496–3501. https://doi.org/10.1073/pnas.0809269106.

Iadecola, C., 2017. The neurovascular unit coming of age: a journey through neurovascular coupling in health and disease. Neuron 96, 17–42. https://doi.org/10.1016/j.neuron.2017.07.030.

Illes, P., Verkhratsky, A., Burnstock, G., Franke, H., 2012. P2X receptors and their roles in astroglia in the central and peripheral nervous system. Neuroscientist 18, 422–438. https://doi.org/10.1177/1073858411418524.

Illes, P., Verkhratsky, A., Burnstock, G., Sperlagh, B., 2016. Purines in neurodegeneration and neuroregeneration. Neuropharmacology 104, 1–3. https://doi.org/10.1016/j.neuropharm.2016.01.020.

Illes, P., Verkhratsky, A., Tang, Y., 2019. Pathological ATPergic Signaling in major depression and bipolar disorder. Front. Mol. Neurosci. 12, 331. https://doi.org/10.3389/fnmol.2019.00331.

Ishibashi, T., Dakin, K.A., Stevens, B., Lee, P.R., Kozlov, S.V., Stewart, C.L., Fields, R.D., 2006. Astrocytes promote myelination in response to electrical impulses. Neuron 49, 823–832. https://doi.org/10.1016/j.neuron.2006.02.006.

Jennings, A., Tyurikova, O., Bard, L., Zheng, K., Semyanov, A., Henneberger, C., Rusakov, D.A., 2017. Dopamine elevates and lowers astroglial Ca2 + through distinct pathways depending on local synaptic circuitry. Glia 65, 447–459. https://doi.org/10.1002/glia.23103.

Jo, A.O., Ryskamp, D.A., Phuong, T.T.T., Verkman, A.S., Yarishkin, O., MacAulay, N., Križaj, D., 2015. TRPV4 and AQP4 channels synergistically regulate cell volume and calcium homeostasis in retinal Müller glia. J. Neurosci. 35, 13525–13537. https://doi.org/10.1523/JNEUROSCI.1987-15.2015.

Joost, E., Jordão, M.J.C., Mages, B., Prinz, M., Bechmann, I., Krueger, M., 2019. Microglia contribute to the glia limitans around arteries, capillaries and veins under physiological conditions, in a model of neuroinflammation and in human brain tissue. Brain Struct. Funct. 224, 1301–1314. https://doi.org/10.1007/s00429-019-01834-8.

Joseph, S.K., Brownell, S., Khan, M.T., 2005. Calcium regulation of inositol 1,4,5-trisphosphate receptors. Cell Calcium 38, 539–546. https://doi.org/10.1016/j.ceca.2005.07.007.

Kabbani, N., Nichols, R.A., 2018. Beyond the channel: metabotropic Signaling by nicotinic receptors. Trends Pharmacol. Sci. 39, 354–366. https://doi.org/10.1016/j.tips.2018.01.002.

Kanemaru, K., Sekiya, H., Xu, M., Satoh, K., Kitajima, N., Yoshida, K., Okubo, Y., Sasaki, T., Moritoh, S., Hasuwa, H., Mimura, M., Horikawa, K., Matsui, K., Nagai, T., Iino, M., Tanaka, K.F., 2014. In vivo visualization of subtle, transient, and local activity of astrocytes using an ultrasensitive ca(2+) indicator. Cell Rep. 8, 311–318. https://doi.org/10.1016/j.celrep.2014.05.056.

Kettenmann, H., Hanisch, U.-K., Noda, M., Verkhratsky, A., 2011. Physiology of microglia. Physiol. Rev. 91, 461–553. https://doi.org/10.1152/physrev.00011.2010.

Kettenmann, H., Verkhratsky, A., 2008. Neuroglia: the 150 years after. Trends Neurosci. 31, 653–659. https://doi.org/10.1016/j.tins.2008.09.003.

Khakh, B.S., Deneen, B., 2019. The emerging nature of astrocyte diversity. Annu. Rev. Neurosci. 42, 187–207. https://doi.org/10.1146/annurev-neuro-070918-050443.

Khakh, B.S., Sofroniew, M.V., 2015. Diversity of astrocyte functions and phenotypes in neural circuits. Nat. Neurosci. 18, 942–952. https://doi.org/10.1038/nn.4043.

King, C.M., Bohmbach, K., Minge, D., Delekate, A., Zheng, K., Reynolds, J., Rakers, C., Zeug, A., Petzold, G.C., Rusakov, D.A., Henneberger, C., 2020. Local resting Ca2 + controls the scale of Astroglial Ca2 + signals. Cell Rep. 30, 3466–3477.e4. https://doi.org/10.1016/j.celrep.2020.02.043.

Kirischuk, S., Kettenmann, H., Verkhratsky, A., 1997. Na+/Ca2 + exchanger modulates kainate-triggered Ca2 + signaling in Bergmann glial cells in situ. FASEB J. 11, 566–572. https://doi.org/10.1096/fasebj.11.7.9212080.

Kirischuk, S., Kettenmann, H., Verkhratsky, A., 2007. Membrane currents and cytoplasmic sodium transients generated by glutamate transport in Bergmann glial cells. Pflugers Arch. 454, 245–252. https://doi.org/10.1007/s00424-007-0207-5.

Kirischuk, S., Kirchhoff, F., Matyash, V., Kettenmann, H., Verkhratsky, A., 1999. Glutamate-triggered calcium signalling in mouse bergmann glial cells in situ: role of inositol-1,4,5-trisphosphate-mediated intracellular calcium release. Neuroscience 92, 1051–1059. https://doi.org/10.1016/s0306-4522(99)00067-6.

Kirischuk, S., Möller, T., Voitenko, N., Kettenmann, H., Verkhratsky, A., 1995a. ATP-induced cytoplasmic calcium mobilization in Bergmann glial cells. J. Neurosci. 15, 7861–7871.

Kirischuk, S., Parpura, V., Verkhratsky, A., 2012. Sodium dynamics: another key to astroglial excitability? Trends Neurosci. 35, 497–506. https://doi.org/10.1016/j.tins.2012.04.003.

Kirischuk, S., Scherer, J., Möller, T., Verkhratsky, A., Kettenmann, H., 1995b. Subcellular heterogeneity of voltage-gated Ca2+ channels in cells of the oligodendrocyte lineage. Glia 13, 1–12. https://doi.org/10.1002/glia.440130102.

Kirischuk, S., Tuschick, S., Verkhratsky, A., Kettenmann, H., 1996. Calcium signalling in mouse Bergmann glial cells mediated by alpha1-adrenoreceptors and H1 histamine receptors. Eur. J. Neurosci. 8, 1198–1208. https://doi.org/10.1111/j.1460-9568.1996.tb01288.x.

Kjaerby, C., Rasmussen, R., Andersen, M., Nedergaard, M., 2017. Does global astrocytic calcium signaling participate in awake brain state transitions and neuronal circuit function? Neurochem. Res. 42, 1810–1822. https://doi.org/10.1007/s11064-017-2195-y.

Klegeris, A., Choi, H.B., McLarnon, J.G., McGeer, P.L., 2007. Functional ryanodine receptors are expressed by human microglia and THP-1 cells: their possible involvement in modulation of neurotoxicity. J. Neurosci. Res. 85, 2207–2215. https://doi.org/10.1002/jnr.21361.

Kofuji, P., Araque, A., 2020. G-protein-coupled receptors in astrocyte-neuron communication. Neuroscience 456, 71–84. https://doi.org/10.1016/j.neuroscience.2020.03.025.

Kubes, M., Cordier, J., Glowinski, J., Girault, J.A., Chneiweiss, H., 1998. Endothelin induces a calcium-dependent phosphorylation of PEA-15 in intact astrocytes: identification of Ser104 and Ser116 phosphorylated, respectively, by protein kinase C and calcium/calmodulin kinase II in vitro. J. Neurochem. 71, 1307–1314. https://doi.org/10.1046/j.1471-4159.1998.71031307.x.

Kuchibhotla, K.V., Lattarulo, C.R., Hyman, B.T., Bacskai, B.J., 2009. Synchronous hyperactivity and intercellular calcium waves in astrocytes in Alzheimer mice. Science 323, 1211–1215. https://doi.org/10.1126/science.1169096.

Kukley, M., Barden, J.A., Steinhäuser, C., Jabs, R., 2001. Distribution of P2X receptors on astrocytes in juvenile rat hippocampus. Glia 36, 11–21. https://doi.org/10.1002/glia.1091.

Kwon, J., An, H., Sa, M., Won, J., Shin, J.I., Lee, C.J., 2017. Orai1 and Orai3 in combination with Stim1 mediate the majority of store-operated calcium entry in astrocytes. Exp. Neurobiol. 26, 42–54. https://doi.org/10.5607/en.2017.26.1.42.

Lalo, U., Pankratov, Y., Kirchhoff, F., North, R.A., Verkhratsky, A., 2006. NMDA receptors mediate neuron-to-glia signaling in mouse cortical astrocytes. J. Neurosci. 26, 2673–2683. https://doi.org/10.1523/JNEUROSCI.4689-05.2006.

Lalo, U., Pankratov, Y., Wichert, S.P., Rossner, M.J., North, R.A., Kirchhoff, F., Verkhratsky, A., 2008. P2X1 and P2X5 subunits form the functional P2X receptor in mouse cortical astrocytes. J. Neurosci. 28, 5473–5480. https://doi.org/10.1523/JNEUROSCI.1149-08.2008.

Lalo, U., Verkhratsky, A., Pankratov, Y., 2011. Ionotropic ATP receptors in neuronal-glial communication. Semin. Cell Dev. Biol. 22, 220–228. https://doi.org/10.1016/j.semcdb.2011.02.012.

Lebedeva, A., Plata, A., Nosova, O., Tyurikova, O., Semyanov, A., 2018. Activity-dependent changes in transporter and potassium currents in hippocampal astrocytes. Brain Res. Bull. 136, 37–43. https://doi.org/10.1016/j.brainresbull.2017.08.015.

Lee, M.-C., Ting, K.K., Adams, S., Brew, B.J., Chung, R., Guillemin, G.J., 2010. Characterisation of the expression of NMDA receptors in human astrocytes. PLoS One 5, e14123. https://doi.org/10.1371/journal.pone.0014123.

Lenhossék, M.V., 1895. Der feinere Bau des Nervensystems im Lichte neuester Forschung, second ed. Fischer's Medicinische Buchhandlung H. Kornfield, Berlin.

Li, G., Liu, Y., Olson, J.E., 2002. Calcium/calmodulin-modulated chloride and taurine conductances in cultured rat astrocytes. Brain Res. 925, 1–8. https://doi.org/10.1016/s0006-8993(01)03235-8.

Lim, D., Iyer, A., Ronco, V., Grolla, A.A., Canonico, P.L., Aronica, E., Genazzani, A.A., 2013. Amyloid beta deregulates astroglial mGluR5-mediated calcium signaling via calcineurin and Nf-kB. Glia 61, 1134–1145. https://doi.org/10.1002/glia.22502.

Lim, D., Mapelli, L., Canonico, P.L., Moccia, F., Genazzani, A.A., 2018. Neuronal activity-dependent activation of Astroglial Calcineurin in mouse primary hippocampal cultures. Int. J. Mol. Sci. 19, 2997. https://doi.org/10.3390/ijms19102997.

Lim, D., Rocchio, F., Lisa, M., Fcancesco, M., 2016a. From pathology to physiology of Calcineurin signalling in astrocytes. Opera Medica Physiol. 2, 122–140. https://doi.org/:10.20388/OMP2016.002.0029.

Lim, D., Rodríguez-Arellano, J.J., Parpura, V., Zorec, R., Zeidán-Chuliá, F., Genazzani, A.A., Verkhratsky, A., 2016b. Calcium signalling toolkits in astrocytes and spatio-temporal progression of Alzheimer's disease. Curr. Alzheimer Res. 13, 359–369.

Lim, D., Ronco, V., Grolla, A.A., Verkhratsky, A., Genazzani, A.A., 2014. Glial calcium signalling in Alzheimer's disease. Rev. Physiol. Biochem. Pharmacol. 167, 45–65. https://doi.org/10.1007/112_2014_19.

Lin, C.-C., Lee, I.-T., Wu, W.-B., Liu, C.-J., Hsieh, H.-L., Hsiao, L.-D., Yang, C.-C., Yang, C.-M., 2013. Thrombin mediates migration of rat brain astrocytes via PLC, Ca^{2+}, CaMKII, PKCα, and AP-1-dependent matrix metalloproteinase-9 expression. Mol. Neurobiol. 48, 616–630. https://doi.org/10.1007/s12035-013-8450-6.

Lind, B.L., Brazhe, A.R., Jessen, S.B., Tan, F.C.C., Lauritzen, M.J., 2013. Rapid stimulus-evoked astrocyte Ca2+ elevations and hemodynamic responses in mouse somatosensory cortex in vivo. Proc. Natl. Acad. Sci. U. S. A. 110, E4678–E4687. https://doi.org/10.1073/pnas.1310065110.

Liu, X., Zhang, Z., Guo, W., Burnstock, G., He, C., Xiang, Z., 2013. The superficial glia limitans of mouse and monkey brain and spinal cord. Anat. Rec. (Hoboken) 296, 995–1007. https://doi.org/10.1002/ar.22717.

Lopes, J.P., Cunha, R.A., 2019. What is the extracellular calcium concentration within brain synapses?: An editorial for "ionized calcium in human cerebrospinal fluid and its influence on intrinsic and synaptic excitability of hippocampal pyramidal neurons in the rat" on page 452. J. Neurochem. 149, 435–437. https://doi.org/10.1111/jnc.14696.

Lugaro, E., 1907. Sulle funzioni della nevroglia. Riv. Patol. Nerv. Ment. 12, 225–233.

Lundgaard, I., Osório, M.J., Kress, B.T., Sanggaard, S., Nedergaard, M., 2014. White matter astrocytes in health and disease. Neuroscience 276, 161–173. https://doi.org/10.1016/j.neuroscience.2013.10.050.

Malarkey, E.B., Ni, Y., Parpura, V., 2008. Ca2+ entry through TRPC1 channels contributes to intracellular Ca2+ dynamics and consequent glutamate release from rat astrocytes. Glia 56, 821–835. https://doi.org/10.1002/glia.20656.

Marina, N., Christie, I.N., Korsak, A., Doronin, M., Brazhe, A., Hosford, P.S., Wells, J.A., Sheikhbahaei, S., Humoud, I., Paton, J.F.R., Lythgoe, M.F., Semyanov, A., Kasparov, S., Gourine, A.V., 2020. Astrocytes monitor cerebral perfusion and control systemic circulation to maintain brain blood flow. Nat. Commun. 11, 131. https://doi.org/10.1038/s41467-019-13956-y.

Martorana, F., Brambilla, L., Valori, C.F., Bergamaschi, C., Roncoroni, C., Aronica, E., Volterra, A., Bezzi, P., Rossi, D., 2012. The BH4 domain of Bcl-X(L) rescues astrocyte degeneration in amyotrophic lateral sclerosis by modulating intracellular calcium signals. Hum. Mol. Genet. 21, 826–840. https://doi.org/10.1093/hmg/ddr513.

Mathiesen, C., Brazhe, A., Thomsen, K., Lauritzen, M., 2013. Spontaneous calcium waves in Bergman glia increase with age and hypoxia and may reduce tissue oxygen. J. Cereb. Blood Flow Metab. 33, 161–169. https://doi.org/10.1038/jcbfm.2012.175.

Mathiisen, T.M., Lehre, K.P., Danbolt, N.C., Ottersen, O.P., 2010. The perivascular astroglial sheath provides a complete covering of the brain microvessels: an electron microscopic 3D reconstruction. Glia 58, 1094–1103. https://doi.org/10.1002/glia.20990.

Matute, C., Alberdi, E., Domercq, M., Sánchez-Gómez, M.-V., Pérez-Samartín, A., Rodríguez-Antigüedad, A., Pérez-Cerdá, F., 2007. Excitotoxic damage to white matter. J. Anat. 210, 693–702. https://doi.org/10.1111/j.1469-7580.2007.00733.x.

Matyash, M., Matyash, V., Nolte, C., Sorrentino, V., Kettenmann, H., 2002. Requirement of functional ryanodine receptor type 3 for astrocyte migration. FASEB J. 16, 84–86. https://doi.org/10.1096/fj.01-0380fje.

Mehina, E.M.F., Murphy-Royal, C., Gordon, G.R., 2017. Steady-state free Ca2 + in astrocytes is decreased by experience and impacts arteriole tone. J. Neurosci. 37, 8150–8165. https://doi.org/10.1523/JNEUROSCI.0239-17.2017.

Metea, M.R., Newman, E.A., 2006. Calcium signaling in specialized glial cells. Glia 54, 650–655. https://doi.org/10.1002/glia.20352.

Michaelis, M., Nieswandt, B., Stegner, D., Eilers, J., Kraft, R., 2015. STIM1, STIM2, and Orai1 regulate store-operated calcium entry and purinergic activation of microglia. Glia 63, 652–663. https://doi.org/10.1002/glia.22775.

Miyake, T., Shirakawa, H., Nakagawa, T., Kaneko, S., 2015. Activation of mitochondrial transient receptor potential vanilloid 1 channel contributes to microglial migration. Glia 63, 1870–1882. https://doi.org/10.1002/glia.22854.

Molnár, T., Yarishkin, O., Iuso, A., Barabas, P., Jones, B., Marc, R.E., Phuong, T.T.T., Križaj, D., 2016. Store-operated calcium entry in Müller glia is controlled by synergistic activation of TRPC and Orai channels. J. Neurosci. 36, 3184–3198. https://doi.org/10.1523/JNEUROSCI.4069-15.2016.

Molotkov, D., Zobova, S., Arcas, J.M., Khiroug, L., 2013. Calcium-induced outgrowth of astrocytic peripheral processes requires actin binding by Profilin-1. Cell Calcium 53, 338–348. https://doi.org/10.1016/j.ceca.2013.03.001.

Mongin, A.A., Cai, Z., Kimelberg, H.K., 1999. Volume-dependent taurine release from cultured astrocytes requires permissive [Ca(2 +)](i) and calmodulin. Am. J. Physiol. 277, C823–C832. https://doi.org/10.1152/ajpcell.1999.277.4.C823.

Morán-Jiménez, M.J., Matute, C., 2000. Immunohistochemical localization of the P2Y(1) purinergic receptor in neurons and glial cells of the central nervous system. Brain Res. Mol. Brain Res. 78, 50–58. https://doi.org/10.1016/s0169-328x(00)00067-x.

Mu, Y., Bennett, D.V., Rubinov, M., Narayan, S., Yang, C.-T., Tanimoto, M., Mensh, B.D., Looger, L.L., Ahrens, M.B., 2019. Glia accumulate evidence that actions are futile and suppress unsuccessful behavior. Cell 178, 27–43.e19. https://doi.org/10.1016/j.cell.2019.05.050.

Muller, M.S., Taylor, C.W., 2017. ATP evokes Ca2 + signals in cultured foetal human cortical astrocytes entirely through G protein-coupled P2Y receptors. J. Neurochem. 142, 876–885. https://doi.org/10.1111/jnc.14119.

Naranjo, J.R., Mellström, B., 2012. Ca2 +–dependent transcriptional control of Ca2 + homeostasis. J. Biol. Chem. 287, 31674–31680. https://doi.org/10.1074/jbc.R112.384982.

Navarrete, M., Araque, A., 2008. Endocannabinoids mediate neuron-astrocyte communication. Neuron 57, 883–893. https://doi.org/10.1016/j.neuron.2008.01.029.

Neary, J.T., Norenberg, L.O., Norenberg, M.D., 1986. Calcium-activated, phospholipid-dependent protein kinase and protein substrates in primary cultures of astrocytes. Brain Res. 385, 420–424. https://doi.org/10.1016/0006-8993(86)91095-4.

Nett, W.J., Oloff, S.H., McCarthy, K.D., 2002. Hippocampal astrocytes in situ exhibit calcium oscillations that occur independent of neuronal activity. J. Neurophysiol. 87, 528–537. https://doi.org/10.1152/jn.00268.2001.

Newman, E.A., 2001. Propagation of intercellular calcium waves in retinal astrocytes and Müller cells. J. Neurosci. 21, 2215–2223.

Newman, E.A., Zahs, K.R., 1997. Calcium waves in retinal glial cells. Science 275, 844–847. https://doi.org/10.1126/science.275.5301.844.

Nimmerjahn, A., Mukamel, E.A., Schnitzer, M.J., 2009. Motor behavior activates Bergmann glial networks. Neuron 62, 400–412. https://doi.org/10.1016/j.neuron.2009.03.019.

Niswender, C.M., Conn, P.J., 2010. Metabotropic glutamate receptors: physiology, pharmacology, and disease. Annu. Rev. Pharmacol. Toxicol. 50, 295–322. https://doi.org/10.1146/annurev.pharmtox.011008.145533.

Nobile, M., Monaldi, I., Alloisio, S., Cugnoli, C., Ferroni, S., 2003. ATP-induced, sustained calcium signalling in cultured rat cortical astrocytes: evidence for a non-capacitative, P2X7-like-mediated calcium entry. FEBS Lett. 538, 71–76. https://doi.org/10.1016/s0014-5793(03)00129-7.

Noda, M., Ifuku, M., Mori, Y., Verkhratsky, A., 2013. Calcium influx through reversed NCX controls migration of microglia. Adv. Exp. Med. Biol. 961, 289–294. https://doi.org/10.1007/978-1-4614-4756-6_24.

North, R.A., Verkhratsky, A., 2006. Purinergic transmission in the central nervous system. Pflugers Arch. 452, 479–485. https://doi.org/10.1007/s00424-006-0060-y.

Nuriya, M., Morita, A., Shinotsuka, T., Yamada, T., Yasui, M., 2018. Norepinephrine induces rapid and long-lasting phosphorylation and redistribution of connexin 43 in cortical astrocytes. Biochem. Biophys. Res. Commun. 504, 690–697. https://doi.org/10.1016/j.bbrc.2018.09.021.

Oberheim, N.A., Takano, T., Han, X., He, W., Lin, J.H.C., Wang, F., Xu, Q., Wyatt, J.D., Pilcher, W., Ojemann, J.G., Ransom, B.R., Goldman, S.A., Nedergaard, M., 2009. Uniquely hominid features of adult human astrocytes. J. Neurosci. 29, 3276–3287. https://doi.org/10.1523/JNEUROSCI.4707-08.2009.

Oe, Y., Wang, X., Patriarchi, T., Konno, A., Ozawa, K., Yahagi, K., Hirai, H., Tsuboi, T., Kitaguchi, T., Tian, L., McHugh, T.J., Hirase, H., 2020. Distinct temporal integration of noradrenaline signaling by astrocytic second messengers during vigilance. Nat. Commun. 11, 471. https://doi.org/10.1038/s41467-020-14378-x.

Ogata, K., Kosaka, T., 2002. Structural and quantitative analysis of astrocytes in the mouse hippocampus. Neuroscience 113, 221–233. https://doi.org/10.1016/s0306-4522(02)00041-6.

Oh-hora, M., Rao, A., 2008. Calcium signaling in lymphocytes. Curr. Opin. Immunol. 20, 250–258. https://doi.org/10.1016/j.coi.2008.04.004.

Oikawa, H., Nakamichi, N., Kambe, Y., Ogura, M., Yoneda, Y., 2005. An increase in intracellular free calcium ions by nicotinic acetylcholine receptors in a single cultured rat cortical astrocyte. J. Neurosci. Res. 79, 535–544. https://doi.org/10.1002/jnr.20398.

Oliveira, J.F., Sardinha, V.M., Guerra-Gomes, S., Araque, A., Sousa, N., 2015. Do stars govern our actions? Astrocyte involvement in rodent behavior. Trends Neurosci. 38, 535–549. https://doi.org/10.1016/j.tins.2015.07.006.

Oliveira, S.R., Figueiredo-Pereira, C., Duarte, C.B., Vieira, H.L.A., 2019. P2X7 receptors mediate CO-induced alterations in gene expression in cultured cortical astrocytes-transcriptomic study. Mol. Neurobiol. 56, 3159–3174. https://doi.org/10.1007/s12035-018-1302-7.

Olson, J.E., Li, G.-Z., Wang, L., Lu, L., 2004. Volume-regulated anion conductance in cultured rat cerebral astrocytes requires calmodulin activity. Glia 46, 391–401. https://doi.org/10.1002/glia.20014.

Orre, M., Kamphuis, W., Osborn, L.M., Melief, J., Kooijman, L., Huitinga, I., Klooster, J., Bossers, K., Hol, E.M., 2014. Acute isolation and transcriptome characterization of cortical astrocytes and microglia from young and aged mice. Neurobiol. Aging 35, 1–14. https://doi.org/10.1016/j.neurobiolaging.2013.07.008.

Otsu, Y., Couchman, K., Lyons, D.G., Collot, M., Agarwal, A., Mallet, J.-M., Pfrieger, F.W., Bergles, D.E., Charpak, S., 2015. Calcium dynamics in astrocyte processes during neurovascular coupling. Nat. Neurosci. 18, 210–218. https://doi.org/10.1038/nn.3906.

Palygin, O., Lalo, U., Verkhratsky, A., Pankratov, Y., 2010. Ionotropic NMDA and P2X1/5 receptors mediate synaptically induced Ca2+ signalling in cortical astrocytes. Cell Calcium 48, 225–231. https://doi.org/10.1016/j.ceca.2010.09.004.

Panatier, A., Vallée, J., Haber, M., Murai, K.K., Lacaille, J.-C., Robitaille, R., 2011. Astrocytes are endogenous regulators of basal transmission at central synapses. Cell 146, 785–798. https://doi.org/10.1016/j.cell.2011.07.022.

Pankratov, Y., Lalo, U., 2015. Role for astroglial α1-adrenoreceptors in gliotransmission and control of synaptic plasticity in the neocortex. Front. Cell. Neurosci. 9, 230. https://doi.org/10.3389/fncel.2015.00230.

Papanikolaou, M., Lewis, A., Butt, A.M., 2017. Store-operated calcium entry is essential for glial calcium signalling in CNS white matter. Brain Struct. Funct. 222, 2993–3005. https://doi.org/10.1007/s00429-017-1380-8.

Patrushev, I., Gavrilov, N., Turlapov, V., Semyanov, A., 2013. Subcellular location of astrocytic calcium stores favors extrasynaptic neuron-astrocyte communication. Cell Calcium 54, 343–349. https://doi.org/10.1016/j.ceca.2013.08.003.

Paukert, M., Agarwal, A., Cha, J., Doze, V.A., Kang, J.U., Bergles, D.E., 2014. Norepinephrine controls astroglial responsiveness to local circuit activity. Neuron 82, 1263–1270. https://doi.org/10.1016/j.neuron.2014.04.038.

Pekny, M., Pekna, M., Messing, A., Steinhäuser, C., Lee, J.-M., Parpura, V., Hol, E.M., Sofroniew, M.V., Verkhratsky, A., 2016. Astrocytes: a central element in neurological diseases. Acta Neuropathol. 131, 323–345. https://doi.org/10.1007/s00401-015-1513-1.

Petravicz, J., Boyt, K.M., McCarthy, K.D., 2014. Astrocyte IP3R2-dependent ca(2+) signaling is not a major modulator of neuronal pathways governing behavior. Front. Behav. Neurosci. 8, 384. https://doi.org/10.3389/fnbeh.2014.00384.

Petzold, G.C., Albeanu, D.F., Sato, T.F., Murthy, V.N., 2008. Coupling of neural activity to blood flow in olfactory glomeruli is mediated by astrocytic pathways. Neuron 58, 897–910. https://doi.org/10.1016/j.neuron.2008.04.029.

Pizzo, P., Burgo, A., Pozzan, T., Fasolato, C., 2001. Role of capacitative calcium entry on glutamate-induced calcium influx in type-I rat cortical astrocytes. J. Neurochem. 79, 98–109. https://doi.org/10.1046/j.1471-4159.2001.00539.x.

Plattner, H., Verkhratsky, A., 2016. Inseparable tandem: evolution chooses ATP and Ca^{2+} to control life, death and cellular signalling. Philos. Trans. R. Soc. Lond. B Biol. Sci. 371, 20150419. https://doi.org/10.1098/rstb.2015.0419.

Plattner, H., Verkhratsky, A., 2018. The remembrance of the things past: conserved signalling pathways link protozoa to mammalian nervous system. Cell Calcium 73, 25–39. https://doi.org/10.1016/j.ceca.2018.04.001.

Pocock, J.M., Kettenmann, H., 2007. Neurotransmitter receptors on microglia. Trends Neurosci. 30, 527–535. https://doi.org/10.1016/j.tins.2007.07.007.

Popov, A., Brazhe, A., Denisov, P., Sutyagina, O., Lazareva, N., Verkhratsky, A., Semyanov, A., 2021. Astrocytes dystrophy in ageing brain parallels impaired synaptic plasticity. Aging Cell 20, e13334. https://doi.org/10.1111/acel.13334.

Porter, J.T., McCarthy, K.D., 1996. Hippocampal astrocytes in situ respond to glutamate released from synaptic terminals. J. Neurosci. 16, 5073–5081.

Quintana, F.J., 2017. Astrocytes to the rescue! Glia limitans astrocytic endfeet control CNS inflammation. J. Clin. Invest. 127, 2897–2899. https://doi.org/10.1172/JCI95769.

Ramón y Cajal, S., 1895. Algunas conjeturas sobre el mechanismoanatomico de la ideacion. asociacion y atencion (Imprenta y Libreria de Nicolas Moya).

Ransohoff, R.M., Perry, V.H., 2009. Microglial physiology: unique stimuli, specialized responses. Annu. Rev. Immunol. 27, 119–145. https://doi.org/10.1146/annurev. immunol.021908.132528.

Reichenbach, A., Bringmann, A., 2016. Role of purines in Müller glia. J. Ocul. Pharmacol. Ther. 32, 518–533. https://doi.org/10.1089/jop.2016.0131.

Reichenbach, A., Bringmann, A., 2017. 2.16 Comparative anatomy of glial cells in mammals. Evol. Nerv. Syst. 2, 309–348. https://doi.org/10.1016/B978-0-12-804042-3. 00050-6.

Reichenbach, A., Derouiche, A., Kirchhoff, F., 2010. Morphology and dynamics of perisynaptic glia. Brain Res. Rev. 63, 11–25. https://doi.org/10.1016/j.brainresrev.2010. 02.003.

Reichenbach, N., Delekate, A., Breithausen, B., Keppler, K., Poll, S., Schulte, T., Peter, J., Plescher, M., Hansen, J.N., Blank, N., Keller, A., Fuhrmann, M., Henneberger, C., Halle, A., Petzold, G.C., 2018. P2Y1 receptor blockade normalizes network dysfunction and cognition in an Alzheimer's disease model. J. Exp. Med. 215, 1649–1663. https:// doi.org/10.1084/jem.20171487.

Reyes, R.C., Verkhratsky, A., Parpura, V., 2012. Plasmalemmal Na^+/Ca^{2+} exchanger modulates Ca^{2+}−dependent exocytotic release of glutamate from rat cortical astrocytes. ASN Neuro 4, e00075. https://doi.org/10.1042/AN20110059.

Reyes, R.C., Verkhratsky, A., Parpura, V., 2013. TRPC1-mediated Ca2 + and Na + signalling in astroglia: differential filtering of extracellular cations. Cell Calcium 54, 120–125. https://doi.org/10.1016/j.ceca.2013.05.005.

Rizzoli, S., Sharma, G., Vijayaraghavan, S., 2002. Calcium rise in cultured neurons from medial septum elicits calcium waves in surrounding glial cells. Brain Res. 957, 287–297. https://doi.org/10.1016/s0006-8993(02)03618-1.

Ronco, V., Grolla, A.A., Glasnov, T.N., Canonico, P.L., Verkhratsky, A., Genazzani, A.A., Lim, D., 2014. Differential deregulation of astrocytic calcium signalling by amyloid-β, TNFα, IL-1β and LPS. Cell Calcium 55, 219–229. https://doi.org/10.1016/j.ceca. 2014.02.016.

Rose, C.R., Verkhratsky, A., 2016. Principles of sodium homeostasis and sodium signalling in astroglia. Glia 64, 1611–1627. https://doi.org/10.1002/glia.22964.

Rose, C.R., Ziemens, D., Verkhratsky, A., 2020. On the special role of NCX in astrocytes: translating Na +−transients into intracellular Ca2 + signals. Cell Calcium 86, 102154. https://doi.org/10.1016/j.ceca.2019.102154.

Rosenegger, D.G., Tran, C.H.T., Wamsteeker Cusulin, J.I., Gordon, G.R., 2015. Tonic local brain blood flow control by astrocytes independent of phasic neurovascular coupling. J. Neurosci. 35, 13463–13474. https://doi.org/10.1523/JNEUROSCI.1780- 15.2015.

Rossi, A., Pizzo, P., Filadi, R., 2019. Calcium, mitochondria and cell metabolism: a functional triangle in bioenergetics. Biochim. Biophys. Acta Mol. Cell Res. 1866, 1068–1078. https://doi.org/10.1016/j.bbamcr.2018.10.016.

Rosso, L., Mienville, J.-M., 2009. Pituicyte modulation of neurohormone output. Glia 57, 235–243. https://doi.org/10.1002/glia.20760.

Rothstein, J.D., Dykes-Hoberg, M., Pardo, C.A., Bristol, L.A., Jin, L., Kuncl, R.W., Kanai, Y., Hediger, M.A., Wang, Y., Schielke, J.P., Welty, D.F., 1996. Knockout of glutamate transporters reveals a major role for astroglial transport in excitotoxicity and clearance of glutamate. Neuron 16, 675–686. https://doi.org/10.1016/s0896-6273 (00)80086-0.

Rouach, N., Glowinski, J., Giaume, C., 2000. Activity-dependent neuronal control of gap-junctional communication in astrocytes. J. Cell Biol. 149, 1513–1526. https:// doi.org/10.1083/jcb.149.7.1513.

Rungta, R.L., Bernier, L.-P., Dissing-Olesen, L., Groten, C.J., LeDue, J.M., Ko, R., Drissler, S., MacVicar, B.A., 2016. Ca2+ transients in astrocyte fine processes occur via Ca2+ influx in the adult mouse hippocampus. Glia 64, 2093–2103. https://doi. org/10.1002/glia.23042.

Rusakov, D.A., Fine, A., 2003. Extracellular Ca2+ depletion contributes to fast activity-dependent modulation of synaptic transmission in the brain. Neuron 37, 287–297. https://doi.org/10.1016/s0896-6273(03)00025-4.

Salas, E., Carrasquero, L.M.G., Olivos-Oré, L.A., Bustillo, D., Artalejo, A.R., Miras-Portugal, M.T., Delicado, E.G., 2013. Purinergic P2X7 receptors mediate cell death in mouse cerebellar astrocytes in culture. J. Pharmacol. Exp. Ther. 347, 802–815. https:// doi.org/10.1124/jpet.113.209452.

Salido, G.M., Sage, S.O., Rosado, J.A., 2009. TRPC channels and store-operated ca(2+) entry. Biochim. Biophys. Acta 1793, 223–230. https://doi.org/10.1016/j.bbamcr. 2008.11.001.

Salman, M.M., Kitchen, P., Woodroofe, M.N., Brown, J.E., Bill, R.M., Conner, A.C., Conner, M.T., 2017. Hypothermia increases aquaporin 4 (AQP4) plasma membrane abundance in human primary cortical astrocytes via a calcium/transient receptor potential vanilloid 4 (TRPV4)- and calmodulin-mediated mechanism. Eur. J. Neurosci. 46, 2542–2547. https://doi.org/10.1111/ejn.13723.

Salter, M.G., Fern, R., 2005. NMDA receptors are expressed in developing oligodendrocyte processes and mediate injury. Nature 438, 1167–1171. https://doi.org/10.1038/ nature04301.

Santello, M., Toni, N., Volterra, A., 2019. Astrocyte function from information processing to cognition and cognitive impairment. Nat. Neurosci. 22, 154–166. https://doi.org/ 10.1038/s41593-018-0325-8.

Savtchouk, I., Volterra, A., 2018. Gliotransmission: beyond black-and-White. J. Neurosci. 38, 14–25. https://doi.org/10.1523/JNEUROSCI.0017-17.2017.

Scemes, E., Giaume, C., 2006. Astrocyte calcium waves: what they are and what they do. Glia 54, 716–725. https://doi.org/10.1002/glia.20374.

Schleich, C.L., 1894. Schmerzlose Operationen: Örtliche Betäubung mit indiffrenten Flüssigkeiten. Psychophysik des natürlichen und künstlichen Schlafes. Julius Springer, Berlin.

Schummers, J., Yu, H., Sur, M., 2008. Tuned responses of astrocytes and their influence on hemodynamic signals in the visual cortex. Science 320, 1638–1643. https://doi.org/10. 1126/science.1156120.

Semyanov, A., 2008. Can diffuse extrasynaptic signaling form a guiding template? Neurochem. Int. 52, 31–33. https://doi.org/10.1016/j.neuint.2007.07.021.

Semyanov, A., 2019. Spatiotemporal pattern of calcium activity in astrocytic network. Cell Calcium 78, 15–25. https://doi.org/10.1016/j.ceca.2018.12.007.

Semyanov, A., Henneberger, C., Agarwal, A., 2020. Making sense of astrocytic calcium signals—from acquisition to interpretation. Nat. Rev. Neurosci. 21, 551–564. https://doi.org/10.1038/s41583-020-0361-8.

Sharma, G., Vijayaraghavan, S., 2001. Nicotinic cholinergic signaling in hippocampal astrocytes involves calcium-induced calcium release from intracellular stores. Proc. Natl. Acad. Sci. U. S. A. 98, 4148–4153. https://doi.org/10.1073/pnas.071540198.

Sharp, A.H., Nucifora, F.C., Blondel, O., Sheppard, C.A., Zhang, C., Snyder, S.H., Russell, J.T., Ryugo, D.K., Ross, C.A., 1999. Differential cellular expression of isoforms of inositol 1,4,5-triphosphate receptors in neurons and glia in brain. J. Comp. Neurol. 406, 207–220.

Sheppard, C.A., Simpson, P.B., Sharp, A.H., Nucifora, F.C., Ross, C.A., Lange, G.D., Russell, J.T., 1997. Comparison of type 2 inositol 1,4,5-trisphosphate receptor distribution and subcellular Ca2+ release sites that support Ca2+ waves in cultured astrocytes. J. Neurochem. 68, 2317–2327. https://doi.org/10.1046/j.1471-4159.1997.68062317.x.

Sherwood, M.W., Arizono, M., Hisatsune, C., Bannai, H., Ebisui, E., Sherwood, J.L., Panatier, A., Oliet, S.H.R., Mikoshiba, K., 2017. Astrocytic IP3 Rs: contribution to Ca2+ signalling and hippocampal LTP. Glia 65, 502–513. https://doi.org/10.1002/glia.23107.

Shideman, C.R., Hu, S., Peterson, P.K., Thayer, S.A., 2006. CCL5 evokes calcium signals in microglia through a kinase-, phosphoinositide-, and nucleotide-dependent mechanism. J. Neurosci. Res. 83, 1471–1484. https://doi.org/10.1002/jnr.20839.

Shigetomi, E., Bushong, E.A., Haustein, M.D., Tong, X., Jackson-Weaver, O., Kracun, S., Xu, J., Sofroniew, M.V., Ellisman, M.H., Khakh, B.S., 2013a. Imaging calcium microdomains within entire astrocyte territories and endfeet with GCaMPs expressed using adeno-associated viruses. J. Gen. Physiol. 141, 633–647. https://doi.org/10.1085/jgp.201210949.

Shigetomi, E., Hirayama, Y.J., Ikenaka, K., Tanaka, K.F., Koizumi, S., 2018. Role of purinergic receptor P2Y1 in spatiotemporal Ca2+ dynamics in astrocytes. J. Neurosci. 38, 1383–1395. https://doi.org/10.1523/JNEUROSCI.2625-17.2017.

Shigetomi, E., Jackson-Weaver, O., Huckstepp, R.T., O'Dell, T.J., Khakh, B.S., 2013b. TRPA1 channels are regulators of astrocyte basal calcium levels and long-term potentiation via constitutive D-serine release. J. Neurosci. 33, 10143–10153. https://doi.org/10.1523/JNEUROSCI.5779-12.2013.

Shigetomi, E., Kracun, S., Sofroniew, M.V., Khakh, B.S., 2010. A genetically targeted optical sensor to monitor calcium signals in astrocyte processes. Nat. Neurosci. 13, 759–766. https://doi.org/10.1038/nn.2557.

Shih, P.-Y., Savtchenko, L.P., Kamasawa, N., Dembitskaya, Y., McHugh, T.J., Rusakov, D.A., Shigemoto, R., Semyanov, A., 2013. Retrograde synaptic signaling mediated by K+ efflux through postsynaptic NMDA receptors. Cell Rep. 5, 941–951. https://doi.org/10.1016/j.celrep.2013.10.026.

Simpson, P.B., Holtzclaw, L.A., Langley, D.B., Russell, J.T., 1998a. Characterization of ryanodine receptors in oligodendrocytes, type 2 astrocytes, and O-2A progenitors. J. Neurosci. Res. 52, 468–482. https://doi.org/10.1002/(SICI)1097-4547(19980515)52:4<468::AID-JNR11>3.0.CO;2-#.

Simpson, P.B., Mehotra, S., Langley, D., Sheppard, C.A., Russell, J.T., 1998b. Specialized distributions of mitochondria and endoplasmic reticulum proteins define Ca2+ wave amplification sites in cultured astrocytes. J. Neurosci. Res. 52, 672–683. https://doi.org/10.1002/(SICI)1097-4547(19980615)52:6<672::AID-JNR6>3.0.CO;2-5.

Solovyova, N., Veselovsky, N., Toescu, E.C., Verkhratsky, A., 2002. Ca(2+) dynamics in the lumen of the endoplasmic reticulum in sensory neurons: direct visualization of ca(2+)-induced ca(2+) release triggered by physiological ca(2+) entry. EMBO J. 21, 622–630. https://doi.org/10.1093/emboj/21.4.622.

Sompol, P., Norris, C.M., 2018. Ca2+, astrocyte activation and Calcineurin/NFAT Signaling in age-related neurodegenerative diseases. Front. Aging Neurosci. 10, 199. https://doi.org/10.3389/fnagi.2018.00199.

Sonoda, K., Matsui, T., Bito, H., Ohki, K., 2018. Astrocytes in the mouse visual cortex reliably respond to visual stimulation. Biochem. Biophys. Res. Commun. 505, 1216–1222. https://doi.org/10.1016/j.bbrc.2018.10.027.

Sosunov, A.A., Wu, X., Tsankova, N.M., Guilfoyle, E., McKhann, G.M., Goldman, J.E., 2014. Phenotypic heterogeneity and plasticity of isocortical and hippocampal astrocytes in the human brain. J. Neurosci. 34, 2285–2298. https://doi.org/10.1523/JNEUROSCI.4037-13.2014.

Srinivasan, R., Huang, B.S., Venugopal, S., Johnston, A.D., Chai, H., Zeng, H., Golshani, P., Khakh, B.S., 2015. Ca(2+) signaling in astrocytes from Ip3r2(−/−) mice in brain slices and during startle responses in vivo. Nat. Neurosci. 18, 708–717. https://doi.org/10.1038/nn.4001.

Stallcup, W.B., 1981. The NG2 antigen, a putative lineage marker: immunofluorescent localization in primary cultures of rat brain. Dev. Biol. 83, 154–165. https://doi.org/10.1016/s0012-1606(81)80018-8.

Stobart, J.L., Ferrari, K.D., Barrett, M.J.P., Glück, C., Stobart, M.J., Zuend, M., Weber, B., 2018a. Cortical circuit activity evokes rapid astrocyte calcium signals on a similar timescale to neurons. Neuron 98, 726–735.e4. https://doi.org/10.1016/j.neuron.2018.03.050.

Stobart, J.L., Ferrari, K.D., Barrett, M.J.P., Stobart, M.J., Looser, Z.J., Saab, A.S., Weber, B., 2018b. Long-term in vivo calcium imaging of astrocytes reveals distinct cellular compartment responses to sensory stimulation. Cereb. Cortex 28, 184–198. https://doi.org/10.1093/cercor/bhw366.

Stratoulias, V., Venero, J.L., Tremblay, M.-È., Joseph, B., 2019. Microglial subtypes: diversity within the microglial community. EMBO J. 38, e101997. https://doi.org/10.15252/embj.2019101997.

Sun, W., McConnell, E., Pare, J.-F., Xu, Q., Chen, M., Peng, W., Lovatt, D., Han, X., Smith, Y., Nedergaard, M., 2013. Glutamate-dependent neuroglial calcium signaling differs between young and adult brain. Science 339, 197–200. https://doi.org/10.1126/science.1226740.

Taheri, M., Handy, G., Borisyuk, A., White, J.A., 2017. Diversity of evoked astrocyte Ca2+ dynamics quantified through experimental measurements and mathematical Modeling. Front. Syst. Neurosci. 11, 79. https://doi.org/10.3389/fnsys.2017.00079.

Takata, N., Hirase, H., 2008. Cortical layer 1 and layer 2/3 astrocytes exhibit distinct calcium dynamics in vivo. PLoS One 3, e2525. https://doi.org/10.1371/journal.pone.0002525.

Takata, N., Mishima, T., Hisatsune, C., Nagai, T., Ebisui, E., Mikoshiba, K., Hirase, H., 2011. Astrocyte calcium signaling transforms cholinergic modulation to cortical plasticity in vivo. J. Neurosci. 31, 18155–18165. https://doi.org/10.1523/JNEUROSCI.5289-11.2011.

Takey, Y., Pearl, G.S., 1984. Ultrastructure of the human neurohypophysis. In: Motta, P.M. (Ed.), Ultrastructure of Endocrine Cells and Tissues, Electron Microscopy in Biology and Medicine. Springer US, Boston, MA, pp. 77–88. https://doi.org/10.1007/978-1-4613-3861-1_7.

Takuma, K., Matsuda, T., Hashimoto, H., Asano, S., Baba, A., 1994. Cultured rat astrocytes possess Na(+)-Ca2+ exchanger. Glia 12, 336–342. https://doi.org/10.1002/glia.440120410.

Talantova, M., Sanz-Blasco, S., Zhang, X., Xia, P., Akhtar, M.W., Okamoto, S., Dziewczapolski, G., Nakamura, T., Cao, G., Pratt, A.E., Kang, Y.-J., Tu, S., Molokanova, E., McKercher, S.R., Hires, S.A., Sason, H., Stouffer, D.G., Buczynski, M.W., Solomon, J.P., Michael, S., Powers, E.T., Kelly, J.W., Roberts, A., Tong, G., Fang-Newmeyer, T., Parker, J., Holland, E.A., Zhang, D., Nakanishi, N., Chen, H.-S.V., Wolosker, H., Wang, Y., Parsons, L.H., Ambasudhan, R., Masliah, E., Heinemann, S.F., Piña-Crespo, J.C., Lipton, S.A.,

2013. Aβ induces astrocytic glutamate release, extrasynaptic NMDA receptor activation, and synaptic loss. Proc. Natl. Acad. Sci. U. S. A. 110, E2518–E2527. https://doi.org/10. 1073/pnas.1306832110.

Tamamushi, S., Nakamura, T., Inoue, T., Ebisui, E., Sugiura, K., Bannai, H., Mikoshiba, K., 2012. Type 2 inositol 1,4,5-trisphosphate receptor is predominantly involved in agonist-induced ca(2+) signaling in Bergmann glia. Neurosci. Res. 74, 32–41. https://doi.org/10.1016/j.neures.2012.06.005.

Tanaka, M., Shih, P.-Y., Gomi, H., Yoshida, T., Nakai, J., Ando, R., Furuichi, T., Mikoshiba, K., Semyanov, A., Itohara, S., 2013. Astrocytic Ca2+ signals are required for the functional integrity of tripartite synapses. Mol. Brain 6, 6. https://doi.org/10. 1186/1756-6606-6-6.

Tapella, L., Cerruti, M., Biocotino, I., Stevano, A., Rocchio, F., Canonico, P.L., Grilli, M., Genazzani, A.A., Lim, D., 2018. TGF-β2 and TGF-β3 from cultured β-amyloid-treated or 3xTg-AD-derived astrocytes may mediate astrocyte-neuron communication. Eur. J. Neurosci. 47, 211–221. https://doi.org/10.1111/ejn.13819.

Tapella, L., Soda, T., Mapelli, L., Bortolotto, V., Bondi, H., Ruffinatti, F.A., Dematteis, G., Stevano, A., Dionisi, M., Ummarino, S., Di Ruscio, A., Distasi, C., Grilli, M., Genazzani, A.A., D'Angelo, E., Moccia, F., Lim, D., 2020a. Deletion of calcineurin from GFAP-expressing astrocytes impairs excitability of cerebellar and hippocampal neurons through astroglial Na+ /K+ ATPase. Glia 68, 543–560. https://doi.org/10. 1002/glia.23737.

Tapella, L., Soda, T., Mapelli, L., Bortolotto, V., Bondi, H., Ruffinatti, F.A., Dematteis, G., Stevano, A., Dionisi, M., Ummarino, S., Di Ruscio, A., Distasi, C., Grilli, M., Genazzani, A.A., D'Angelo, E., Moccia, F., Lim, D., 2020b. Deletion of calcineurin from GFAP-expressing astrocytes impairs excitability of cerebellar and hippocampal neurons through astroglial Na+ /K+ ATPase. Glia 68, 543–560. https://doi.org/10. 1002/glia.23737.

Teaktong, T., Graham, A., Court, J., Perry, R., Jaros, E., Johnson, M., Hall, R., Perry, E., 2003. Alzheimer's disease is associated with a selective increase in alpha7 nicotinic ace-tylcholine receptor immunoreactivity in astrocytes. Glia 41, 207–211. https://doi.org/ 10.1002/glia.10132.

Toth, A.B., Hori, K., Novakovic, M.M., Bernstein, N.G., Lambot, L., Prakriya, M., 2019. CRAC channels regulate astrocyte Ca^{2+} signaling and gliotransmitter release to modu-late hippocampal GABAergic transmission. Sci. Signal. 12, eaaw5450. https://doi.org/ 10.1126/scisignal.aaw5450.

Traaseth, N., Elfering, S., Solien, J., Haynes, V., Giulivi, C., 2004. Role of calcium signaling in the activation of mitochondrial nitric oxide synthase and citric acid cycle. Biochim. Biophys. Acta 1658, 64–71. https://doi.org/10.1016/j.bbabio.2004.04.015.

Tuschick, S., Kirischuk, S., Kirchhoff, F., Liefeldt, L., Paul, M., Verkhratsky, A., Kettenmann, H., 1997. Bergmann glial cells in situ express endothelinB receptors linked to cytoplasmic calcium signals. Cell Calcium 21, 409–419. https://doi.org/10. 1016/s0143-4160(97)90052-x.

Underhill, S.M., Wheeler, D.S., Amara, S.G., 2015. Differential regulation of two isoforms of the glial glutamate transporter EAAT2 by DLG1 and CaMKII. J. Neurosci. 35, 5260–5270. https://doi.org/10.1523/JNEUROSCI.4365-14.2015.

Van Bockstaele, E.J., Colago, E.E., 1996. Selective distribution of the NMDA-R1 glutamate receptor in astrocytes and presynaptic axon terminals in the nucleus locus coeruleus of the rat brain: an immunoelectron microscopic study. J. Comp. Neurol. 369, 483–496. https://doi.org/10.1002/(SICI)1096-9861(19960610)369:4<483::AID-CNE1>3.0.CO;2-0.

Vardjan, N., Parpura, V., Verkhratsky, A., Zorec, R., 2019. Gliocrine system: Astroglia as secretory cells of the CNS. Adv. Exp. Med. Biol. 1175, 93–115. https://doi.org/10. 1007/978-981-13-9913-8_4.

Vardjan, N., Parpura, V., Zorec, R., 2016. Loose excitation-secretion coupling in astrocytes. Glia 64, 655–667. https://doi.org/10.1002/glia.22920.

Vardjan, N., Zorec, R., 2015. Excitable astrocytes: Ca(2+)- and cAMP-regulated exocytosis. Neurochem. Res. 40, 2414–2424. https://doi.org/10.1007/s11064-015-1545-x.

Verkhratsky, A., Butt, A.M., 2013. Glial Physiology and Pathophysiology. Wiley-Blackwell, Chichester.

Verkhratsky, A., Chvátal, A., 2020. NMDA receptors in astrocytes. Neurochem. Res. 45, 122–133. https://doi.org/10.1007/s11064-019-02750-3.

Verkhratsky, A., Kettenmann, H., 1996. Calcium signalling in glial cells. Trends Neurosci. 19, 346–352. https://doi.org/10.1016/0166-2236(96)10048-5.

Verkhratsky, A., Nedergaard, M., 2014. Astroglial cradle in the life of the synapse. Philos. Trans. R. Soc. Lond. B Biol. Sci. 369, 20130595. https://doi.org/10.1098/rstb.2013.0595.

Verkhratsky, A., Nedergaard, M., 2016. The homeostatic astroglia emerges from evolutionary specialization of neural cells. Philos. Trans. R. Soc. Lond. B Biol. Sci. 371, 20150428. https://doi.org/10.1098/rstb.2015.0428.

Verkhratsky, A., Nedergaard, M., 2018. Physiology of astroglia. Physiol. Rev. 98, 239–389. https://doi.org/10.1152/physrev.00042.2016.

Verkhratsky, A., Noda, M., Parpura, V., 2015. Microglia: structure and function. In: Anatomy and Physiology, Systems. vol. 2, pp. 109–113. https://doi.org/10.1016/B978-0-12-397025-1.00356-0.

Verkhratsky, A., Parpura, V., 2014. Store-operated calcium entry in neuroglia. Neurosci. Bull. 30, 125–133. https://doi.org/10.1007/s12264-013-1343-x.

Verkhratsky, A., Reyes, R.C., Parpura, V., 2014. TRP channels coordinate ion signalling in astroglia. Rev. Physiol. Biochem. Pharmacol. 166, 1–22. https://doi.org/10.1007/112_2013_15.

Verkhratsky, A., Rodrigues, J.J., Pivoriunas, A., Zorec, R., Semyanov, A., 2019. Astroglial atrophy in Alzheimer's disease. Pflugers Arch. 471, 1247–1261. https://doi.org/10.1007/s00424-019-02310-2.

Verkhratsky, A., Rodríguez, J.J., Parpura, V., 2012. Calcium signalling in astroglia. Mol. Cell. Endocrinol. 353, 45–56. https://doi.org/10.1016/j.mce.2011.08.039.

Verkhratsky, A., Semyanov, A., Zorec, R., 2020. Physiology of astroglial excitability. Function 1 (2), zqaa016. https://doi.org/10.1093/function/zqaa016.

Verkhratsky, A., Solovyova, N., Toescu, E.C., 2002. Calcium excitability of glial cells. The Tripartite Synapse: Glia in Synaptic Transmission. Oxford University Press, pp. 99–109.

Verkhratsky, A., Trebak, M., Perocchi, F., Khananshvili, D., Sekler, I., 2018. Crosslink between calcium and sodium signalling. Exp. Physiol. 103, 157–169. https://doi.org/10.1113/EP086534.

Verkhratsky, A., Untiet, V., Rose, C.R., 2020b. Ionic signalling in astroglia beyond calcium. J. Physiol. (Lond.) 598, 1655–1670. https://doi.org/10.1113/JP277478.

Verkhratsky, A., Zorec, R., Parpura, V., 2017. Stratification of astrocytes in healthy and diseased brain. Brain Pathol. 27, 629–644. https://doi.org/10.1111/bpa.12537.

Verkhratsky, A.N., Trotter, J., Kettenmann, H., 1990. Cultured glial precursor cells from mouse cortex express two types of calcium currents. Neurosci. Lett. 112, 194–198. https://doi.org/10.1016/0304-3940(90)90202-k.

Virchow, R., 1856. Ueber das granulirte Ansehen der Wandungen der Gehirnventrikel. In: Virchow, R. (Ed.), Gesammelte Abhandlungen zur wissenschaftlichen Medicin. Frankfurt A.M.: Meidinger Sohn & Comp, pp. 885–891.

Virchow, R., 1858. Die Cellularpathologie in ihrer Begründung auf physiologische und pathologische Gewebelehre 20 Vorlesungen, gehalten während d. Monate Febr., März u. April 1858 im Patholog. Inst. zu Berlin. August Hirschwald, Berlin.

Volterra, A., Liaudet, N., Savtchouk, I., 2014. Astrocyte Ca^{2+} signalling: an unexpected complexity. Nat. Rev. Neurosci. 15, 327–335. https://doi.org/10.1038/nrn3725.

Walsh, C., Barrow, S., Voronina, S., Chvanov, M., Petersen, O.H., Tepikin, A., 2009. Modulation of calcium signalling by mitochondria. Biochim. Biophys. Acta 1787, 1374–1382. https://doi.org/10.1016/j.bbabio.2009.01.007.

Wang, C., Pralong, W.F., Schulz, M.F., Rougon, G., Aubry, J.M., Pagliusi, S., Robert, A., Kiss, J.Z., 1996. Functional N-methyl-D-aspartate receptors in O-2A glial precursor cells: a critical role in regulating polysialic acid-neural cell adhesion molecule expression and cell migration. J. Cell Biol. 135, 1565–1581. https://doi.org/10.1083/jcb. 135.6.1565.

Wang, F., Smith, N.A., Xu, Q., Fujita, T., Baba, A., Matsuda, T., Takano, T., Bekar, L., Nedergaard, M., 2012a. Astrocytes modulate neural network activity by Ca^{2+}-dependent uptake of extracellular K+. Sci. Signal. 5, ra26. https://doi.org/10.1126/scisignal. 2002334.

Wang, X., Lou, N., Xu, Q., Tian, G.-F., Peng, W.G., Han, X., Kang, J., Takano, T., Nedergaard, M., 2006. Astrocytic Ca^{2+} signaling evoked by sensory stimulation in vivo. Nat. Neurosci. 9, 816–823. https://doi.org/10.1038/nn1703.

Wang, X., Saegusa, H., Huntula, S., Tanabe, T., 2019. Blockade of microglial Cav1.2 Ca^{2+} channel exacerbates the symptoms in a Parkinson's disease model. Sci. Rep. 9, 9138. https://doi.org/10.1038/s41598-019-45681-3.

Wang, Y., Zhu, N., Wang, K., Zhang, Z., Wang, Y., 2012b. Identification of α7 nicotinic acetylcholine receptor on hippocampal astrocytes cultured in vitro and its role on inflammatory mediator secretion. Neural Regen. Res. 7, 1709–1714. https://doi.org/10. 3969/j.issn.1673-5374.2012.22.005.

Watterson, D.M., Mirzoeva, S., Guo, L., Whyte, A., Bourguignon, J.J., Hibert, M., Haiech, J., Van Eldik, L.J., 2001. Ligand modulation of glial activation: cell permeable, small molecule inhibitors of serine-threonine protein kinases can block induction of interleukin 1 beta and nitric oxide synthase II. Neurochem. Int. 39, 459–468. https://doi.org/10.1016/s0197-0186(01)00053-5.

Weisman, G.A., Camden, J.M., Peterson, T.S., Ajit, D., Woods, L.T., Erb, L., 2012. P2 receptors for extracellular nucleotides in the central nervous system: role of P2X7 and $P2Y_2$ receptor interactions in neuroinflammation. Mol. Neurobiol. 46, 96–113. https://doi.org/10.1007/s12035-012-8263-z.

Winship, I.R., Plaa, N., Murphy, T.H., 2007. Rapid astrocyte calcium signals correlate with neuronal activity and onset of the hemodynamic response in vivo. J. Neurosci. 27, 6268–6272. https://doi.org/10.1523/JNEUROSCI.4801-06.2007.

Wu, Y.-W., Gordleeva, S., Tang, X., Shih, P.-Y., Dembitskaya, Y., Semyanov, A., 2019. Morphological profile determines the frequency of spontaneous calcium events in astrocytic processes. Glia 67, 246–262. https://doi.org/10.1002/glia.23537.

Yamaguchi, N., 2020. Molecular insights into calcium dependent regulation of ryanodine receptor calcium release channels. Adv. Exp. Med. Biol. 1131, 321–336. https://doi. org/10.1007/978-3-030-12457-1_13.

Young, J.K., Baker, J.H., Muller, T., 1996. Immunoreactivity for brain-fatty acid binding protein in gomori-positive astrocytes. Glia 16, 218–226. https://doi.org/10.1002/ (SICI)1098-1136(199603)16:3<218::AID-GLIA4>3.0.CO;2-Y.

Young, J.K., McKenzie, J.C., Baker, J.H., 1990. Association of iron-containing astrocytes with dopaminergic neurons of the arcuate nucleus. J. Neurosci. Res. 25, 204–213. https://doi.org/10.1002/jnr.490250208.

Zhang, Y., Chen, K., Sloan, S.A., Bennett, M.L., Scholze, A.R., O'Keeffe, S., Phatnani, H.P., Guarnieri, P., Caneda, C., Ruderisch, N., Deng, S., Liddelow, S.A., Zhang, C., Daneman, R., Maniatis, T., Barres, B.A., Wu, J.Q., 2014. An RNA-sequencing transcriptome and splicing database of glia, neurons, and vascular cells of the cerebral cortex. J. Neurosci. 34, 11929–11947. https://doi.org/10.1523/ JNEUROSCI.1860-14.2014.

Zheng, K., Bard, L., Reynolds, J.P., King, C., Jensen, T.P., Gourine, A.V., Rusakov, D.A., 2015. Time-resolved imaging reveals heterogeneous landscapes of Nanomolar Ca(2+) in neurons and Astroglia. Neuron 88, 277–288. https://doi.org/10.1016/j.neuron. 2015.09.043.

Ziemens, D., Oschmann, F., Gerkau, N.J., Rose, C.R., 2019. Heterogeneity of activity-induced sodium transients between astrocytes of the mouse hippocampus and neocortex: mechanisms and consequences. J. Neurosci. 39, 2620–2634. https://doi. org/10.1523/JNEUROSCI.2029-18.2019.

CHAPTER TWO

Ca^{2+} homeostasis in brain microvascular endothelial cells

Roberta Stoica^{a,b,†}, Călin Mircea Rusu^{a,b,†}, Cristina Elena Staicu^{a,c,†}, Andreea Elena Burlacu^{a}, Mihai Radu^{b}, and Beatrice Mihaela Radu^{a,d,*}

^{a}Department of Anatomy, Animal Physiology and Biophysics, Faculty of Biology, University of Bucharest, Bucharest, Romania
^{b}Department of Life and Environmental Physics, 'Horia Hulubei' National Institute for Physics and Nuclear Engineering, Măgurele, Romania
^{c}Center for Advanced Laser Technologies, National Institute for Laser, Plasma and Radiation Physics, Măgurele, Romania
^{d}Life, Environmental and Earth Sciences Division, Research Institute of the University of Bucharest, Bucharest, Romania
*Corresponding author: e-mail address: beatrice.radu@bio.unibuc.ro

Contents

---

† R Stoica, CM Rusu and CE Staicu had equal contributions.

International Review of Cell and Molecular Biology, Volume 362
ISSN 1937-6448
https://doi.org/10.1016/bs.ircmb.2021.01.001

Abstract

Blood brain barrier (BBB) is formed by the brain microvascular endothelial cells (BMVECs) lining the wall of brain capillaries. Its integrity is regulated by multiple mechanisms, including up/downregulation of tight junction proteins or adhesion molecules, altered Ca^{2+} homeostasis, remodeling of cytoskeleton, that are confined at the level of BMVECs. Beside the contribution of BMVECs to BBB permeability changes, other cells, such as pericytes, astrocytes, microglia, leukocytes or neurons, etc. are also exerting direct or indirect modulatory effects on BBB. Alterations in BBB integrity play a key role in multiple brain pathologies, including neurological (e.g. epilepsy) and neurodegenerative disorders (e.g. Alzheimer's disease, Parkinson's disease, amyotrophic lateral sclerosis etc.). In this review, the principal Ca^{2+} signaling pathways in brain microvascular endothelial cells are discussed and their contribution to BBB integrity is emphasized. Improving the knowledge of Ca^{2+} homeostasis alterations in BMVECa is fundamental to identify new possible drug targets that diminish/prevent BBB permeabilization in neurological and neurodegenerative disorders.

Abbreviations

Ach	acetylcholine
AD	Alzheimer's disease
ADP	adenosine diphosphate
AJ	adherens junctions
ALS	amyotrophic lateral sclerosis
ATP	adenosine triphosphate
BBB	blood brain barrier
BKR	bradykinin receptors
BMVECs	brain microvascular endothelial cells
cAMP	cyclic adenosine monophosphate
CNS	central nervous system
CRAC	calcium release-activated calcium channel protein
CREB	cAMP response element-binding protein
DAG	diacylglycerol
ER	endoplasmic reticulum
ICAM-1	intercellular Adhesion Molecule 1
IP3	inositol 1,4,5-trisphosphate
IP3R	inositol 1,4,5-trisphosphate receptor
LFA-1	lymphocyte function-associated antigen 1
MAC-1	macrophage-1 antigen

MLC	myosin light chain
MLCK	myosin light chain kinase
NO	nitric oxide
PD	Parkinson's disease
PIP2	phosphatidylinositol 4,5-bisphosphate
PKC	protein kinase C
PLCβ	phospholipase Cβ
PM2.5	particles smaller than 2.5 μM
PMCA	plasma membrane Ca^{2+}-ATPase
PMNs	polymorphonuclear neutrophils
ROS	reactive oxygen species
SEM	scanning electron microscopy
SERCA	sarco/endoplasmic reticulum Ca^{2+}-ATPase
siRNA	small interfering RNA
SOC	store-operated channels
SOCE	store-operated Ca^{2+} entry
SPECT	single-photon emission computerized tomography
SSH	slingshot homolog
SSH1L	slingshot protein phosphatase 1
STIM	stromal interaction molecule
TcGH	99mTc-glucoheptonate
TEER	trans-endothelial electrical resistance
TJ	tight junctions
TRP	transient receptor potential channels
TRPC	canonical TRP channels
UTP	uridine triphosphate
VEGF	vascular endothelial growth factor
ZO	zonula occludens

1. Introduction

Calcium is a key ion in the central nervous system (CNS). According to the compartmental model of brain calcium, calcium is distributed in CNS among five compartments: free Ca^{2+} in the extracellular space ($\sim$10%), Ca^{2+} closely associated with the extracellular surface of plasma membrane ($\sim$55%), Ca^{2+} with moderate binding affinity in intracellular compartment ($\sim$17%), tightly bound Ca^{2+} in the non-exchangeable intracellular compartment ($\sim$15%), and free Ca^{2+} in the cytoplasmic compartment ($<$0.01%) (Kass and Lipton, 1986; Moriarty, 1980; Newman et al., 1995; Yarlagadda et al., 2007). In the whole brain, the steady-state values of calcium were estimated to be approximately 3.5–5 µmol/g (Anghileri et al., 1994; Dienel et al., 1995; Mies et al., 1993; Murphy et al., 1988).

Endothelial cells line the inner lumen of blood vessels and are, therefore, exposed to a multitude of chemical and physical stimuli from the bloodstream, including chemokines, hormones, growth factors, mechanical stress, etc. (Banks, 2012; Moccia et al., 2012, 2019). When exposed to endogenous or exogenous stimuli various proteins expressed in the brain endothelium are activated, further triggering calcium signaling pathways, tight junctions loosening and/or cytoskeleton remodeling, that mediate a myriad of physiological processes, including vascular relaxation/constriction, extravasation of solutes, fluid, hormones, and macromolecules, and extravasation of red blood cells and leukocytes (Fig. 1).

Multiple calcium signaling pathways involved in calcium homeostasis have been characterized in brain endothelium. However, the ion channels/receptors that activate these signaling pathways have been described

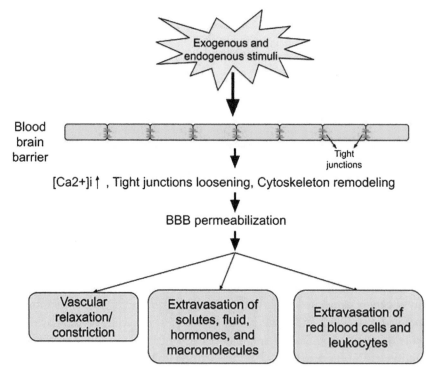

Fig. 1 Calcium signaling in brain endothelium regulates multiple physiological processes.

to vary in expression across the vascular tree and in different animal species. In particular, their expression in *in vitro* endothelial cells cultures may vary between different preparations of the same type, for example due to serial passages (Earley and Brayden, 2015; Moccia et al., 2012; Smani et al., 2018; Thakore and Earley, 2019), or may be distinct between *in vivo* and *in vitro* samples. Therefore, in this review we have focused our attention only on the studies containing data on the expression/functional expression of the ion channels/receptors controlling calcium signaling pathways in brain microvascular endothelial cells and on their contribution to calcium homeostasis.

2. Calcium measurements in brain microvascular endothelial cells

2.1 Calcium measurements in cell cultures

Calcium is an important second messenger and it is involved in a variety of cellular processes and functions, such as cell proliferation, metabolism, migration, fertilization and gene transcription, in various types of cells (Berridge et al., 2003). Changes in the intracellular calcium concentration, generated as response to an extracellular agent, can modulate a number of vascular mechanisms (inflammation, coagulation, regulation of vascular permeability). In particular, in brain microvascular endothelial cells, calcium ions' intracellular concentration plays an important role in disrupting tight junctions between endothelial cells and also affects the permeability of the blood–brain barrier (BBB) (Brown and Davis, 2002; Dalal et al., 2020).

In order to demonstrate that certain vasoactive agents (histamine, brady-kinin, endothelin) and nucleotides (adenosine triphosphate (ATP), adenosine diphosphate (ADP) and uridine triphosphate (UTP)) increase the BBB permeability along with an elevation in the intracellular calcium concentration, different studies were conducted on cultured brain endothelial cells (Abbott, 1998, 2000; Revest et al., 1991).

In order to measure the agonist-induced intracellular calcium changes, standard commercially available fluorescent calcium indicators are used (the most usual fluorescent dye is the ratiometric FURA-2AM probe). The fluorescence signal, recorded as the 340/380 nm excitation ratio, allows a quantitative measurement of the intracellular calcium concentration and as the technique is becoming very popular, it has began being used by many scientific groups (Radu et al., 2017; Tsai et al., 2019; Wu et al., 2019). There are a lot of studies using this method on the *in vitro* murine brain microvascular endothelial cells (bEnd.3), having as target the calcium concentration changes after different treatments (Leong et al., 2018; Park et al., 2010; Radu et al., 2017; Tsai et al., 2019; Wu et al., 2019). In order to exemplify, in the Figs. 2 and 3 we presented one typical protocol used in our laboratory. In these particular recordings, the calcium transients which appear in the endothelial cells after the stimulation with different endogenous agents such as ATP and acetylcholine can be observed. Either if it is about the cellular response to stimuli such as Ach (Fig. 2) or ATP (Fig. 3), the endothelial cell has the capacity to refill its internal deposits after a certain period and also to react to a new stimuli by increasing its intracellular calcium concentration (the second transient, almost comparable to the first one). Such double stimulus application protocol reveals the effect of certain drugs and the cellular response to the stimulus by intercalating a certain period of time exposure to the drug just before the second application of the stimulus. Thus the effect can be observed on the same cell. For these experiments, the cell line bEnd.3 (ATCC, CRL-2299™) which represents endothelial cells from mice cerebral microvasculature was used.

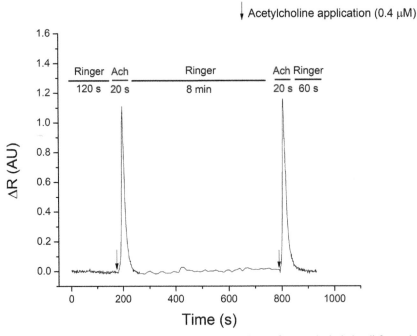

Fig. 2 Original ratiometric calcium imaging recording of an endothelial cell from the mouse brain microvasculature after acetylcholine stimulation of the muscarinic receptors.

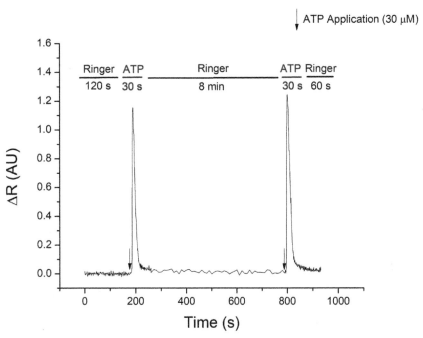

Fig. 3 Original ratiometric calcium imaging recording of endothelial cells from the mouse brain microvasculature after ATP stimulation of the purinergic receptors.

In the literature reports, in almost all the articles which present calcium imaging recordings and their analysis, the single parameter of interest is the amplitude of the calcium transient, and sometimes the duration. This way all the assumptions made over the capacity of certain antagonists or partial agonists to inhibit the cellular signaling, are developed based on the calcium transient amplitude modifications. On the other hand, the shape of the calcium transient could be characterized by a large set of parameters as: amplitude, area, rising velocity, decreasing velocity, latency and duration (Radu et al., 2015). These parameters, which are illustrated in Fig. 4, are strongly correlated with the molecular processes initiated by endogenous agent stimulation. Quantitative analysis of the transient shape using the abovementioned parameters is an appropriate solution to provide a more complex description of the results obtained from such kind of experiments.

Such complex analysis of the data resulting from intracellular calcium concentration recordings could be necessary in order to observe the alteration degree of the molecular pathways after certain treatments. For example, in Fig. 5, there are illustrated two ratiometric calcium imaging recordings

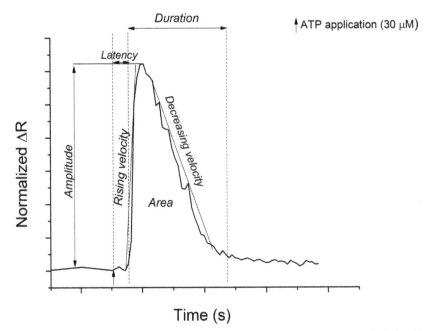

Fig. 4 Calcium transient triggered by Ach in brain microvascular endothelial cells. Parameters that characterize the calcium transient are indicated, i.e., amplitudine, latency, rising velocity, decreasing velocity, duration, area (for more details regarding the protocol of Ca^{2+} signal analysis see Radu et al., 2015).

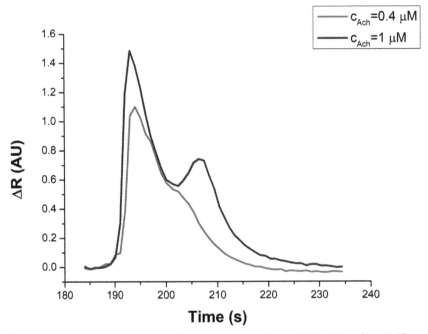

Fig. 5 Calcium transients triggered by two concentrations of ACh, 0.4 and 1 μM. These calcium signals are examples that can be analyzed using the parameters indicated in Fig. 4. One can observe the complexity of the calcium signal triggered at higher ACh concentrations that requires an extensive analysis, and cannot be limited to the measurement of the transient amplitude.

after the application of two concentrations of acetylcholine: 0.4 and 1 µM on untreated bEnd.3 endothelial cells. It can be observed that once the endogenous agent concentration is high, the molecular processes are activated differently. Although the application was simultaneous, the analyzed cells, according to their Ach concentration, respond differently. At small concentrations of endogenous agents, in this particular case acetylcholine, the cell response is diminished in comparison with the case for higher concentrations. Also, it can be observed that after one certain threshold value for the acetylcholine concentration, the return to the normal physiological concentration of intracellular calcium will be attained through different processes (i.e. SOCE mechanism is strongly involved). It has become obvious that such type of parametrization could offer information about the molecular processes activated by the stimulation with an endogenous agent.

In this particular case of Ach induced calcium transients, the amplitude and area values show the capacity of the endothelial cell to mobilize the calcium ions as response to the interaction with extracellular acetylcholine molecules. The latency of the transients can be correlated with the initial phase of the Ach triggered signaling pathways (activation of the muscarinic receptors, phospholipase C hydrolysis, scinding PIP$_2$ in DAG and IP$_3$ and also the activation of the IP$_3$ specific receptors found on the endoplasmic reticulum membrane). Also, the rising velocity, which represents the maximal slope of the ascendent phase, can be associated with the cellular capacity of calcium ions release from the internal deposits. For the descendent part, the decreasing velocity gives the cell the ability to restore the calcium deposits. Here, it can be observed a difference between the two descending parts of the calcium transients (Fig. 5): for higher concentration of endogenous agent, the process through which the endothelial cell returns to physiological calcium concentrations involves other mechanisms (i.e. SOCE mechanism). On the endothelial cells' surface, there are expressed a variety of ion channels which mediate the calcium ions influx from the extracellular media and which will also determine a calcium mobilization from internal stores (especially from endoplasmic reticulum) (Filippini et al., 2019). From this category, it is important to mention voltage-dependent calcium channels, transient receptor potential channels (TRP) (Thakore and Earley, 2019), ORAI family of calcium channels and store-operated calcium channels (SOC) (Shinde et al., 2013).

Not only the exogen or endogen chemical factors can be used as stimuli to induce intracellular calcium transients. Also, physical factors such as shear stress (Jafarnejad et al., 2015), ultrasounds (Park et al., 2010), ionizing radiation

(Baselet et al., 2019; Kabacik and Raj, 2017) are able to trigger changes of intracellular free calcium concentration. A relevant issue revealed by the experiments done with endothelial cell cultures under shear stress, pointed out the importance of the conditions regarding the movement of the culture medium during the experiments (Ando and Yamamoto, 2013; Scheitlin et al., 2016). In most of the experiments the static condition is used, but in order to better mimic the *in vivo* case a dynamic condition is more appropriate. It was proved by several reports that not only the movement of the medium can modulate the endothelial cell physiology, in particular the intracellular calcium level, but also the shear stress profile during time is important (Jafarnejad et al., 2015). The shear stress produced by the flow of the external medium on the cells in culture modulates the expression and function of several cellular components like membrane ion channels and receptors, cellular cytoskeleton (Ando and Yamamoto, 2013).

However, a more relevant approach with respect to clinical translation of results is given by studies made on the *in vitro* human models of BBB (hCMEC/D3 or HBMEC cell lines) (Al Suleimani and Hiley, 2016; Rakkar and Bayraktutan, 2016).

At temporal level, the changes of intracellular calcium concentration can be oscillatory due to the repetitive occurrence of calcium transients. This continuous oscillation allows the maintenance of the temporal control of the cellular functions, but also the avoidance of possible toxic effects caused by an elevated permanent concentration. At spatial level, the calcium signals can be initially generated through local fast stimulation (puffs) as result of activation of one or more specific receptors, followed by the appearance of a global intracellular calcium wave. These waves are not specific for a single cell, as they become intercellular events. They can be seen as an increase of the intracellular calcium concentration which starts from one cell and travels along the neighboring cells. To assure the propagation of the calcium intercellular wave, two mechanisms are used: direct intercellular communication through gap junction and paracrine signaling (Leybaert and Sanderson, 2012).

2.2 Calcium measurements in freshly isolated brain endothelium

Throughout the broader literature, various studies have reported the association of vascular endothelial dysfunction with the appearance of chronic diseases such as hypertension, diabetes or even heart failure (Endemann and Schiffrin, 2004). This justifies the significance of studying endothelial function in physiological as well as in pathological conditions. Vascular endothelium is involved in the production of hyperpolarization, vasodilation and tissue perfusion and thus the study of its native cellular properties is crucial.

In contrast with the freshly isolated brain endothelium, commonly used cell culture induces significant alterations in morphology (Simmers et al., 2007) and ion channel expression (Sandow and Grayson, 2009). Thus, the comparison between the physiological observations determined *ex vivo* and *in vivo* can be affected.

A relevant study in this field demonstrated the isolation of fresh endothelium from posterior cerebral arteries and simultaneous measurements of endothelial intracellular calcium concentration and membrane voltage using FURA-2 photometry and intracellular sharp electrodes (Hakim and Behringer, 2019). In 1985, a mechanical isolation procedure was established for cerebral microvessels, but unfortunately the endothelial cells which form the blood-brain barrier, had the membrane integrity disturbed (Bradbury, 1985). However, 4 years later, Hess and his team presented a new, simplified modification of the collagenase digestion technique for isolation of microvessels from rat hippocampus (Hess et al., 1989). They used a fluorescence microscope adapted to an image analyzer; the equivalent of the calcium imaging devices used nowadays. Their system had the sufficient spatial and temporal resolution for detection of the changes in the intracellular calcium concentration in isolated cerebral microvessels and also for the study of calcium-mobilizing receptors found on the blood-brain barrier membrane.

Nowadays, despite the importance of the blood-brain barrier involved in various neurological disorders, the molecular and cellular mechanisms that lead to BBB dysfunctionality remain mostly unknown. Therefore, it is necessary to develop a method to isolate brain microvasculature with preserved structural integrity. A recent article (Paraiso et al., 2020) describes an optimized protocol of isolation of the brain microvessels that closely mimic *in vivo* structures suitable for characterization of molecular pathways of BBB component cells. Also, there are several reports in the literature which describe different protocols for isolation of the microvessels from the mouse cerebral cortex (Hartz et al., 2004, 2018; Miller et al., 2000; Wang et al., 2010, 2014b).

2.3 Measurements of *in vivo* calcium waves

Intercellular calcium waves consist of increases in the intracellular calcium ion concentration that are communicated between the cells. These calcium waves spread radially from an initiating cell. The speed and size of the calcium waves depend on the nature and strength of the stimulus, but also on the propagation mechanism. The appearance of these calcium waves is confirmed in different *in vitro* cell cultures (Leybaert and Sanderson, 2012; Paemeleire et al., 1999) and also in the two brain endothelial cells lines: RBE4 and GP8 (Vandamme et al., 2004).

This type of calcium waves obey to two fundamental mechanisms of propagation: diffusion of IP_3 through the gap junctions and the release of extracellular ATP, which will stimulate the neighboring cells *via* the calcium ions release from the endoplasmic reticulum or the indirect release of calcium ions by stimulating the purinergic receptors (Suadicani et al., 2004). The intercellular direct communication through gap junctions is done by calcium ions or inositol triphosphate (IP_3) diffusion through the gap junction channels. These channels are formed from the interaction of two connexin hemichannels, each hemichannel belonging to one of the two neighboring cells. Paracrine signaling involves a messenger release, one of the most used being ATP, in the extracellular space of the nearby cells, which will determine the activation of a signaling cascade having as purpose an increase of intracellular calcium concentration (Wang et al., 2013).

In the case of observing the *in vivo* intercellular calcium waves, the experiments can be achieved by two-photon microscopy. This technique allows the noninvasive monitoring of the intracellular calcium concentration changes, which can be detected by loading a calcium indicator dye intro cells. In order to avoid the mechanical stimulation of the neural tissue associated with the dye injection, cells are genetically modified to express an encoded GCaMP2 calcium ion indicator (Hoogland et al., 2009). Also, another study used this method in order to test the hypothesis that the conduction of vasodilation along arteriolar networks, *in vivo*, involves a wave of calcium ions traveling along the endothelium (Tallini et al., 2007).

The intercellular calcium waves also appear to be involved in pathological conditions like brain ischemia (Ding et al., 2009) and Alzheimer's disease (Kuchibhotla et al., 2009).

3. Ion channels/receptors regulating Ca^{2+} homeostasis in brain microvascular endothelial cells

3.1 TRP channels

Transient receptor potential (TRP) channels are a family of ion channels located mainly in the plasma membrane, that were classified into several subgroups, including Transient Receptor Potential-Canonical (TRPC) channels, Transient receptor potential channels, of the vanilloid subtype (TRPV), Transient Receptor Potential Melastatin (TRPM) channels, Transient receptor potential channel, subfamily N (TRPN; "N" for no mechanoreceptor potential C), Transient receptor potential channel, subfamily A (TRPA; "A" for ankyrin), Transient receptor potential polycystic (TRPP) channel, and Transient receptor potential channel, mucolipin subfamily (TRPML).

TRP channels are expressed in brain microvascular endothelial cells (Table 1). qRT-PCR analysis demonstrated that murine brain microvascular endothelial cells, either freshly isolated or immortalized, express multiple TRP channels, belonging to TRPC, TRPV, TRPM, TRPP, TRPML subfamilies (Berrout, 2012; Brown et al., 2008; Chang et al., 2018). As expected, there are differences in TRP channels expression between freshly isolated and immortalized murine BMVECs preparations (Brown et al., 2008), depending on the cell line producer (Brown et al., 2008; Chang et al., 2018), or depending on the method employed (i.e. qRT-PCR, immuno-fluorescence, Western-blot, calcium imaging etc.) (Berrout, 2012; Brown et al., 2008; Chang et al., 2018).

Table 1 TRP channels expression in brain microvascular endothelial cells.

TRP channel	Level	Sample	Specie	Method	References
TRPC1	–	hCMEC/D3 cells	Human	RT-PCR	Zuccolo et al. (2019)
TRPC3	–				
TRPC4	+				
TRPC5	–				
TRPC6	+				
TRPC7	++				
TRPV2	+	HBCE	Human	qRT-PCR	Hatano et al. (2013)
TRPV4	++				
TRPV4	++	Brain endothelial cells in hematoma	Rat	Immunofluorescence	Zhao et al. (2018)
				Western blot	
				Functional testing with antagonists	

Continued

Table 1 TRP channels expression in brain microvascular endothelial cells.—cont'd

TRP channel	Level	Sample	Specie	Method	References
TRPC1	+	bEnd.3 cells (ATCC)	Mouse	qRT-PCR	Brown et al. (2008)
TRPC2	+				
TRPC4	++				
TRPC7	++				
TRPV2	++				
TRPV4	++				
TRPM2	+				
TRPM3	+				
TRPM4	++				
TRPM7	++				
TRPV2	+++	bEnd.3 cells (ATCC)	Mouse	qRT-PCR	Berrout (2012)
TRPV4	+				
TRPC1	++				
TRPC4	+++				
TRPC6	++				
TRPM2	+				
TRPM4	+++				
TRPM7	++				
TRPP2	+++				
TRPP2	N/A	bEnd.3 cells (ATCC)	Mouse	Immunofluorescence	Berrout (2012)
TRPC1				Western-blot	
TRPP2	N/A	bEnd.3 cells (ATCC)	Mouse	Calcium imaging	Berrout (2012)
TRPC					
TRPC1	+	Freshly isolated mouse brain microvessels	Mouse	RT-PCR	Brown et al. (2008)
TRPC2	+				
TRPC3	+				
TRPC4	++				

Table 1 TRP channels expression in brain microvascular endothelial cells.—cont'd

TRP channel	Level	Sample	Specie	Method	References
TRPC5	++				
TRPC6	++				
TRPC7	++				
TRPV2	++				
TRPV4	+				
TRPV6	+				
TRPM2	++				
TRPM3	+				
TRPM4	++				
TRPM7	+				
TRPM2	+	BMVECs (Cell Biologics)	Mouse	qRT-PCR	Chang et al. (2018)
TRPM4	++				
TRPM5	+				
TRPM7	++				
TRPML1	++				
TRPML2	++				
TRPP2	++				
TRPP5	+				
TRPC1	+				
TRPC2	+				
TRPC5	+				
TRPC7	+				
TRPV2	+				
TRPV4	+				

Scale (−) no expression, (+/−) no expression, very low expression, (+) low expression; (++) medium expression; (+++) high expression.

Human brain microvascular endothelial cells (hCMEC/D3) have a medium expression of TRPC7 and low expression of TRPC4 and TRPC6 (Zuccolo et al., 2019). Moreover, functional testing using 1-oleoyl-2-acetyl-sn-glycerol (OAG; 100 µM), a membrane permeable DAG analogue which was described as an agonist for TRPC3 and TRPC6 in endothelial cells and endothelial colony forming cells (Dragoni et al., 2013; Hamdollah Zadeh et al., 2008), did not activate any intracellular Ca^{2+} increase in hCMEC/D3 cells (Zuccolo et al., 2019).

Murine brain microvascular endothelial cell lines express a wide range of TRP channels, including TRPC1, TRPC2, TRPC4, TRPC6, TRPC7, TRPV2, TRPV4, TRPM2, TRPM3, TRPM4, TRPM7, TRPML1, TRPML2, TRPP2, TRPP5 (Berrout, 2012; Brown et al., 2008; Chang et al., 2018). Among these, the functional expression of TRPC and TRPP2 was demonstrated by calcium imaging recordings, applying specific antagonists SKF96365, LOE908, and respectively, amiloride (Berrout, 2012). Thus, it was also proved that TRPP2 and TRPC1 channels play an essential role in stretch-induced intracellular Ca^{2+} responses (Berrout, 2012). Moreover, the authors showed that the mechanical injury of brain endothelial cells mediated through TRPC and TRPP2 channels, triggers Ca^{2+} release followed by NO production and actin cytoskeleton remodeling (Berrout, 2012).

TRP channels are activated by several vasoactive agents, such as thrombin, ATP, angiotensin II or bradykinin (Bishara and Ding, 2010; Sundivakkam et al., 2013) and have been described to play essential roles in a multitude of physiological processes, including endothelial-dependent hyperpolarization, angiogenesis and arteriogenesis, vascular permeability etc.

TRP channels were documented to trigger endothelial-dependent hyperpolarization in endothelial cells from the cerebral vasculature (Guerra et al., 2018), thus inducing vasodilation in cerebral arteries. To detail, endothelial-dependent hyperpolarization can be initiated by two pathways: (i) neurotransmitters (i.e., glutamate and acetylcholine) and neuromodulators (e.g., ATP) $\rightarrow$ dilation of cerebral vessels $\rightarrow$ endothelial $\uparrow [Ca^{2+}]i \rightarrow$ NO or prostaglandin E2 (PGE) release $\rightarrow$ activation of intermediate and small-conductance Ca^{2+}-activated K^+ channels, (ii) activation of several TRP channels, such as TRPC3, TRPV3, TRPV4, TRPA1 etc. (Guerra et al., 2018).

Multiple studies described the role played by TRP channels in regulating angiogenesis and arteriogenesis. Indeed, studies have demonstrated that angiogenic growth factors activate TRP channels that trigger intracellular Ca^{2+} release and further initiate angiogenesis (Kwan et al., 2007). TRPC channels' role in angiogenesis was extensively described. Inhibition and/silencing studies indicated the critical role played by TRPC1 (Du et al., 2018; Moccia et al., 2014), TRPC3 (Antigny et al., 2012), TRPC4 (Antigny et al., 2012; Qin et al., 2016; Song et al., 2015), TRPC5 (Antigny et al., 2012), TRPC6 (Ding et al., 2014; Ge et al., 2009; Hamdollah Zadeh et al., 2008) in controlling endothelial cell migration and proliferation, tubulogenesis or in some studies the generation of spontaneous [Ca^{2+}]i oscillations. Moreover, *in vitro* and *in vivo* studies demonstrated that some TRPV (i.e. TRPV1 and TRPV4) channels activation stimulates abnormal angiogenesis, while their knockout reduces angiogenesis and tube formation (Adapala et al., 2016; Ching et al., 2011; Su et al., 2014; Thoppil et al., 2016). TRPM channels are also involved in angiogenesis. To be precise, activation of TRPM2 mediates the VEGF-induced endothelial cells migration, and its knockout determined a reduction in vessel formation (Mittal et al., 2015). Additionally, blocking or silencing TRPM4 enhanced *in vitro* tube formation and improved *in vivo* capillary integrity (Loh et al., 2014), while silencing TRPM7 has similar effects with Mg^{2+} deficiency in microvascular endothelial cells growth and migration (Baldoli and Maier, 2012). TRPA1 also contributes to angiogenesis, where the *in vitro* inhibition decreased angiogenesis tube formation and the *in vivo* knockout partially reduced simvastatin-induced angiogenesis (Su et al., 2014).

It was shown by immunofluorescence staining, Western blot and functional testing that TRPV4 is present in human brain microvascular endothelial cells (Hatano et al., 2013; Zhao et al., 2018). TRPV4 is involved in the regulation of BBB integrity *via* the PKCα/RhoA/MLC2 pathway (Zhao et al., 2018). In pathological conditions, antagonist-induced blockage of TRPV4 was shown to preserve BBB integrity in a rat model of intracerebral hemorrhage (Zhao et al., 2018), while no contribution to the BBB permeability was evidenced in inflammation (Rosenkranz et al., 2020).

TRPM2 is a sensor for reactive oxygen species (ROS) and is known to be involved in the modulation of multiple signaling pathways (Sita et al., 2018), and also the apoptosis of endothelial cells (Hecquet et al., 2014). As previously described in β cells, this channel is located in the lysosomal compartment, where it regulates the usage of calcium from intracellular compartments and it also contributes to H$_2$O$_2$ induced apoptosis (Lange et al., 2009). Activation of this channel is highly dependent on the presence of intracellular and/or extracellular Ca^{2+}. The calcium ions inside the cell sensitize TRPM2 to ADPR (adenosine diphosphate ribose) using a calcium dependent interaction of calmodulin (CaM) with the N-terminal IQ-like motif. Mutations to this motif or expression of a mutant CaM protein that is unable to bind Ca^{2+}, reduce the calcium conductance rate induced by H_2O_2 in TRPM2 channels (Du et al., 2009; Tong et al., 2006). TRPM2 also plays a role in trans endothelial migration of PMNs (polymorphonuclear neutrophils). Depletion of this channel *via* siRNA led to a decrease in an adhesion molecule that mediates the opening of adherens junctions (AJ) and facilitates the migration of the PMNs through the BBB—phosphorylated VE-cadherin (Alcaide et al., 2008; Mittal et al., 2017).

3.2 Purinergic receptors

Purinergic receptors are divided in two groups: P1 (metabotropic, adenosine receptors) and P2 (it recognizes the extracellular ATP and ADP). The P2 class is also divided in two subcategories: P2X and P2Y.

P2X is a subcategory of Ca^{2+} permeable ionic channels and it can constitute homo- and heteromeric channels whose preferential physiologic agonist is ATP. The members of this family (P2X1-7) have the same subunitar topology: N and C intracellular terminal tales to which protein kinases can bind, two transmembranar regions, TM1 (implicated in the channel gate) and TM2 (implicated in ionic pores), but also an extracellular loop. The binding of ATP to this type of receptors allows a flux of cations (Na^+ and Ca^{2+}) across the membrane.

P2Y category contains G-protein coupled receptors which includes 8 subtypes, divided in 2 groups based on associated G-protein selectivity and on sequence similarity. In particular, P2Y1, P2Y2, P2Y4, P2Y6 and P2Y11 are associated with Gq protein and it activates phospholipase C beta (PLCbeta), while P2Y12, P2Y13 and P2Y14 are coupled with Gi/o, thus inhibiting the adenylate cyclase activity (Scarpellino et al., 2019). In terms of molecular structure, these types of receptors have a characteristic subunitary topology: an extracellular terminal N tale and an intracellular C tale, to which protein kinases can bind, 7 transmembrane regions, from which TM3, TM6 and TM7 represent high-level sequence homology. Due to intracellular loops, but also to C terminal tale, the diversity of subtypes in P2Y category is imposed and as consequence, the way in which proteins Gq/11, Gs and Gi couple is modified (Burnstock and Knight, 2004).

In particular, P2Y1, P2Y11, P2Y12 and P2Y13 receptors are activated mainly by adenine nucleotides, P2Y2 and P2Y4 receptors respond to both adenine and uracil nucleotides, while P2Y6 is activated by uracil nucleotides and P2Y14 receptor by uridine diphosphate (UDP)-glucose. There are a couple of studies which confirm their expression in different brain microvascular endothelial cells, summarized in Table 2.

Table 2 Purinergic receptors expression in brain microvascular endothelial cells.

Purinergic receptor	Level	Sample	Specie	Method	References
P2Y2	N/A	bEND.3 cells	Mouse	Calcium imaging	Wu et al. (2019)
P2Y2	+++	Cultures of primary BBB endothelial cells	Rat	RT-PCR	Anwar et al. (1999)
P2Y4	+++				
P2Y6	+++				
P2Y2	+++	hCMEC/D3 cells	Human	RT-PCR	Bintig et al. (2012)
P2Y11	+++				
P2Y6	+				
P2X4	+				
P2X5	+				
P2X7	+				
P2X7	N/A	hCMEC/D3	Human	Immunofluorescence Western Blot	Yang et al. (2016)

Scale (+) low intensity; (++) medium intensity; (+++) high intensity.

One of the most important signaling involved in the normal function of the neurovascular unit is definitely the purinergic signaling. Vasodilatation and inflammatory reactions are modulated by purine receptors (Iadecola and Nedergaard, 2007). On the abluminal side of the endothelial cells (brain-vessel interface), the release of purines or pyrimidines (ATP/UTP) from astrocytes can stimulate the purinergic receptors found on endothelial cells' membranes. On the luminal side (vessel-blood interface), the blood cells release agonists which can stimulate the endothelial cells by purinergic signaling. In inflammatory conditions, the ATP release on both sides of the endothelial cells can be affected.

By binding of a hormone, a growth factor or adenosines on G-protein coupled receptor, P2Y, it is determined phospholipase C hydrolysis, which will transform phosphatidyl-inositol 4,5-bisphosphate in diacylglycerol and 1.4.5-inositol triphosphate which will follow different signaling pathways.

IP$_3$ is an important intracellular secondary messenger, which by binding with the specific receptors for IP$_3$ found on the endoplasmic reticulum membrane, will determine an increase in the intracellular calcium concentration caused by a fast passage of calcium ions from the reticulum lumen in cytosol. Calcium release from the endoplasmic reticulum (ER) and its decrease from ER lumen activate the SOCE mechanism (store-operated Ca^{2+} entry). This mechanism modulates a series of cellular functions like restorage of calcium internal deposits from ER, gene expression, cellular cycle regulation, cytoskeleton remodeling, and the release of nitric oxide. Also, it is modulated by the physical interaction between STIM1 and Orai1. At the ER level, there are expressed STIM proteins (Stromal Interaction Molecule) which act as sensors for keeping at physiological level the calcium concentration. These proteins detect the calcium change, squeeze ER membrane surface and activate various channels operated by stores. Orai represents a channel of small conductance, very selective for calcium ions, modulated by STIM protein (Moccia et al., 2019).

All specific receptors for IP_3 are modulated by IP_3 and also by calcium ions from cytosol, both ligands being necessary for receptors' activation. But, if there are high calcium cytosolic concentrations, these receptors are inhibited. This suggests that there exists a biphasic dependence between activation of IP_3 receptors and calcium cytosolic concentration: small calcium concentrations are stimulative while high concentrations are inhibitors (Thillaiappan et al., 2019). At the same time, the presence of a high calcium quantity in cytosol will activate other calcium channels through a process called calcium induced calcium released (CICR). This process consists in an increase of opening probability of IP_3 specific receptors determined by any local increase of calcium concentration. These receptors have also the possibility to amplify the evoked signals from neighboring receptors or other calcium channels.

Once the internal calcium deposits are empty, the endothelial cell activates different mechanisms through which the intracellular calcium concentration has physiological values and the internal deposits are filled.

In calcium imaging experiments, the purinergic stimulation determines the increase of the cytosolic calcium concentration which consists in an initial transient caused by calcium release from internal deposits (event dependent on IP_3), followed by a sustained plateau dependent on the calcium influx from the extracellular medium (Scarpellino et al., 2019).

3.3 Bradykinin receptors

The bradykinin receptors (BKR) are G-protein coupled receptors, whose principal ligand is the protein bradykinin. Although bradykinin is a pro-inflammatory mediator, it is recognized as a neuromediator and it also modulates several vascular and renal functions (Bascands et al., 2003). There are two members in the BKR family: B1 and B2 receptors. Both types of kinin receptors are primarily linked to the activation of phospholipase C, which will determine calcium mobilization by inositol 1,4,5-triphosphate and additional intracellular effects. The calcium ions regulate the permeability of the cell membrane *via* Ca^{2+}-sensitive ion channels (England et al., 2001) and also diacyl glycerol activation of protein kinase C followed by activation of phospholipase A$_2$ and phospholipase D (Han et al., 2002; Zubakova et al., 2008).

Bradykinin is expressed at low concentrations in the normal brain, but it was demonstrated that it increases significantly in brain injury (Tadahiro et al., 1973). By injection of exogenous bradykinin in mammalian tissue causes the four signs of inflammation: redness, local heat, swelling and pain. Local endothelium-dependent vasodilation determines the redness and local heat and by stimulation of endothelial cells which determines an increase in the microvascular permeability, it is also promoted the swelling process (exudation of protein-rich fluid from circulation). The kinin receptors can be found also on non-myelinated afferent neurons' membranes and that represents the explanation why a simple peptide has the ability to produce pain (Marceau and Regoli, 2004).

Most of the times, the blood-brain barrier breakdown is associated with the production and release of bradykinin due to the fact that the BK antagonist CP-0597 had an inhibitor effect in the appearance of ischemic brain injury (Mayhan, 2001; Relton et al., 1997). The *in vitro* studies demonstrated that bradykinin stimulation reduces trans-endothelial electrical resistance (TEER) across the blood-brain barrier (Butt, 1995; Olesen and Crone, 1986).

Various studies show that B1 receptors are generally reduced or even absent in normal tissues, but they are rapidly induced after a certain type of injury of the tissue (Christiansen et al., 2002; Souza et al., 2004). Also, it is important to mention that B1 receptors are up-regulated in the presence of cytokines and endotoxins. They have a dual role in some pathologies in which they exert either a protective (septic shock, multiple sclerosis) or harmful effect (pain and inflammation) (Gabra et al., 2003). This type of receptors is stimulated almost entirely by Des-Arg-9-BK, a metabolic by-product of bradykinin (Dobrivojević et al., 2015).

On the other hand, B2 receptors have been reported in many tissues like neurons, thalamus, basal nuclei, cerebral cortex, hypothalamus (Raidoo and Bhoola, 1997) and they are activated by bradykinin itself. The expression of the bradykinin receptors in the brain endothelial cells can be seen in Table 3.

A relevant study in this domain demonstrated that bradykinin induces Ca^{2+} influx in rat primary cultured aortic endothelial cells and in the H5V cell line (mouse microvessel endothelial cells) (Leung et al., 2006). The authors treated both cell cultures with thapsigargin in order to empty

Table 3 Bradykinin receptors expression in brain microvascular endothelial cells.

Bradykinin receptor	Level	Sample	Specie	Method	References
B1	N/A	hCMVEC cell line	Human	Immunofluorescence	Mugisho et al. (2019)
B2	++	pCMVEC (porcine primary cell from cerebral microvessels)	Porcine	RT-PCR	Bovenzi et al. (2010)
B1	+/−	HBEC cell line	Human	RT-PCR	Prat et al. (2000)
B2	+++				
B1	N/A	HBEC primary culture	Human	Western Blot	Prat et al. (2000)

Scale (−) no expression, (+/−) no expression, very low expression, (+) low intensity; (++) medium intensity; (+++) high intensity.

the intracellular calcium stores and then added bradykinin to observe the induced calcium influx. The difference between the two cell cultures was that bradykinin failed to induce a calcium influx in the rat aortic endothelial cells, but not in the H5V cell line (Leung et al., 2006). This result suggests that bradykinin signaling and effects depend on the expression of other ion channels or enzymes in various types of cells.

In brain microvascular endothelial cells, the bradykinin receptor B2 through subunit $alpha_i$ of the protein G inhibits adenylyl cyclase decreasing the expression of cAMP and protein kinase A. Also, through subunit $alpha_q$, it stimulates phospholipase C to increase intracellular calcium. These changes influence the perijunctional actin cytoskeleton and down-regulate the expression of tight junctions proteins: ZO-1, occludin and claudin-5 (Liu et al., 2008).

The activation of both receptors is known to contribute to the appearance of acute and chronic pathological disorders, including hypotension, broncho-constriction, pain and inflammation. Despite much research performed until now, the importance of the B1 and B2 receptors in BBB damage following a traumatic or ischemic brain insult is still not fully understood.

3.4 Muscarinic receptors

Several studies have investigated the expression of muscarinic receptors in brain microvascular endothelial cells, analyzing the gene and/or protein expression, and in some cases demonstrating the functional expression (Table 4).

Table 4 Muscarinic receptors expression in brain microvascular endothelial cells.

Muscarinic receptor	Level	Sample	Specie	Method	References
M1	+/−	Cultures of brain intracortical microvessels	Human (biopsies from patients undergoing surgery for the treatment of temporal lobe epilepsy)	RT-PCR	Elhusseiny et al. (1999)
M2	−				
M3	++				
M4	−				
M5	−				
M1	+/−	Cultures of brain intracortical capillaries			
M2	++				
M3	++				
M4	−				
M5	+/−				
M1	+/−	Cultures of brain endothelial cells			
M2	++				
M3	−				
M4	−				
M5	++				

Continued

Table 4 Muscarinic receptors expression in brain microvascular endothelial cells.—cont'd

Muscarinic receptor	Level	Sample	Specie	Method	References
M1	+++	Brain endothelial cell cultures	Human (fetal brain, 10–18 weeks of gestation)	RT-PCR	Elhusseiny et al. (1999)
M2	+++				
M3	+++				
M4	++				
M5	+++				
M1	+	hCMEC/D3 cells	Human	qRT-PCR	Zuccolo et al. (2019)
M2	+				
M3	+				
M4	+++				
M5	+++				
M5	N/A	hCMEC/D3 cells	Human	Calcium imaging / Western blot	Zuccolo et al. (2019)
M1	++	bEnd.3 cells (ATCC)	Mouse	qRT-PCR	Radu et al. (2017)
M2	+				
M3	+++	BMVEC (PELOBiotech)			
M4	++				
M5	++				

			Method	Reference
M1	+++	bEnd.3 cells (ATCC)	Mouse	Western blot	Radu et al. (2017)
M2	+	BMVEC (PELOBiotech)			
M3	+++				
M4	+++				
M5	+				
M1	N/A	bEnd.3 cells (ATCC)	Mouse	Immunofluorescence	Radu et al. (2017)
M2					
M3					
M4					
M5					
M1	N/A	bEnd.3 cells (ATCC)	Mouse	Calcium imaging	Radu et al. (2017)
M3					

Human brain microvascular endothelial cells (hCMEC/D3) express all five muscarinic receptors, the most expressed being M4 and M5 receptors (Zuccolo et al., 2019). Functional testing after siRNA silencing demonstrated the complete blockage of acetylcholine-induced Ca^{2+} increase indicating that M5 is the only isoform coupled to phospholipase Cβ (PLCβ) present in hCMEC/D3 cells (Zuccolo et al., 2019).

In a recent study, it was also demonstrated that all five muscarinic receptors are expressed in murine BMVECs, and that there are differences in expression between primary cultures and cell lines (Radu et al., 2017). Additionally, it showed that the ranking in expression of genes encoding for muscarinic receptors is different with respect to the protein expression (Radu et al., 2017). Functional testing by calcium imaging using specific antagonists indicated that M3 receptor has a major contribution to the Ach-induced Ca^{2+} transients and that M1, despite the high total protein level, has a low functional expression at the plasma membrane level (Radu et al., 2017).

4. Ca^{2+} homeostasis and cytoskeleton organization in the blood brain barrier

4.1 The cytoskeleton of brain microvascular endothelial cells

Brain microvascular endothelial cells are the main constituents of BBB and act as the primary defending unit against pathogens (Chen et al., 2020). They also act as a securing element for the signaling between neurons and glial cells as well as a protective barrier against circulating toxins (De Bock et al., 2013). The protective role of the BBB is guaranteed by the properties of the brain endothelial cells. Beside the absence of fenestrae and the formation of tight junctions between adjacent cells of the brain endothelium, the brain endothelial cells distinguish themselves through the presence solute carriers for small molecules transport, efflux transporters (P-glycoprotein, Multidrug Resistance Protein, etc.) and several receptor-mediated macromolecules uptake processes (Abbott et al., 2010; Gastfriend et al., 2018; Hawkins, 2005; Zlokovic, 2008).

The physiological response of endothelial cells under different stimuli is often induced by the active remodeling of the cytoskeleton. The dynamics of the cytoskeleton in brain microvascular cells were previously described in fluorescence microscopy studies and characterized against specific changes following exposure to different physical or chemical stimuli (Abbott et al., 2010; Osborn et al., 2006; Shi et al., 2016).

Among the cytoskeleton components there are three important categories, namely microtubules, microfilaments (or actin filaments) and a group of polymers called intermediate filaments (Alberts et al., 2002; Fletcher and Mullins, 2010). The main differences between them, beside the nature of the molecular processes they associate with, are their mechanical rigidity, assembly dynamics and polarity (Brestcher, 2000). These differences not only distinguish the cytoskeleton components in terms of function and network architecture but are also active factors in the cellular motility (Brestcher, 2000; Tang and Gerlach, 2017). All three types of filaments are organized in deformation-resistant networks but can be reorganized in response to applied external stimuli and play important roles in arranging and maintaining the integrity of intracellular compartments (Fife et al., 2014). Following the exposure to different factors such as oxidative stress, hypoxia, cytokines and chemokines, the cytoskeleton responds by reorganizing itself through assembly and disassembly (Abbott et al., 2010). These two processes can induce changes in cellular morphology in response to nocive stimuli or disease, thus making the BBB remodeling a primary characteristic of neurological disease (Shi et al., 2016).

Under normal physiological conditions, brain endothelial cells are tied by intercellular tight junctions (TJs) and adherens junctions (AJs). The zonula occludens proteins (ZO-1, ZO-2, ZO-3) are anchoring the TJ proteins (occludin and claudin-3, claudin-5, claudin-12, etc.), as well as the AJ proteins (cadherin) to the actin filaments of the cytoskeleton (Shi et al., 2016; Stamatovic et al., 2008; Wallez and Huber, 2008).

Actin is the most common protein of the cytoskeleton (Fig. 6). Structurally, the microfilaments are composed of a series of monomers that polymerize to form actin fibers. These are helical shaped flexible fibers of 5 to 9 nm in diameter (Suetsugu and Takenawa, 2003). Microfilaments are dispersed throughout the entire cell but more concentrated in the vicinity of the plasma membrane (Alberts et al., 2002). Actin fibers are much less rigid than microtubules (Fletcher and Mullins, 2010). However, the presence of high concentrations of crosslinking agents that bind actin filaments, promote the assembly of rigid, highly organized structures, including isotropic networks, grouped networks and branched networks (Fallqvist et al., 2014).

Bundles of aligned actin filaments support the filopodial protuberances, which are involved in the directed cell movement along a chemical gradient and in intercellular communication (Rougerie et al., 2008). They elongate

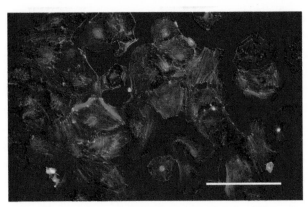

Fig. 6 Original fluorescence microscopy image of endothelial cells from the mouse brain microvasculature (bEnd.3). The actin microfilaments were stained with phalloidin-conjugated with Alexa Fluor 488 (green), the adhesion points were stained with anti-vinculin antibody and secondary antibody conjugated with Alexa Fluor 594 (red), the nuclei were stained with Hoechst 33342 (blue). Scale 100 μm.

constantly in the presence of nucleotide-bound monomers, allowing them to generate the forces necessary to advance the leading edges of a motile cell (Insall and Machesky, 2009). The assembly of contractile actin filament bundles, known as stress fibers, is triggered locally when cell surface adhesion receptors, called integrins, interact with their ligands (Tojkander et al., 2012).

It has been reported that the brain endothelial cells' cytoskeletal elements have a key role in the formation and maintenance of the junctional complexes. The actin filaments network is focally binded by several adhesion proteins such as cadherin and occludin as well as by functional proteins such as ZO-1, ZO-2, ZO-3, catenins or by focal adhesion complexes (Hawkins and Davis, 2005; Kuwabara et al., 2001; Stamatovic et al., 2008). These associations of proteins bound to the F-actin radially distributed bands are forming the actin–rich adhesion belt. The actin–rich adhesion belt is a belt-like structure that forms around endothelial cells in which a bundle of contractile microfilaments is linked to the plasmalemma (Shi et al., 2016; Stamatovic et al., 2016).

The microtubules system of the brain endothelial cells consists of a network of hollow tubules distributed in a polarized manner from the nucleus to the membrane. These hollow structures are polymers of α and β-tubulin (Stamatovic et al., 2008; Wade, 2009). They are oriented along the axis of the cell, usually from the basal to the apical surface (Nag, 1995). They possess an outer diameter of ~25 nm (Alberts et al., 2002). Along with the highest level of rigidity among all the filaments of the cytoskeleton, microtubules have the most complex dynamic processes of assembly and disassembly. Due to their rigidity and high flexibility, microtubules can describe shapes that are almost linear, extending throughout the entire length of the cell (Igaev and Grubmüller, 2018; Kurz and Williams, 1995). Microtubules bend under the compressive force of the cell (Wade, 2009). During the interphase, the cell takes advantage of the microtubules rigidity, assembling radial networks that play the role of intracellular trafficking pathways (Maizels and Gerlitz, 2015). A microtube can switch between a state of constant growth and one of rapid decrease. This instability in its dynamic assembly/disassembly processes, allow for the rapid reorganization of networks of microtubules (Dos Remedios et al., 2003; Insall and Machesky, 2009; Pollard, 2017).

It has been reported that several of the microtubule's functions are intermediated by their interaction with microfilaments (Fife et al., 2014; Verin et al., 2001; Wade, 2009). Among the specified functions we can name the cellular contraction and the increasing of transendothelial leukocyte migration, as well as the contribution to the fast assembly of the microfilaments and to focal adhesion (Pegoraro et al., 2017; Stamatovic et al., 2008). It has been also reported that the process of microtubules disassembly is important in permeability changes in endothelial cells (Verin et al., 2001). Adhesion points of brain microvascular endothelial cells can be visualized by immunostaining (Fig. 6).

The intermediate filaments are fibers of around 10 nm in diameter, made up of specific proteins forming a heterogeneous family (Pegoraro et al., 2017). They form a network called the nuclear lamina, located below the inner nuclear membrane (Aebi et al., 1986). Some types of intermediate filaments are extended along the cytoplasm providing mechanical strength (Alfaro-Aco and Petry, 2015; Costigliola et al., 2017; Pegoraro et al., 2017). The intermediate filaments are the least rigid components of the cytoskeleton and have a greater resistance to extension forces, compared to compressive forces (Pegoraro et al., 2017). The proteins called plectins contribute to the interconnection and crosslinking of intermediate filaments to microfilaments and microtubules. Therefore, several intermediate filament structures are predominantly determined by the interaction with microtubules or actin filaments (Schliwa and Van Blerkom, 1981).

The intermediate filament protein vimentin plays a major role in endothelial brain endothelial cells. It determines cellular shape and provides integrity. Studies report that cellular motility can be linked to vimentin expression, as vimentin deficiency indicated slower migrations (Ivaska, 2012; Tang and Gerlach, 2017). Recent studies highlighted that vimentin and the actin/tubulin cytoskeletal systems are linked. Besides offering structural stability and rigidity to the more dynamic systems of actin and tubulin, vimentin appears to be following their initial structural cues (Costigliola et al., 2017; Gan et al., 2016; Hookway et al., 2015; Jiu et al., 2015; Pan et al., 2019).

4.2 Relation between Ca^{2+} homeostasis and cytoskeleton remodeling and function in brain microvascular endothelial cells

It has been reported before that Ca^{2+} ions modulate the permeability of the BBB through variations in their intra- and extracellular concentrations concentration (De Bock et al., 2013). Several studies emphasized on the fact that Ca^{2+} is an important secondary messenger, involved in the regulatory mechanisms of the BBB functions (De Bock et al., 2013; Olesen, 1985, 1987). An important example is the role of Ca^{2+} in the maintenance of the various intercellular junctions (Brown and Davis, 2002). Studies reported that endothelial cells exposed to low-calcium extracellular media induced a significant decrease in the transendothelial resistance, which was further restored with the restoration of the calcium level (Balda et al., 1991; Brown and Davis, 2002; Shasby et al., 1997). However, the strong decrease/increase in intracellular Ca^{2+} is a well-known factor for the disruption of TJs in brain endothelial cells by lowering the expression of the proteins or disrupting their interaction (Balda et al., 1991; Bauer et al., 2014; Hawkins and Davis, 2005; Rubin and Staddon, 1999). It is well known that the activation of PKC (Protein kinase C) can diminish the effects of low extracellular Ca^{2+} (Balda et al., 1991). Additionally, the calcium transport inside and outside of intracellular stores contributes to the activation of several kinase signaling cascades, regulating the TJ protein expression through a number of transcription factors such as the CREB (cAMP response-element binding protein) or the nuclear factor-κB (Brown and Davis, 2002; Hawkins and Davis, 2005).

A low intracellular calcium concentration is maintained by the transport at the level of membrane pumps, such as the calcium ATPase channels. Additionally, the transmembrane channels present in the membrane of the ER (endoplasmic reticulum) contribute to the maintenance of the low-level intracellular calcium concentration (Berridge et al., 2003; Dalal et al., 2020). The access of calcium from the extracellular space is also allowed by the receptor-operated cation channels. These channels can be activated without the depletion of the ER inner calcium stores (Dalal et al., 2020; Rosado and Sage, 2000; Tran et al., 2000).

During injury, the BBB function often becomes altered (Munji et al., 2019; Stamatovic et al., 2016). This alteration usually occurs once inflammatory mediators are released by the affected tissue. Thus, intracellular calcium influx takes place under the influence of inflammatory mediators such as histamine, bradykinin, or eicosanoids (Dalal et al., 2020; De Bock et al., 2013; Tran et al., 2000). One of the early events in the case of brain inflammation is the extravasation of leukocytes. Their transport over the BBB, as well as their attachment to the plasmalemma occurs with the aid of various adhesion molecules. For example, a study by Greenwood et al., suggested that the lymphocytes that migrate *via* LFA-1 (Lymphocyte function-associated antigen 1) and MAC-1 (Macrophage-1 antigen) integrins are interacting with the (ICAM)-1 adhesion molecule, expressed at different levels in healthy and dysfunctional brain endothelial cells. Among the dysfunctions at which the level of (ICAM)-1 is significantly elevated, multiple sclerosis and Alzheimer's disease stroke are indicated (De Bock et al., 2011, 2013; Etienne-Manneville et al., 2000; Greenwood et al., 2002). Following this mechanism, it has been observed that the crosslinking of ICAM-1 to the surface of rat brain endothelial cells induced, along with other processes such as the production of InsP$_3$, the release of calcium from the ER stores (De Bock et al., 2013; Etienne-Manneville et al., 2000).

The calcium influx that occurs following inflammatory response induces the disassembly of the adherens junctions, as well as the reorganization of the cytoskeleton in order to ensure the retraction and increase in permeability of brain endothelial cells (Dalal et al., 2020). It has been reported that the inflammatory response of the endothelial brain cells induces, among other effects, a significant reorganization of the cytoskeletal filaments. Inflammatory agonists have an increasing effect on the cytosolic calcium, as well as a decreasing one on the cAMP. In addition, they activate the RhoA/Rho kinase, initiating actin reorganization at the level of actin fibers (Prasain and Stevens, 2009).

Studies suggest that F-actin reorganization accompanies the TJ and AJ remodeling (De Bock et al., 2013; Ivanov et al., 2005). The microfilaments network interacts with the motor protein myosin II. This interaction leads to the formation of stress fibers, *via* the regulatory myosin light chain (MLC). The level of myosin light chain phosphorylation is regulated by the myosin light chain kinase (MLCK) in a Ca^{2+}-dependent manner (Choi et al., 2018; De Bock et al., 2013; Goeckeler and Wysolmerski, 1995; Polte et al., 2004; Verin et al., 2001). More precisely, Ca^{2+} locally present between lamellipodia and lamella, induces the contraction of myosin by activating MLCK which further phosphorylates MLC. Moreover, just a slight increase in the local concentration of Ca^{2+} is necessary for the contraction of myosin II to take place, as the affinity between MLCK and myosin-calmodulin is high (Tsai and Meyer, 2012; Tsai et al., 2015).

Ca^{2+} also assists the actin dynamics in brain endothelial cells, by modulating the activity of several related regulators (Mostafavi et al., 2014; Tsai et al., 2015; Wang et al., 2014a). PKC and calmodulin dependent kinases are activated by Ca^{2+}. They interact with actin, regulating its dynamics and network formation properties (Yang et al., 2013). The Rho GTPase is necessary for the formation of actin bundles in lamellipodia, filopodia and at the level of focal adhesion complexes, thus being a key factor for cell motility (Evans and Falke, 2007; Fogh et al., 2014; Tsai and Meyer, 2012; Tsai et al., 2015). The cofilin protein, having the role of F-actin severing, is also modulated by the cytosolic Ca^{2+}. The dephosphorylation of cofilin is Ca^{2+} induced and mediated by the calcineurin-dependent activation of the SSH1L (Slingshot 1L) protein from the SHH family of protein phosphatases (Tsai et al., 2015; Wang et al., 2005).

In a study by Pan et al., it has been indicated that vimentin has a faster degradation mechanism in response to external stimuli. To detail, cells were exposed to hypotonic stress, revealing a more significant degradation at the level of vimentin filaments, rather than microfilaments or microtubules (Pan et al., 2019). It is well known that hypotonic stress increases the intracellular calcium concentration (Desai and Mitchison, 1997; Pan et al., 2019; Pollard, 2017; Xu et al., 1998). Thus, along with the fact that the key pathway of vimentin degradation is the Ca^{2+}-dependent proteolysis, a Ca^{2+}-based interaction classification for the three cytoskeletal components can be established (Pan et al., 2019). Therefore, Ca^{2+} induces the remodeling of endothelial cells morphology, both directly and indirectly, in hypotonic stress conditions, due to the faster degradation mechanism of vimentin.

The store-operated Ca^{2+} entry is a modulator of cytoskeleton organization and cellular motility. The Ca^{2+} entry pathway facilitates the Ca^{2+} transport between microtubules and microfilaments (Martin-Romero et al., 2017). STIM1 protein is a store-operated Ca^{2+} entry modulator that resides inside the ER. It has the role of an intraluminal Ca^{2+} sensing mechanism, as well as a modulator of the Ca^{2+} channels at the level of the plasma membrane. It has been observed that a high expression of STIM1 leads to the extension of ER. This effect was explained by the STIM1 stimulated direct attachment between ER and the microtubule ends (Martin-Romero et al., 2017; Sampieri et al., 2009).

It has been hypothesized that Ca^{2+} is a regulator of the integrin signaling of the cytoskeleton in interaction with plectin (Kostan et al., 2009; Tsai et al., 2015). Evidence suggests that the binding between Ca^{2+} and plectin decreases the interaction with F-actin and integrin β, in a way that causes the decoupling of cell extracellular matrix adhesion with the cytoskeletal structures (Kostan et al., 2009; Song et al., 2015; Tsai et al., 2015).

BBB permeability is linked to Ca^{2+} homeostasis (De Bock et al., 2011, 2013). Several studies confirmed the fact that Ca^{2+} transport as well as intra- and extracellular concentrations, modulate the cytoskeletal components as well as junctional proteins. Therefore, following certain factors that are known to cause BBB disruption, Ca^{2+} acts as an utmost important second messenger, leading to the reorganization of the cytoskeleton and consequently of the remodeling of TJ and AJ.

5. Alterations of Ca^{2+} homeostasis in the blood brain barrier associated with neuropathologies

5.1 Alterations of Ca^{2+} homeostasis in BBB associated with epilepsy

Epilepsy is a brain condition that has recurrent seizures as a hallmark (Beghi et al., 2015) without other specific signs, being a polymorphic disease which can have many forms. This is a disadvantage in epilepsy diagnosis (Thijs et al., 2019).

BBB disruption in epilepsy is a well-known phenomenon that was described both in clinical and experimental epilepsy (Bertini et al., 2013; Marchi et al., 2012). There are two theories linking BBB permeabilization to epileptogenesis, one supporting the idea that BBB leakage contributes to seizures while another that considers BBB opening as a consequence of convulsive episodes (Mendes et al., 2019).

Epilepsy is seen as a combination of seizures and a number of processes such as inflammation, oxidative stress, glycation, methylation (Yuen et al., 2018). Additionally, an equally important factor is Ca^{2+}, playing an indispensable role in the physiological function of the brain, and changes in calcium levels being involved in tissue neurodegeneration, leading to various diseases, including epilepsy. Ca^{2+} is considered to contribute to the onset of epilepsy by changes in its intracellular level that trigger an imbalance in the regulation of neuronal excitability activity (Xu and Tang, 2018).

Moreover, up/down-regulation of ion channels permeable for Ca^{2+} expressed in BBB is considered to be directly involved in the processes of epilepsy. Thus, overexpression or hyperactivation of TRPC3 (non-selectively ion channel permeable to cations, with a prevalence of Ca^{2+} over Na^+ ions) in brain microvascular endothelial cells was associated with BBB permeabilization and with the vasogenic edema formation in rat piriform cortex in response to *status epilepticus* (Ryu et al., 2013). Moreover, TRPV1 was demonstrated to be upregulated in the hippocampus of mice and patients with temporal lobe epilepsy (Nazıroğlu, 2015), while its inhibition with AM404 or capsazepine reduced Ca^{2+} entry in an *in vitro* 4-aminopyridine (4-AP) seizure model-induced in hippocampal and glioblastoma neurons (Nazıroğlu et al., 2019).

It has been hypothesized that calcium fluctuations trigger BBB leakage, by regulating the expression levels of the proteins that maintain tight junctions (e.g. ZO-1, occludin, claudin-1, claudin-5), and finally contributing to epilepsy-specific seizures. This process is part of the activation of calcium-dependent cytosolic phospholipase A2 (cPLA2) signaling cascade. cPLA2 was described to play a major role in maintaining the integrity of the blood-brain barrier during seizures (Rempe et al., 2018). BBB plays a key role in protecting brain parenchyma in terms of calcium fluctuations (Inamura et al., 1990), thus protecting against excess glutathione. In turn, glutathione is important in the activation of cPLA2 (Rempe et al., 2018), but its accumulation and activation in the endothelial cells of BBB causes an increase in intracellular calcium and oxidative stress, both involved in the deterioration of BBB integrity and implicitly, resulting in neuropathologies such as epilepsy (Swissa et al., 2019).

5.2 Alterations of Ca^{2+} homeostasis in BBB associated with Alzheimer's disease

Dementia is a term associated with disrupting cognitive activities that affect daily life and it is recognized by the World Health Organization as a priority for global health issues. Alzheimer's disease (AD) is one of the most common cases of dementia (Briggs et al., 2016). In 2015, AD was reported to be the main cause for dementia in 70% of patients, and the number of cases is increasing every year (Magi et al., 2016). Since the first case was reported in 1907, attempts have been made to understand the pathology, but also to develop an effective treatment, but still without significant results (Lane et al., 2018). The main cause of this pathology is the accumulation in the brain of insoluble aggregates, such as Amyloid β or tau protein (Verri et al., 2012). Although the main cause is linked with the accumulation of Amyloid β plaques, there are also genetic causes (Lane et al., 2018) or aberrant calcium signaling (Magi et al., 2016) that contribute to AD pathology.

The hypothesis that calcium could be involved in Alzheimer's disease was formulated >30 years ago by Khachaturian, 1989. The author realized that although the genetic component is very important and relevant in terms of neurodegeneration, the same processes that regulate the homeostasis of cytosolic calcium ions could play a role in the AD onset (Kachaturian, 1989). A possible mechanism by which Ca^{2+} plays a role in triggering AD is through accumulation of Aβ deposits on the surface of brain endothelial cell membranes, which is an important factor in disrupting Ca^{2+} homeostasis (Mantzavinos and Alexiou, 2017). It should be also highlighted that Aβ interaction with Fe^{3+} and Cu^{2+} induces oxidative stress, and its oligomerization leading to the formation of calcium channels, while in turn these channels affect Ca^{2+} homeostasis triggering oxidative stress (Mantzavinos and Alexiou, 2017).

One hypothesis is that AD is due to the blood-brain barrier permeabilization, followed by the accumulation of amyloid β plaques (Yamazaki and Kanekiyo, 2017). Oppositely, another hypothesis considers that β-amyloid produces irreversible changes in brain endothelial cells contributing to BBB permeabilization. Fonseca and her collaborators demonstrated Aβ1-40 (2.5 μM) induced changes of calcium levels in rat brain cells RBE4 (Fonseca et al., 2015). The same concentration of amyloid β was demonstrated to cause death of brain endothelial cells (Fonseca et al., 2013) and to induce major changes of calcium levels in the endoplasmic reticulum, and through a series of cascade reactions also in mitochondria and cytosol. These Ca^{2+} changes begin in the proteins involved in maintaining calcium homeostasis in brain endothelial cells. In response, endothelial cells became apoptotic, failing to repair the damage caused by amyloid (Fonseca et al., 2013, 2015). Additionally, Nakagawa and colleagues demonstrated that amyloid β induces the adhesion of erythrocytes to endothelial cells, and consequently affecting the endothelial viability and functionality by the generation of oxidative and inflammatory stress (Nakagawa et al., 2011).

The ways in which amyloid β can induce changes of Ca^{2+} levels in brain endothelial cells are varied. Exposure of brain endothelial cells to amyloid β induces calcium transients that are mediated by TRPM2 channels (Park et al., 2014). Moreover, ROS generation in Alzheimer also activates TRPM2 channels triggering intracellular calcium increases, BBB dysfunction and finally death of the brain endothelial cells (Jiang et al., 2018). The accumulation of amyloid β on the surface of these cells can be facilitated by the intervention of external factors such as PM2.5 particles (particulate matter $<2.5\,\mu M$) or combustion-derived nanoparticles, which in turn, can stimulate the generation of this protein (Wang et al., 2020). In addition, exposure of brain endothelial cells to PM2.5 can generate ROS by inducing high levels of Aβ, while leading to increased calcium levels (Wang et al., 2020).

5.3 Alterations of Ca^{2+} homeostasis in BBB associated with Parkinson's disease

Parkinson's is the second most common type of neurodegenerative disease (Poewe et al., 2017). Although similar capillary damage has been observed in Alzheimer's and Parkinson's disease (Farkas et al., 2000), the accumulation of α-synuclein aggregates is characteristic for Parkinson's disease (Poewe et al., 2017). However, Kazmierczak et al. demonstrated that extracellular α-synuclein is also involved in the release and toxicity of β-amyloid peptides, whose effects can be observed in the mitochondria dysfunction or in the caspase-dependent cellular apoptosis (Kazmierczak et al., 2008).

Parkinson's disease is generally characterized by the loss of dopaminergic neurons in the substantia nigra and the formation of intraneuronal protein aggregates called Lewy bodies, and could be classified as a multifactorial disease, due to the fact that environmental factors may contribute to its genetic component (Calì et al., 2012). Lewy bodies are largely made up of α-synuclein, a protein capable of compromising cell membrane integrity (Ingelsson, 2016). Therefore, it is necessary for the α-synuclein level to remain within physiological limits, otherwise the morphological integrity of the cell is affected. The mitochondrial function is particularly affected, due to the endoplasmic reticulum-mitochondrial interactions, and because it supports the activity of these organelles by transferring Ca^{2+} between them. When the protein is below physiological values, mitochondrial fragmentation occurs, and otherwise, when α-synuclein is overexpressed, Ca^{2+} transients help a high interaction between the mitochondria and the endoplasmic reticulum, a process whose purpose is cellular autophagy (Calì et al., 2012).

5.4 Alterations of Ca^{2+} homeostasis in BBB associated with amyotrophic lateral sclerosis

Amyotrophic lateral sclerosis (ALS) is a syndrome that is noted for its neuropathological characteristics, including degeneration of upper and lower motor neurons, that induces muscle weakness, eventually leading to paralysis (Hardiman et al., 2017). Judging by the causes of this syndrome, we can consider it as a multifactorial disease, given the involvement of mutations in several genes such as SOD1, FUS, C9orf72 (Saberi et al., 2015), but also the contribution of environmental factors such as military service, even if their role has not been widely described (Oskarsson et al., 2018).

Although the pathogenicity of ALS is quite difficult to detail, all components of the neurovascular unit have been considered to be involved in the development of this disease, including changes in the blood-brain barrier permeability (Garbuzova-Davis et al., 2011). Among the evidence that support the involvement of these structures in the appearance of the disease are also the alterations of circulating cytokines (Lam et al., 2016; Moreno-Martinez et al., 2019) or in the activity of microglia (Geloso et al., 2017) and astrocytes (Pehar et al., 2017). Moreover, factors such as mitochondrial degeneration, increased levels of reactive oxygen species or intracellular calcium may also be involved in ALS (Evans et al., 2013). The change in intracellular calcium levels occurs due to neurodegeneration in ALS that also involves free radical damage, impaired neuronal transport, organelle fragmentation, and mitochondrial Ca^{2+} overload (Grosskreutz et al., 2010).

Alterations of calcium levels play a key role in the development of ALS, actively contributing to the onset of this disease. The whole process starts from the absorption of glutamate by astrocytes, which causes the increase of calcium levels in the motor neurons (Bonafede and Mariotti, 2017). As calcium homeostasis can no longer be maintained due to mitochondrial degeneration in ALS (Evans et al., 2013), ions accumulate in the cytoplasm and activate calcium-dependent enzymatic pathways, which will cause tissue neurodegeneration (Bonafede and Mariotti, 2017).

Garbuzova-Davis et al. demonstrated that BBB breakdown plays a key role in the development of ALS (Garbuzova-Davis et al., 2007, 2011). As the first step towards triggering the neurodegenerative process is the disturbance of BBB function, by permeabilizing it, resulting in the loss of motor neuron function which is characteristic to this pathology (Bataveljic et al., 2014). We can consider the hypothesis that an important role in ALS is played by the levels of calcium ions at the BBB, and a possible scenario of BBB permeabilization could involve changes in Ca^{2+} homeostasis similar to the general mechanism of BBB breakdown previously described (Nag, 2003). In this mechanism, both intracellular and extracellular Ca^{2+} changes contribute to BBB breakdown; thus, while intracellular Ca^{2+} ensures the cell-to-cell contact by regulating the migration of ZO-1 proteins from the intracellular areas to the plasma membrane, extracellular Ca^{2+} is responsible in maintaining BBB integrity and any decrease in its levels induces a decrease in the BBB electrical resistance and permeability (Nag, 2003). To complete this scenario, previous studies show that BBB permeabilization enables the infiltration of leukocytes to brain parenchyma in animal models of ALS (Lopes Pinheiro et al., 2016), and in turn, the transendothelial leukocyte trafficking requires calcium signaling in brain endothelial cells (Greenwood et al., 2002).

5.5 Alterations of Ca^{2+} homeostasis in BBB associated with ischemic stroke

Stroke is the second leading cause of death, with an increased incidence, especially ischemic stroke that is caused by vessel occlusions (Woodruff et al., 2011). Among the most common signs of ischemic stroke are locomotor problems such as loss of balance and leg numbness, and in some cases other symptoms might include lack of coordination, hands and face numbness, headaches, problems in talking or understanding others, sudden blurred vision unilateral or bilateral. All of these signs may be short-lived episodes, but indicate severe health problems (Randolph, 2016). The processes that can cause this disease are numerous and among them are energy failure, loss of cellular ion homeostasis, increased intracellular calcium levels, free radical-mediated toxicity, cytokine mediated cytotoxicity, activation of glial cells, disruption of the BBB (Woodruff et al., 2011). Therapies that can be applied in ischemic stroke are mechanical removal of the occlusion (Stankowski and Gupta, 2011) or by thrombolysis, but these procedures have the major disadvantage of transforming the ischemia into hemorrhage and early BBB disruption (Warach and Latour, 2004).

The compromise of BBB integrity in ischemic stroke may be due to calcium variation in this structure (Casas et al., 2019). Several mechanisms involving ion transporter dysfunction have been described to contribute in the alterations of BBB that cause cerebral edema upon ischemia, such as Na^+/H^+ exchangers, Na^+-K^+-Cl^- cotransporters, or the calcium-activated potassium channel KCa3.1 that increases the transcellular transport of Na^+ and Cl^- ions from the bloodstream into the brain parenchyma (Chen et al., 2015; O'Donnell, 2014). Another important factor that induces BBB breakdown upon ischemia is the cellular overload with calcium that activates caspases 3/7 triggering apoptosis and necrosis, and the BBB permeabilization can be reduced by inhibiting caspase activity (Rakkar and Bayraktutan, 2016).

In ischemic stroke, the processes involved in BBB permeabilization, by affecting the tight junctions at the endothelial cells, are complex, and in order to understand and clarify the information about these processes, it is ideal to clearly identify the permeabilization phases (Sandoval and Witt, 2008). Intracellular Ca^{2+} fluctuations are a key factor in triggering BBB permeabilization, being considered to regulate the constituent proteins of tight junctions, but also to act as a signal molecule that is responsible for disseminating information to neighboring cells (De Bock et al., 2013).

Alterations of the TRP channels function also play an important role in regulating calcium levels in BBB and their expression changes in traumatic brain injury or stroke. RT-PCR and Western Blot studies demonstrated that the levels of TRPV1, a nonselective cation channel that is important for Ca^{2+} mediated cell signaling, rise after traumatic brain injury. Moreover, if the TRPV1 channels are inhibited, the disruption of the BBB is attenuated (Yang et al., 2019). A study on rats revealed that intracerebral hemorrhage increases the expression of TRPV4, a mechanosensitive calcium permeable channel, which causes an influx of intracellular calcium resulting in edema and apoptosis of brain cells. If the TRPV4 gene is repressed by knockout, the cellular damage is ameliorated. In addition, TRPV4 blockade preserved the tight and adherens junctions as well as the BBB integrity in a rat model of intracerebral hemorrhage (Zhao et al., 2018).

6. Opening of the blood brain barrier and its role in CNS chemotherapy

6.1 Osmotic opening of the blood brain barrier

The BBB permeability can be controlled by osmotic opening and this process is highly relevant for increasing drug delivery to the CNS, in particular in chemotherapy. BBB can be opened *in vivo* by injecting a hypertonic solution of arabinose or mannitol into the carotid artery for 30 s (Rapoport and Robinson, 1986). The BBB permeabilization can be demonstrated by intravenous injection of Evans blue-albumin. This BBB opening is directly correlated with the levels of intracellular Ca^{2+} in brain microvascular endothelial cells (Rapoport, 2000) and lasts about 10 min after osmotic exposure. Indeed, osmotic stress at the BBB level induces increased fluid diffusion from blood into brain parenchyma, and triggers a series of events including tight junctions loosening, brain endothelial cells shrinkage associated with endothelial cytoskeleton remodeling, and increased levels of intracellular Ca^{2+} inside endothelial cells (Rapoport, 2001). The widening of the interendothelial tight junctions in response to BBB osmotic induced opening was estimated to a radius of 200 Å (Rapoport, 2000). Further studies have demonstrated that the BBB opening can be experimentally extended beyond 30 min, if rats were pretreated with KB-R7943 (3 mg/kg), a Na^+/Ca^{2+} exchanger blocker (Bhattacharjee et al., 2001), that reinforces the evidence that Ca^{2+} mediates tight junctions opening.

However, the reversible osmotic BBB disruption is not only a simple mechanical shrinkage of brain microvascular endothelial cells but is rather due to an intracellular Ca^{2+}-activated complex mechanism. Indeed, Nagashima and colleagues demonstrated by *in vitro* experiments that exposure of cultured rat brain capillary endothelial cells to 1.4 M mannitol for 30 s determined an increase of the intracellular Ca^{2+} within 10 s after the mannitol perfusion, which recovered to initial values after 200 s (Nagashima et al., 1997). To understand the mechanism, the authors demonstrated that nifedipine (100 pM, 10 pM), an L-type calcium channel antagonist, did not block the Ca^{2+} increase, while KB-R7943, an Na^+/Ca^{2+} exchanger antagonist, did not affect rising phase of the Ca^{2+} increase but completely abolished its return phase (Nagashima et al., 1997).

In vitro studies have also brought details regarding the mechanisms by which Ca^{2+} signaling in brain microvascular endothelial cells is involved in the regulation of the BBB osmotic opening. To detail, exposure to hypo-osmolar solution (225 mOsm/L) of bEnd3 monolayers determined a rapid and transient increase of the monolayer permeability, while Ruthenium Red (1 μM), a TRPV family inhibitor, partly blocked the BBB osmotic permeabilization (Brown et al., 2008). This hypo-osmolar induced mono-layer permeabilization is presumably mediated by Ca^{2+} influx through TRPV channels (i.e. TRPV2, TRPV4 channels) (Brown et al., 2008).

In experimental animals, the BBB osmotic disruption technique was used to increase delivery to brain parenchyma of water-soluble drugs (Kiviniemi et al., 2017; Rapoport, 2001), peptides (Bors and Erdő, 2019; Oller-Salvia et al., 2016; Upadhyay, 2014), antibodies (Bickel, 1995; Chacko et al., 2014; Stanimirovic et al., 2018), boron compounds for neutron capture therapy (Barth et al., 1997; Yang et al., 1996), viral vectors for gene therapy (Bors and Erdő, 2019; Fu and McCarty, 2016; Stanimirovic et al., 2018) and enzymes (Fredericks and Rapoport, 1988; Neuwelt et al., 1981; Rapoport, 2001).

In human patients with brain cancer (i.e. primary central nervous system lymphoma or primitive neuroectodermal tumors) that were monthly subjected to combined chemotherapy with BBB permeabilization, the window of the osmotically-induced BBB disruption was measured by

99mTc-glucoheptonate (TcGH) single-photon emission computerized tomography (SPECT) scanning (Siegal et al., 2000). Thus, it was demonstrated that the BBB was widely open during the first 40 min after mannitol injection and recovered to baseline values after 6 to 8 h following the induction (Siegal et al., 2000). A subsequent study in patients with high grade malignant gliomas demonstrated that combined chemotherapy (e.g. methotrexate, cyclophosphamide, and procarbazine) with BBB osmotic disrupture demonstrated that 16% of patients remained in complete remission while 65% of patients had partial or temporary remission, and improved survival for all patients (Gumerlock et al., 1992).

6.2 Opening of the blood brain barrier by the activation of bradykinin B2 receptors

RMP-7 (Cereport$^{®}$ from Alkermes Inc.) is a nine amino acid peptide that selectively activates bradykinin B2 receptors and therefore triggers bradykinin-like second messenger systems, including increases in intracellular Ca^{2+} and phosphatidylinositol turnover (Emerich, 2002). RMP-7 increases the BBB permeability from 400 Da size molecules to 1 kDa-size molecules (Borlongan and Emerich, 2003). While bradykinin has a very short circulating half-life (only seconds) due to its rapid degradation, Cereport resists degradation having a half-life 2–3 times greater than bradykinin (Borlongan and Emerich, 2003).

Cereport is used to increase the delivery of chemotherapeutic agents into brain tumors in both animal models and humans. Dean and colleagues demonstrated that Cereport increased carboplatin delivery in rat brain tumors, while pretreatment with dexamethasone (1.5 mg/kg/day, twice a day) slightly reduced tumor carboplatin levels (Dean et al., 1999), and therefore authors concluded that dexamethasone should be administered with caution in patients with brain tumors undergoing Cereport-induced BBB permeabilization procedures. Beside brain tumor therapy, Cereport was also used to permeabilize BBB for other drugs, such as loperamide or cyclosporin-A (Borlongan et al., 2002).

The mechanism by which Cereport permeabilizes BBB was analyzed by electron microscopy. Thus, SEM studies indicated that Cereport administered intravenously increased BBB permeability by loosing the tight junctions (Emerich, 2002). Additionally, Cereport permeabilizes BBB by triggering intracellular calcium increase. The amplitude and concentration dependence of Cereport-induced Ca^{2+} signaling in rat brain endothelial cells was similar to that induced by bradykinin in the same cells or other endothelial cells (Doctrow et al., 1994).

6.3 Ultrasound opening of the blood brain barrier

An alternative technique for BBB opening is based on systemic injection of ultrasound contrast agents. This method was successfully used to increase the delivery of Herceptin to brain parenchyma in a mice model (Hynynen et al., 2001, 2006; Kinoshita et al., 2006).

In vitro studies demonstrated that exposure of bEnd.3 cells to ultrasound stimulated microbubbles (1.25 MHz, 10 cycles, 0.24 MPa peak negative pressure) caused changes in intracellular Ca^{2+} and affected Ca^{2+} even in the cells that had no direct contact with the microbubbles (Park et al., 2010). Very recent, the first clinical trial for testing the utility of BBB opening with focused ultrasound has been announced (Abrahao et al., 2019).

Interestingly, in a recent study was performed the sonoselective transfection of cerebral vasculature by employing low-pressure focused ultrasound technique and demonstrated that the method can be used for drug delivery without triggering BBB opening or activating the inflammatory and immune pathways that commonly were present in high pressure focused ultrasound (Gorick et al., 2020).

Acknowledgments

This work was funded by the Romanian Ministry of Education and Research, CCCDI-UEFISCDI, project number PN-III-P2-2.1-PED-2019-4657, within PNCDI III and from Competitiveness Operational Programme 2014–2020 project P_37_675 (contract no. 146/2016), Priority Axis 1, Action 1.1.4, co-financed by the European Funds for Regional Development and Romanian Government funds. The content of this publication does not necessarily reflect the official position of the European Union or Romanian Government.

References

Abbott, N.J., 1998. Role of intracellular calcium in regulation of brain endothelial permeability. In: Introduction to the Blood-Brain Barrier. Cambridge University Press, pp. 345–353.

Abbott, N.J., 2000. Inflammatory mediators and modulation of blood–brain barrier permeability. Cell. Mol. Neurobiol. 20, 131–147.

Abbott, N.J., Patabendige, A.A.K., Dolman, D.E.M., Yusof, S.R., Begley, D.J., 2010. Structure and function of the blood-brain barrier. Neurobiol. Dis. 37, 13–25.

Abrahao, A., Meng, Y., Llinas, M., Huang, Y., Hamani, C., Mainprize, T., Aubert, I., Heyn, C., Black, S.E., Hynynen, K., Lipsman, N., Zinman, L., 2019. First-in-human trial of blood–brain barrier opening in amyotrophic lateral sclerosis using MR-guided focused ultrasound. Nat. Commun. 10, 4373.

Adapala, R.K., Thoppil, R.J., Ghosh, K., Cappelli, H.C., Dudley, A.C., Paruchuri, S., Keshamouni, V., Klagsbrun, M., Meszaros, J.G., Chilian, W.M., Ingber, D.E., Thodeti, C.K., 2016. Activation of mechanosensitive ion channel TRPV4 normalizes tumor vasculature and improves cancer therapy. Oncogene 35, 314–322.

Aebi, U., Cohn, J., Buhle, L., Gerace, L., 1986. The nuclear lamina is a meshwork of intermediate-type filaments. Nature 323, 560–564.

Al Suleimani, Y.M., Hiley, C.R., 2016. Characterization of calcium signals provoked by lysophosphatidylinositol in human microvascular endothelial cells. Physiol. Res. 65, 53–62.

Alberts, B., Johnson, A., Lewis, J., Raff, M., Roberts, K., Walter, P., 2002. Molecular Biology of The Cell, fifth ed. Garland Science, Taylor & Francis Group, LLC.

Alcaide, P., Newton, G., Auerbach, S., Sehrawat, S., Mayadas, T.N., Golan, D.E., Yacono, P., Vincent, P., Kowalczyk, A., Luscinskas, F.W., 2008. p120-Catenin regulates leukocyte transmigration through an effect on VE-cadherin phosphorylation. Blood 112, 2770–2779.

Alfaro-Aco, R., Petry, S., 2015. Building the microtubule cytoskeleton piece by piece. J. Biol. Chem. 290, 17154–17162.

Ando, J., Yamamoto, K., 2013. Flow detection and calcium signalling in vascular endothelial cells. Cardiovasc. Res. 99, 260–268.

Anghileri, L.J., Maincent, P., Thouvenot, P., 1994. Long-term oral administration of aluminum in mice. Aluminum distribution in tissues and effects on calcium metabolism. Ann. Clin. Lab. Sci. 24, 22–26.

Antigny, F., Girardin, N., Frieden, M., 2012. Transient receptor potential canonical channels are required for in vitro endothelial tube formation. J. Biol. Chem. 287, 5917–5927.

Anwar, Z., Albert, J.L., Gubby, S.E., Boyle, J.P., Roberts, J.A., Webb, T.E., Boarder, M.R., 1999. Regulation of cyclic AMP by extracellular ATP in cultured brain capillary endothelial cells. Br. J. Pharmacol. 128, 465–471.

Balda, M.S., González-Mariscal, L., Contreras, R.G., Macias-Silva, M., Torres-Marquez, M.E., Sáinz, J.A.G., Cereijido, M., 1991. Assembly and sealing of tight junctions: possible participation of G-proteins, phospholipase C, protein kinase C and calmodulin. J. Membr. Biol. 122, 193–202.

Baldoli, E., Maier, J.A.M., 2012. Silencing TRPM7 mimics the effects of magnesium deficiency in human microvascular endothelial cells. Angiogenesis 15, 47–57.

Banks, W.A., 2012. Brain meets body: the blood-brain barrier as an endocrine interface. Endocrinology 153, 4111–4119.

Barth, R.F., Yang, W., Rotaru, J.H., Moeschberger, M.L., Joel, D.D., Nawrocky, M.M., Goodman, J.H., Soloway, A.H., 1997. Boron neutron capture therapy of brain tumors: enhanced survival following intracarotid injection of either sodium borocaptate or boronophenylalanine with or without blood-brain barrier disruption. Cancer Res. 57, 1129 LP–1136.

Bascands, J.-L., Schanstra, J.P., Couture, R., Girolami, J.-P., 2003. Bradykinin receptors: towards new pathophysiological roles. Med. Sci. (Paris) 19, 1093–1100.

Baselet, B., Sonveaux, P., Baatout, S., Aerts, A., 2019. Pathological effects of ionizing radiation: endothelial activation and dysfunction. Cell. Mol. Life Sci. 76, 699–728.

Batavelijc, D., Milosevic, M., Radenovic, L., Andjus, P., 2014. Novel molecular biomarkers at the blood-brain barrier in ALS. Biomed. Res. Int. 2014, 907545.

Bauer, H.C., Krizbai, I.A., Bauer, H., Traweger, A., 2014. "You shall not pass"-tight junctions of the blood brain barrier. Front. Neurosci. 8, 1–21.

Beghi, E., Giussani, G., Sander, J.W., 2015. The natural history and prognosis of epilepsy. Epileptic Disord. 17, 243–253.

Berridge, M.J., Bootman, M.D., Roderick, H.L., 2003. Calcium signalling: dynamics, homeostasis and remodelling. Nat. Rev. Mol. Cell Biol. 4, 517–529.

Berrout, J., 2012. Role of TRP Channels in Mediating the Calcium Signaling Response of Brain Endothelial Cells to Mechanical Stretch. Ph.D. Thesis, The University of Texas MD Anderson Cancer Center UTHealth Graduate School of Biomedical Sciences.

Bertini, G., Bramanti, P., Constantin, G., Pellitteri, M., Radu, B.M., Radu, M., Fabene, P.F., 2013. New players in the neurovascular unit: insights from experimental and clinical epilepsy. Neurochem. Int. 63, 652–659.

Bhattacharjee, A.K., Nagashima, T., Kondoh, T., Tamaki, N., 2001. The effects of the Na+/Ca++ exchange blocker on osmotic blood–brain barrier disruption. Brain Res. 900, 157–162.

Bickel, U., 1995. Antibody delivery through the blood-brain barrier. Adv. Drug Deliv. Rev. 15, 53–72.

Bintig, W., Begandt, D., Schlingmann, B., Gerhard, L., Pangalos, M., Dreyer, L., Hohnjec, N., Couraud, P.-O., Romero, I.A., Weksler, B.B., Ngezahayo, A., 2012. Purine receptors and Ca(2+) signalling in the human blood-brain barrier endothelial cell line hCMEC/D3. Purinergic Signal. 8, 71–80.

Bishara, N.B., Ding, H., 2010. Glucose enhances expression of TRPC1 and calcium entry in endothelial cells. Am. J. Physiol. Heart Circ. Physiol. 298, H171–H178.

Bonafede, R., Mariotti, R., 2017. ALS pathogenesis and therapeutic approaches: the role of mesenchymal stem cells and extracellular vesicles. Front. Cell. Neurosci. 11, 80.

Borlongan, C.V., Emerich, D.F., 2003. Facilitation of drug entry into the CNS via transient permeation of blood brain barrier: laboratory and preliminary clinical evidence from bradykinin receptor agonist, Cereport. Brain Res. Bull. 60, 297–306.

Borlongan, C.V., Emerich, D.F., Hoffer, B.J., Bartus, R.T., 2002. Bradykinin receptor agonist facilitates low-dose cyclosporine-A protection against 6-hydroxydopamine neurotoxicity. Brain Res. 956, 211–220.

Bors, L., Erdő, F., 2019. Overcoming the blood–brain barrier. Challenges and tricks for CNS drug delivery. Sci. Pharm. 87, 6.

Bovenzi, V., Savard, M., Morin, J., Cuerrier, C.M., Grandbois, M., Gobeil, F.J., 2010. Bradykinin protects against brain microvascular endothelial cell death induced by pathophysiological stimuli. J. Cell. Physiol. 222, 168–176.

Bradbury, M.W., 1985. The blood-brain barrier. Transport across the cerebral endothelium. Circ. Res. 57, 213–222.

Brestcher, A., 2000. The cytoskeleton: from regulation to function. EMBO Rep. 1, 473–476.

Briggs, R., Kennelly, S.P., O'Neill, D., 2016. Drug treatments in Alzheimer's disease. Clin. Med. 16, 247–253.

Brown, R.C., Davis, T.P., 2002. Calcium modulation of adherens and tight junction function: a potential mechanism for blood-brain barrier disruption after stroke. Stroke 33, 1706–1711.

Brown, R.C., Wu, L., Hicks, K., O'neil, R.G., 2008. Regulation of blood-brain barrier permeability by transient receptor potential type C and type v calcium-permeable channels. Microcirculation 15, 359–371.

Burnstock, G., Knight, G.E., 2004. Cellular distribution and functions of P2 receptor subtypes in different systems. Int. Rev. Cytol. 240, 31–304.

Butt, A.M., 1995. Effect of inflammatory agents on electrical resistance across the blood-brain barrier in pial microvessels of anaesthetized rats. Brain Res. 696, 145–150.

Calì, T., Ottolini, D., Negro, A., Brini, M., 2012. α-Synuclein controls mitochondrial calcium homeostasis by enhancing endoplasmic reticulum-mitochondria interactions. J. Biol. Chem. 287, 17914–17929.

Casas, A.I., Kleikers, P.W., Geuss, E., Langhauser, F., Adler, T., Busch, D.H., Gailus-Durner, V., de Angelis, M.H., Egea, J., Lopez, M.G., Kleinschnitz, C., Schmidt, H.H., 2019. Calcium-dependent blood-brain barrier breakdown by NOX5 limits postreperfusion benefit in stroke. J. Clin. Invest. 129, 1772–1778.

Chacko, A.-M., Li, C., Pryma, D.A., Brem, S., Coukos, G., Muzykantov, V.R., 2014. Targeted delivery of antibody-based therapeutic and imaging agents to CNS tumors: crossing the blood-brain-barrier divide. Expert Opin. Drug Deliv. 10, 907–926.

Chang, S.L., Huang, W., Mao, X., Mack, M.L., 2018. Ethanol's effects on transient receptor potential channel expression in brain microvascular endothelial cells. J. Neuroimmune Pharmacol. 13, 498–508.

Chen, Y.-J., Wallace, B.K., Yuen, N., Jenkins, D.P., Wulff, H., O'Donnell, M.E., 2015. Blood-brain barrier KCa3.1 channels: evidence for a role in brain Na uptake and edema in ischemic stroke. Stroke 46, 237–244.

Chen, M.B., Yang, A.C., Yousef, H., Lee, D., Chen, W., Schaum, N., Lehallier, B., Quake, S.R., Wyss-Coray, T., 2020. Brain endothelial cells are exquisite sensors of age-related circulatory cues. Cell Rep. 30, 4418–4432.e4.

Ching, L.-C., Kou, Y.R., Shyue, S.-K., Su, K.-H., Wei, J., Cheng, L.-C., Yu, Y.-B., Pan, C.-C., Lee, T.-S., 2011. Molecular mechanisms of activation of endothelial nitric oxide synthase mediated by transient receptor potential vanilloid type 1. Cardiovasc. Res. 91, 492–501.

Choi, H.J., Kim, N.E., Kim, J., An, S., Yang, S.H., Ha, J., Cho, S., Kwon, I., Kim, Y.D., Nam, H.S., Heo, J.H., 2018. Dabigatran reduces endothelial permeability through inhibition of thrombin-induced cytoskeleton reorganization. Thromb. Res. 167, 165–171.

Christiansen, S.C., Eddleston, J., Woessner, K.M., Chambers, S.S., Ye, R., Pan, Z.K., Zuraw, B.L., 2002. Up-regulation of functional kinin B1 receptors in allergic airway inflammation. J. Immunol. 169, 2054–2060.

Costigliola, N., Ding, L., Burckhardt, C.J., Han, S.J., Gutierrez, E., Mota, A., Groisman, A., Mitchison, T.J., Danuser, G., 2017. Vimentin fibers orient traction stress. Proc. Natl. Acad. Sci. U. S. A. 114, 5195–5200.

Dalal, P.J., Muller, W.A., Sullivan, D.P., 2020. Endothelial cell calcium signaling during barrier function and inflammation. Am. J. Pathol. 190, 535–542.

De Bock, M., Culot, M., Wang, N., Bol, M., Decrock, E., De Vuyst, E., Da Costa, A., Dauwe, I., Vinken, M., Simon, A.M., Rogiers, V., De Ley, G., Evans, W.H., Bultynck, G., Dupont, G., Cecchelli, R., Leybaert, L., 2011. Connexin channels provide a target to manipulate brain endothelial calcium dynamics and blood-brain barrier permeability. J. Cereb. Blood Flow Metab. 31, 1942–1957.

De Bock, M., Wang, N., Decrock, E., Bol, M., Gadicherla, A.K., Culot, M., Cecchelli, R., Bultynck, G., Leybaert, L., 2013. Endothelial calcium dynamics, connexin channels and blood–brain barrier function. Prog. Neurobiol. 108, 1–20.

Dean, R.L., Emerich, D.F., Hasler, B.P., Bartus, R.T., 1999. Cereport (RMP-7) increases carboplatin levels in brain tumors after pretreatment with dexamethasone. Neuro Oncol. 1, 268–274.

Desai, A., Mitchison, T.J., 1997. Microtubule polymerization dynamics. Annu. Rev. Cell Dev. Biol. 13, 83–117.

Dienel, G.A., Tofel-Grehl, B., Cruz, C.C., Luludis, K., Pettigrew, K., Sokoloff, L., Gibson, G.E., 1995. Determination of local rates of 45Ca influx into rat brain by quantitative autoradiography: studies of aging. Am. J. Physiol. Integr. Comp. Physiol. 269, R453–R462.

Ding, S., Wang, T., Cui, W., Haydon, P.G., 2009. Photothrombosis ischemia stimulates a sustained astrocytic Ca^{2+} signaling in vivo. Glia 57, 767–776.

Ding, Y., Frömel, T., Popp, R., Falck, J.R., Schunck, W.-H., Fleming, I., 2014. The biological actions of 11,12-epoxyeicosatrienoic acid in endothelial cells are specific to the R/S-enantiomer and require the G(s) protein. J. Pharmacol. Exp. Ther. 350, 14–21.

Dobrivojević, M., Špiranec, K., Sinđić, A., 2015. Involvement of bradykinin in brain edema development after ischemic stroke. Pflugers Arch. 467, 201–212.

Doctrow, S.R., Abelleira, S.M., Curry, L.A., Heller-Harrison, R., Kozarich, J.W., Malfroy, B., McCarroll, L.A., Morgan, K.G., Morrow, A.R., Musso, G.F., 1994. The bradykinin analog RMP-7 increases intracellular free calcium levels in rat brain microvascular endothelial cells. J. Pharmacol. Exp. Ther. 271, 229–237.

Dos Remedios, C.G., Chhabra, D., Kekic, M., Dedova, I.V., Tsubakihara, M., Berry, D.A., Nosworthy, N.J., 2003. Actin binding proteins: regulation of cytoskeletal microfilaments. Physiol. Rev. 83, 433–473.

Dragoni, S., Laforenza, U., Bonetti, E., Lodola, F., Bottino, C., Guerra, G., Borghesi, A., Stronati, M., Rosti, V., Tanzi, F., Moccia, F., 2013. Canonical transient receptor potential 3 channel triggers vascular endothelial growth factor-induced intracellular Ca^{2+} oscillations in endothelial progenitor cells isolated from umbilical cord blood. Stem Cells Dev. 22, 2561–2580.

Du, J., Xie, J., Yue, L., 2009. Intracellular calcium activates TRPM2 and its alternative spliced isoforms. Proc. Natl. Acad. Sci. U. S. A. 106, 7239–7244.

Du, L.-L., Shen, Z., Li, Z., Ye, X., Wu, M., Hong, L., Zhao, Y., 2018. TRPC1 deficiency impairs the endothelial progenitor cell function via inhibition of calmodulin/eNOS pathway. J. Cardiovasc. Transl. Res. 11, 339–345.

Earley, S., Brayden, J.E., 2015. Transient receptor potential channels in the vasculature. Physiol. Rev. 95, 645–690.

Elhusseiny, A., Cohen, Z., Olivier, A., Stanimirović, D.B., Hamel, E., 1999. Functional acetylcholine muscarinic receptor subtypes in human brain microcirculation: identification and cellular localization. J. Cereb. Blood Flow Metab. 19, 794–802.

Emerich, D.F., 2002. Use of the bradykinin agonist, cereport as a pharmacological means of increasing drug delivery to the CNS. Curr. Med. Chem.: Immunol., Endocr. Metab. Agents 2, 109–123.

Endemann, D.H., Schiffrin, E.L., 2004. Endothelial dysfunction. J. Am. Soc. Nephrol. 15, 1983–1992.

England, S., Heblich, F., James, I.F., Robbins, J., Docherty, R.J., 2001. Bradykinin evokes a Ca^{2+}-activated chloride current in non-neuronal cells isolated from neonatal rat dorsal root ganglia. J. Physiol. 530, 395–403.

Etienne-Manneville, S., Manneville, J.-B., Adamson, P., Wilbourn, B., Greenwood, J., Couraud, P.-O., 2000. ICAM-1-coupled cytoskeletal rearrangements and trans-endothelial lymphocyte migration involve intracellular calcium signaling in brain endothelial cell lines. J. Immunol. 165, 3375–3383.

Evans, J.H., Falke, J.J., 2007. Ca^{2+} influx is an essential component of the positive-feedback loop that maintains leading-edge structure and activity in macrophages. Proc. Natl. Acad. Sci. U. S. A. 104, 16176–16181.

Evans, M.C., Couch, Y., Sibson, N., Turner, M.R., 2013. Inflammation and neurovascular changes in amyotrophic lateral sclerosis. Mol. Cell. Neurosci. 53, 34–41.

Fallqvist, B., Kulachenko, A., Kroon, M., 2014. Modelling of cross-linked actin networks— influence of geometrical parameters and cross-link compliance. J. Theor. Biol. 350, 57–69.

Farkas, E., De Jong, G.I., Apró, E., De Vos, R.A., Steur, E.N., Luiten, P.G., 2000. Similar ultrastructural breakdown of cerebrocortical capillaries in Alzheimer's disease, Parkinson's disease, and experimental hypertension. What is the functional link? Ann. N. Y. Acad. Sci. 903, 72–82.

Fife, C.M., McCarroll, J.A., Kavallaris, M., 2014. Movers and shakers: cell cytoskeleton in cancer metastasis. Br. J. Pharmacol. 171, 5507–5523.

Filippini, A., D'Amore, A., D'Alessio, A., 2019. Calcium mobilization in endothelial cell functions. Int. J. Mol. Sci. 20, 4525.

Fletcher, D.A., Mullins, R.D., 2010. Cell mechanisms and cytoskeleton. Nature 463, 485–492.

Fogh, B.S., Multhaupt, H.A.B., Couchman, J.R., 2014. Protein kinase C, focal adhesions and the regulation of cell migration. J. Histochem. Cytochem. 62, 172–184.

Fonseca, A.C.R.G., Ferreiro, E., Oliveira, C.R., Cardoso, S.M., Pereira, C.F., 2013. Activation of the endoplasmic reticulum stress response by the amyloid-beta 1-40 peptide in brain endothelial cells. Biochim. Biophys. Acta 1832, 2191–2203.

Fonseca, A.C.R.G., Moreira, P.I., Oliveira, C.R., Cardoso, S.M., Pinton, P., Pereira, C.F., 2015. Amyloid-beta disrupts calcium and redox homeostasis in brain endothelial cells. Mol. Neurobiol. 51, 610–622.

Fredericks, W.R., Rapoport, S.I., 1988. Reversible osmotic opening of the blood-brain barrier in mice. Stroke 19, 266–268.

Fu, H., McCarty, D.M., 2016. Crossing the blood–brain-barrier with viral vectors. Curr. Opin. Virol. 21, 87–92.

Gabra, B.H., Couture, R., Sirois, P., 2003. Functional duality of kinin receptors in pathophysiology. Med. Sci. (Paris) 19, 1101–1110.

Gan, Z., Ding, L., Burckhardt, C.J., Lowery, J., Zaritsky, A., Sitterley, K., Mota, A., Costigliola, N., Starker, C.G., Voytas, D.F., Tytell, J., Goldman, R.D., Danuser, G., 2016. Vimentin intermediate filaments template microtubule networks to enhance persistence in cell polarity and directed migration. Cell Syst. 3, 252–263.e8.

Garbuzova-Davis, S., Haller, E., Saporta, S., Kolomey, I., Nicosia, S.V., Sanberg, P.R., 2007. Ultrastructure of blood-brain barrier and blood-spinal cord barrier in SOD1 mice modeling ALS. Brain Res. 1157, 126–137.

Garbuzova-Davis, S., Rodrigues, M.C.O., Hernandez-Ontiveros, D.G., Louis, M.K., Willing, A.E., Borlongan, C.V., Sanberg, P.R., 2011. Amyotrophic lateral sclerosis: a neurovascular disease. Brain Res. 1398, 113–125.

Gastfriend, B.D., Palecek, S.P., Shusta, E.V., 2018. Modeling the blood–brain barrier: beyond the endothelial cells. Curr. Opin. Biomed. Eng. 5, 6–12.

Ge, R., Tai, Y., Sun, Y., Zhou, K., Yang, S., Cheng, T., Zou, Q., Shen, F., Wang, Y., 2009. Critical role of TRPC6 channels in VEGF-mediated angiogenesis. Cancer Lett. 283, 43–51.

Geloso, M.C., Corvino, V., Marchese, E., Serrano, A., Michetti, F., D'Ambrosi, N., 2017. The dual role of microglia in ALS: mechanisms and therapeutic approaches. Front. Aging Neurosci. 9, 242.

Goeckeler, Z.M., Wysolmerski, R.B., 1995. Myosin light chain kinase-regulated endothelial cell contraction: the relationship between isometric tension, actin polymerization, and myosin phosphorylation. J. Cell Biol. 130, 613–627.

Gorick, C.M., Mathew, A.S., Garrison, W.J., Thim, E.A., Fisher, D.G., Copeland, C.A., Song, J., Klibanov, A.L., Miller, G.W., Price, R.J., 2020. Sonoselective transfection of cerebral vasculature without blood–brain barrier disruption. Proc. Natl. Acad. Sci. U. S. A. 117, 5644 LP–5654.

Greenwood, J., Etienne-manneville, S., Adamson, P., Couraud, P., 2002. Lymphocyte migration into the central nervous system: implication of ICAM-1 signalling at the blood–brain barrier. Vascul. Pharmacol. 38, 315–322.

Grosskreutz, J., Van Den Bosch, L., Keller, B.U., 2010. Calcium dysregulation in amyotrophic lateral sclerosis. Cell Calcium 47, 165–174.

Guerra, G., Lucariello, A., Perna, A., Botta, L., De Luca, A., Moccia, F., 2018. The role of endothelial Ca(2 +) signaling in neurovascular coupling: a view from the lumen. Int. J. Mol. Sci. 19, 938.

Gumerlock, M.K., Belshe, B.D., Madsen, R., Watts, C., 1992. Osmotic blood-brain barrier disruption and chemotherapy in the treatment of high grade malignant glioma: patient series and literature review. J. Neurooncol 12, 33–46.

Hakim, M.A., Behringer, E.J., 2019. Simultaneous measurements of intracellular calcium and membrane potential in freshly isolated and intact mouse cerebral endothelium. J. Vis. Exp. 143, e58832.

Hamdollah Zadeh, M.A., Glass, C.A., Magnussen, A., Hancox, J.C., Bates, D.O., 2008. VEGF-mediated elevated intracellular calcium and angiogenesis in human microvascular endothelial cells in vitro are inhibited by dominant negative TRPC6. Microcirculation 15, 605–614.

Han, J.M., Kim, J.H., Lee, B.D., Do Lee, S., Kim, Y., Jung, Y.W., Lee, S., Cho, W., Ohba, M., Kuroki, T., Suh, P.-G., Ryu, S.H., 2002. Phosphorylation-dependent regulation of phospholipase D2 by protein kinase C delta in rat Pheochromocytoma PC12 cells. J. Biol. Chem. 277, 8290–8297.

Hardiman, O., Al-Chalabi, A., Chio, A., Corr, E.M., Logroscino, G., Robberecht, W., Shaw, P.J., Simmons, Z., van den Berg, L.H., 2017. Amyotrophic lateral sclerosis. Nat. Rev. Dis. Primers. 3, 17071.

Hartz, A.M.S., Bauer, B., Fricker, G., Miller, D.S., 2004. Rapid regulation of P-glycoprotein at the blood-brain barrier by endothelin-1. Mol. Pharmacol. 66, 387–394.

Hartz, A.M.S., Zhong, Y., Shen, A.N., Abner, E.L., Bauer, B., 2018. Preventing P-gp ubiquitination lowers Aβ brain levels in an Alzheimer's disease mouse model. Front. Aging Neurosci. 10, 186.

Hatano, N., Suzuki, H., Itoh, Y., Muraki, K., 2013. TRPV4 partially participates in proliferation of human brain capillary endothelial cells. Life Sci. 92, 317–324.

Hawkins, B.T., 2005. The blood-brain barrier/neurovascular unit in health and disease. Pharmacol. Rev. 57, 173–185.

Hecquet, C.M., Zhang, M., Mittal, M., Vogel, S.M., Di, A., Gao, X., Bonini, M.G., Malik, A.B., 2014. Cooperative interaction of trp melastatin channel transient receptor potential (TRPM2) with its splice variant TRPM2 short variant is essential for endothelial cell apoptosis. Circ. Res. 114, 469–479.

Hess, J., Jensen, C.V., Diemer, N.H., 1989. Calcium-imaging with Fura-2 in isolated cerebral microvessels. Acta Histochem. 87, 107–114.

Hoogland, T.M., Kuhn, B., Göbel, W., Huang, W., Nakai, J., Helmchen, F., Flint, J., Wang, S.S.-H., 2009. Radially expanding transglial calcium waves in the intact cerebellum. Proc. Natl. Acad. Sci. U. S. A. 106, 3496 LP–3501.

Hookway, C., Ding, L., Davidson, M.W., Rappoport, J.Z., Danuser, G., Gelfand, V.I., 2015. Microtubule-dependent transport and dynamics of vimentin intermediate filaments. Mol. Biol. Cell 26, 1675–1686.

Hynynen, K., McDannold, N., Vykhodtseva, N., Jolesz, F.A., 2001. Noninvasive MR imaging-guided focal opening of the blood-brain barrier in rabbits. Radiology 220, 640–646.

Hynynen, K., McDannold, N., Vykhodtseva, N., Raymond, S., Weissleder, R., Jolesz, F.A., Sheikov, N., 2006. Focal disruption of the blood-brain barrier due to 260-kHz ultrasound bursts: a method for molecular imaging and targeted drug delivery. J. Neurosurg. 105, 445–454.

Iadecola, C., Nedergaard, M., 2007. Glial regulation of the cerebral microvasculature. Nat. Neurosci. 10, 1369–1376.

Igaev, M., Grubmüller, H., 2018. Microtubule assembly governed by tubulin allosteric gain in flexibility and lattice induced fit. Elife 7, 1–21.

Inamura, K., Martins, E., Themner, K., Tapper, S., Pallon, J., Lövestam, G., Malmqvist, K.G., Siesjö, B.K., 1990. Accumulation of calcium in substantia nigra lesions induced by status epilepticus. A microprobe analysis. Brain Res. 514, 49–54.

Ingelsson, M., 2016. Alpha-synuclein oligomers–neurotoxic molecules in Parkinson's disease and other lewy body disorders. Front. Neurosci. 10, 408.

Insall, R.H., Machesky, L.M., 2009. Actin dynamics at the leading edge: from simple machinery to complex networks. Dev. Cell 17, 310–322.

Ivanov, A.I., Hunt, D., Utech, M., Nusrat, A., Parkos, C.A., 2005. Differential roles for actin polymerization and a myosin ii motor in assembly of the epithelial apical junctional complex. Mol. Biol. Cell 16, 5356–5372.

Ivaska, J., 2012. Unanchoring integrins in focal adhesions. Nat. Cell Biol. 14, 981–983.

Jafarnejad, M., Cromer, W.E., Kaunas, R.R., Zhang, S.L., Zawieja, D.C., Moore, J.E.J., 2015. Measurement of shear stress-mediated intracellular calcium dynamics in human dermal lymphatic endothelial cells. Am. J. Physiol. Heart Circ. Physiol. 308, H697–H706.

Jiang, L.-H., Li, X., Syed Mortadza, S.A., Lovatt, M., Yang, W., 2018. The TRPM2 channel nexus from oxidative damage to Alzheimer's pathologies: an emerging novel intervention target for age-related dementia. Ageing Res. Rev. 47, 67–79.

Jiu, Y., Lehtimäki, J., Tojkander, S., Cheng, F., Jäälinoja, H., Liu, X., Varjosalo, M., Eriksson, J.E., Lappalainen, P., 2015. Bidirectional interplay between vimentin intermediate filaments and contractile actin stress fibers. Cell Rep. 11, 1511–1518.

Kabacik, S., Raj, K., 2017. Ionising radiation increases permeability of endothelium through ADAM10-mediated cleavage of VE-cadherin. Oncotarget 8, 82049–82063.

Kachaturian, Z.S., 1989. Introduction and overview. Ann. N. Y. Acad. Sci. 568, 1–4.

Kass, I.S., Lipton, P., 1986. Calcium and long-term transmission damage following anoxia in dentate gyrus and CA1 regions of the rat hippocampal slice. J. Physiol. 378, 313–334.

Kazmierczak, A., Strosznajder, J.B., Adamczyk, A., 2008. α-Synuclein enhances secretion and toxicity of amyloid beta peptides in PC12 cells. Neurochem. Int. 53, 263–269.

Kinoshita, M., McDannold, N., Jolesz, F.A., Hynynen, K., 2006. Targeted delivery of antibodies through the blood-brain barrier by MRI-guided focused ultrasound. Biochem. Biophys. Res. Commun. 340, 1085–1090.

Kiviniemi, V., Korhonen, V., Kortelainen, J., Rytky, S., Keinänen, T., Tuovinen, T., Isokangas, M., Sonkajärvi, E., Siniluoto, T., Nikkinen, J., Alahuhta, S., Tervonen, O., Turpeenniemi-Hujanen, T., Myllylä, T., Kuittinen, O., Voipio, J., 2017. Real-time monitoring of human blood-brain barrier disruption. PLoS One 12, 1–16.

Kostan, J., Gregor, M., Walko, G., Wiche, G., 2009. Plectin isoform-dependent regulation of keratin-integrin α6β4 anchorage via Ca^{2+}/calmodulin. J. Biol. Chem. 284, 18525–18536.

Kuchibhotla, K.V., Lattarulo, C.R., Hyman, B.T., Bacskai, B.J., 2009. Synchronous hyperactivity and intercellular calcium waves in astrocytes in Alzheimer mice. Science 323, 1211–1215.

Kurz, J.C., Williams, R.C., 1995. Microtubule-associated proteins and the flexibility of microtubules. Biochemistry 34, 13374–13380.

Kuwabara, H., Kokai, Y., Kojima, T., Takakuwa, R., Mori, M., Sawada, N., 2001. Occludin regulates actin cytoskeleton in endothelial cells. Cell Struct. Funct. 26, 109–116.

Kwan, H.-Y., Huang, Y., Yao, X., 2007. TRP channels in endothelial function and dysfunction. Biochim. Biophys. Acta, Mol. Basis Dis. 1772, 907–914.

Lam, L., Chin, L., Halder, R.C., Sagong, B., Famenini, S., Sayre, J., Montoya, D., Rubbi, L., Pellegrini, M., Fiala, M., 2016. Epigenetic changes in T-cell and monocyte signatures and production of neurotoxic cytokines in ALS patients. FASEB J. 30, 3461–3473.

Lane, C.A., Hardy, J., Schott, J.M., 2018. Alzheimer's disease. Eur. J. Neurol. 25, 59–70.

Lange, I., Yamamoto, S., Partida-Sanchez, S., Mori, Y., Fleig, A., Penner, R., 2009. TRPM2 functions as a lysosomal Ca^{2+}-release channel in β cells. Sci. Signal. 2, ra23.

Leong, I.-L., Tsai, T.-Y., Wong, K.-L., Shiao, L.-R., Cheng, K.-S., Chan, P., Leung, Y.-M., 2018. Valproic acid inhibits ATP-triggered Ca(2 +) release via a p38-dependent mechanism in bEND.3 endothelial cells. Fundam. Clin. Pharmacol. 32, 499–506.

Leung, P.-C., Cheng, K.-T., Liu, C., Cheung, W.-T., Kwan, H.-Y., Lau, K.-L., Huang, Y., Yao, X., 2006. Mechanism of non-capacitative Ca^{2+} influx in response to bradykinin in vascular endothelial cells. J. Vasc. Res. 43, 367–376.

Leybaert, L., Sanderson, M.J., 2012. Intercellular Ca(2 +) waves: mechanisms and function. Physiol. Rev. 92, 1359–1392.

Liu, L.-B., Xue, Y.-X., Liu, Y.-H., Wang, Y.-B., 2008. Bradykinin increases blood-tumor barrier permeability by down-regulating the expression levels of ZO-1, occludin, and claudin-5 and rearranging actin cytoskeleton. J. Neurosci. Res. 86, 1153–1168.

Loh, K.P., Ng, G., Yu, C.Y., Fhu, C.K., Yu, D., Vennekens, R., Nilius, B., Soong, T.W., Liao, P., 2014. TRPM4 inhibition promotes angiogenesis after ischemic stroke. Pflugers Arch. 466, 563–576.

Lopes Pinheiro, M.A., Kooij, G., Mizee, M.R., Kamermans, A., Enzmann, G., Lyck, R., Schwaninger, M., Engelhardt, B., de Vries, H.E., 2016. Immune cell trafficking across the barriers of the central nervous system in multiple sclerosis and stroke. Biochim. Biophys. Acta 1862, 461–471.

Magi, S., Castaldo, P., Macrì, M.L., Maiolino, M., Matteucci, A., Bastioli, G., Gratteri, S., Amoroso, S., Lariccia, V., 2016. Intracellular calcium dysregulation: implications for Alzheimer's disease. Biomed. Res. Int. 2016, 6701324.

Maizels, Y., Gerlitz, G., 2015. Shaping of interphase chromosomes by the microtubule network. FEBS J. 282, 3500–3524.

Mantzavinos, V., Alexiou, A., 2017. Biomarkers for Alzheimer's disease diagnosis. Curr. Alzheimer Res. 14, 1149–1154.

Marceau, F., Regoli, D., 2004. Bradykinin receptor ligands: therapeutic perspectives. Nat. Rev. Drug Discov. 3, 845–852.

Marchi, N., Granata, T., Ghosh, C., Janigro, D., 2012. Blood-brain barrier dysfunction and epilepsy: pathophysiologic role and therapeutic approaches. Epilepsia 53, 1877–1886.

Martin-Romero, F.J., Lopez-Guerrero, A.M., Pascual-Caro, C., Pozo-Guisado, E., 2017. The interplay between cytoskeleton and calcium dynamics. In: Jimenez-Lopez, J.C. (Ed.), Cytoskeleton—Structure, Dynamics, Function and Disease. IntechOpen.

Mayhan, W.G., 2001. Regulation of blood—brain barrier permeability. Microcirculation 8, 89–104.

Mendes, N.F., Pansani, A.P., Carmanhães, E.R.F., Tange, P., Meireles, J.V., Ochikubo, M., Chagas, J.R., da Silva, A.V., Monteiro de Castro, G., Le Sueur-Maluf, L., 2019. The blood-brain barrier breakdown during acute phase of the pilocarpine model of epilepsy is dynamic and time-dependent. Front. Neurol. 10, 382.

Mies, G., Kawai, K., Saito, N., Nagashima, G., Nowak, T.S., Ruetzler, C.A., Klatzo, I., 1993. Cardiac arrest-induced complete cerebral ischaemia in the rat: dynamics of post-ischaemic in vivo calcium uptake and protein synthesis. Neurol. Res. 15, 253–263.

Miller, D.S., Nobmann, S.N., Gutmann, H., Toeroek, M., Drewe, J., Fricker, G., 2000. Xenobiotic transport across isolated brain microvessels studied by confocal microscopy. Mol. Pharmacol. 58, 1357–1367.

Mittal, M., Urao, N., Hecquet, C.M., Zhang, M., Sudhahar, V., Gao, X.-P., Komarova, Y., Ushio-Fukai, M., Malik, A.B., 2015. Novel role of reactive oxygen species-activated Trp melastatin channel-2 in mediating angiogenesis and postischemic neovascularization. Arterioscler. Thromb. Vasc. Biol. 35, 877–887.

Mittal, M., Nepal, S., Tsukasaki, Y., Hecquet, C.M., Soni, D., Rehman, J., Tiruppathi, C., Malik, A.B., 2017. Neutrophil activation of endothelial cell-expressed TRPM2 mediates transendothelial neutrophil migration and vascular injury. Circ. Res. 121, 1081–1091.

Moccia, F., Berra-Romani, R., Tanzi, F., 2012. Update on vascular endothelial Ca(2+) signalling: a tale of ion channels, pumps and transporters. World J. Biol. Chem. 3, 127–158.

Moccia, F., Dragoni, S., Poletto, V., Rosti, V., Tanzi, F., Ganini, C., Porta, C., 2014. Orai1 and transient receptor potential channels as novel molecular targets to impair tumor neovascularization in renal cell carcinoma and other malignancies. Anticancer Agents Med. Chem. 14, 296–312.

Moccia, F., Negri, S., Shekha, M., Faris, P., Guerra, G., 2019. Endothelial Ca(2+) signaling, angiogenesis and vasculogenesis: just what it takes to make a blood vessel. Int. J. Mol. Sci. 20, 3962.

Moreno-Martinez, L., Calvo, A.C., Muñoz, M.J., Osta, R., 2019. Are circulating cytokines reliable biomarkers for amyotrophic lateral sclerosis? Int. J. Mol. Sci. 20, 2759.

Moriarty, C.M., 1980. Kinetic analysis of calcium distribution in rat anterior pituitary slices. Am. J. Physiol. Metab. 238, E167–E173.

Mostafavi, E., Nargesi, A.A., Ghazizadeh, Z., Larry, M., Farahani, R.H., Morteza, A., Esteghamati, A., Vigneron, C., Nakhjavani, M., 2014. The degree of resistance of erythrocyte membrane cytoskeletal proteins to supra-physiologic concentrations of calcium: an in vitro study. J. Membr. Biol. 247, 695–701.

Mugisho, O.O., Robilliard, L.D., Nicholson, L.F.B., Graham, E.S., O'Carroll, S.J., 2019. Bradykinin receptor-1 activation induces inflammation and increases the permeability of human brain microvascular endothelial cells. Cell Biol. Int. 44, 343–351.

Munji, R.N., Soung, A.L., Weiner, G.A., Sohet, F., Semple, B.D., Trivedi, A., Gimlin, K., Kotoda, M., Korai, M., Aydin, S., Batugal, A., Cabangcala, A.C., Schupp, P.G., Oldham, M.C., Hashimoto, T., Noble-Haeusslein, L.J., Daneman, R., 2019. Profiling the mouse brain endothelial transcriptome in health and disease models reveals a core blood–brain barrier dysfunction module. Nat. Neurosci. 22, 1892–1902.

Murphy, V.A., Smith, Q.R., Rapoport, S.I., 1988. Regulation of brain and cerebrospinal fluid calcium by brain barrier membranes following vitamin D-related chronic hypo- and hypercalcemia in rats. J. Neurochem. 51, 1777–1782.

Nag, S., 1995. Role of the endothelial cytoskeleton in blood-brain-barrier permeability to protein. Acta Neuropathol. 90, 454–460.

Nag, S., 2003. Patophysiology of blood-brain barier breakdown. Methods Mol. Med. 89, 97–119.

Nagashima, T., Ikeda, K., Wu, S., Kondo, T., Yamaguehi, M., Tamaki, N., 1997. In: James, H.E., Marshall, L.F., Raulen, H.J., Baethmann, A., Marmarou, A., Ito, U., Czernicki, Z. (Eds.), The Mechanism of Reversible Osmotic Opening of the Blood-Brain Barrier: Role of Intracellular Calcium Ion in Capillary Endothelial Cells BT—Brain Edema X. Springer Vienna, Vienna, pp. 231–233.

Nakagawa, K., Kiko, T., Kuriwada, S., Miyazawa, T., Kimura, F., Miyazawa, T., 2011. Amyloid β induces adhesion of erythrocytes to endothelial cells and affects endothelial viability and functionality. Biosci. Biotechnol. Biochem. 75, 2030–2033.

Nazıroğlu, M., 2015. TRPV1 channel: a potential drug target for treating epilepsy. Curr. Neuropharmacol. 13, 239–247.

Nazıroğlu, M., Taner, A.N., Balbay, E., Çiğ, B., 2019. Inhibitions of anandamide transport and FAAH synthesis decrease apoptosis and oxidative stress through inhibition of TRPV1 channel in an in vitro seizure model. Mol. Cell. Biochem. 453, 143–155.

Neuwelt, E.A., Barranger, J.A., Brady, R.O., Pagel, M., Furbish, F.S., Quirk, J.M., Mook, G.E., Frenkel, E., 1981. Delivery of hexosaminidase A to the cerebrum after osmotic modification of the blood-brain barrier. Proc. Natl. Acad. Sci. U. S. A. 78, 5838–5841.

Newman, G.C., Hospod, F.E., Qi, H., Patel, H., 1995. Effects of dextran on hippocampal brain slice water, extracellular space, calcium kinetics and histology. J. Neurosci. Methods 61, 33–46.

O'Donnell, M.E., 2014. Blood-brain barrier Na transporters in ischemic stroke. Adv. Pharmacol. 71, 113–146.

Olesen, S.-P., 1985. A calcium-dependent reversible permeability increase in microvessels in frog brain, induced by serotonin. J. Physiol. 361, 103–113.

Olesen, S.-P., 1987. Regulation of ion permeability in frog brain venules. significance of calcium, cyclic nucleotides and protein kinase C. J. Physiol. 387, 59–68.

Olesen, S.P., Crone, C., 1986. Substances that rapidly augment ionic conductance of endothelium in cerebral venules. Acta Physiol. Scand. 127, 233–241.

Oller-Salvia, B., Sánchez-Navarro, M., Giralt, E., Teixidó, M., 2016. Blood-brain barrier shuttle peptides: an emerging paradigm for brain delivery. Chem. Soc. Rev. 45, 4690–4707.

Osborn, E.A., Rabodzey, A., Dewey, C.F., Hartwig, J.H., 2006. Endothelial actin cytoskeleton remodeling during mechanostimulation with fluid shear stress. Am. J. Physiol. Cell Physiol. 290, 444–452.

Oskarsson, B., Gendron, T.F., Staff, N.P., 2018. Amyotrophic lateral sclerosis: an update for 2018. Mayo Clin. Proc. 93, 1617–1628.

Paemeleire, K., de Hemptinne, A., Leybaert, L., 1999. Chemically, mechanically, and hyperosmolarity-induced calcium responses of rat cortical capillary endothelial cells in culture. Exp. Brain Res. 126, 473–481.

Pan, L., Zhang, P., Hu, F., Yan, R., He, M., Li, W., Xu, J., Xu, K., 2019. Hypotonic stress induces fast, reversible degradation of the vimentin cytoskeleton via intracellular calcium release. Adv. Sci. 6, 1–8.

Paraiso, H.C., Wang, X., Kuo, P.-C., Furnas, D., Scofield, B.A., Chang, F.-L., Yen, J.-H., Yu, I.-C., 2020. Isolation of mouse cerebral microvasculature for molecular and single-cell analysis. Front. Cell. Neurosci. 14, 84.

Park, J., Fan, Z., Kumon, R.E., El-Sayed, M.E.H., Deng, C.X., 2010. Modulation of intra-cellular Ca^{2+} concentration in brain microvascular endothelial cells in vitro by acoustic cavitation. Ultrasound Med. Biol. 36, 1176–1187.

Park, L., Wang, G., Moore, J., Girouard, H., Zhou, P., Anrather, J., Iadecola, C., 2014. The key role of transient receptor potential melastatin-2 channels in amyloid-β-induced neurovascular dysfunction. Nat. Commun. 5, 5318.

Pegoraro, A.F., Janmey, P., Weitz, D.A., 2017. Mechanical properties of the cytoskeleton and cells. Cold Spring Harb. Perspect. Biol. 9, 1–12.

Pehar, M., Harlan, B.A., Killoy, K.M., Vargas, M.R., 2017. Role and therapeutic potential of astrocytes in amyotrophic lateral sclerosis. Curr. Pharm. Des. 23, 5010–5021.

Poewe, W., Seppi, K., Tanner, C.M., Halliday, G.M., Brundin, P., Volkmann, J., Schrag, A.-E., Lang, A.E., 2017. Parkinson disease. Nat. Rev. Dis. Primers. 3, 17013.

Pollard, T.D., 2017. Actin and actin-binding proteins. Cold Spring Harb. Perspect. Biol. 8 (8), a018226.

Polte, T.R., Eichler, G.S., Wang, N., Ingber, D.E., 2004. Extracellular matrix controls myosin light chain phosphorylation and cell contractility through modulation of cell shape and cytoskeletal prestress. Am. J. Physiol. Cell Physiol. 286, 518–528.

Prasain, N., Stevens, T., 2009. The actin cytoskeleton in endothelial cell phenotypes. Microvasc. Res. 77, 53–63.

Prat, A., Biernacki, K., Pouly, S., Nalbantoglu, J., Couture, R., Antel, J.P., 2000. Kinin B1 receptor expression and function on human brain endothelial cells. J. Neuropathol. Exp. Neurol. 59, 896–906.

Qin, W., Xie, W., Xia, N., He, Q., Sun, T., 2016. Silencing of transient receptor potential channel 4 alleviates oxLDL-induced angiogenesis in human coronary artery endothelial cells by inhibition of VEGF and NF-κB. Med. Sci. Monit. 22, 930–936.

Radu, B.M., Radu, M., Tognoli, C., Benati, D., Merigo, F., Assfalg, M., Solani, E., Stranieri, C., Ceccon, A., Fratta Pasini, A.M., Cominacini, L., Bramanti, P., Osculati, F., Bertini, G., Fabene, P.F., 2015. Are they in or out? The elusive interaction between Qtracker®800 vascular labels and brain endothelial cells. Nanomedicine 10, 3329–3342.

Radu, B.M., Osculati, A.M.M., Suku, E., Banciu, A., Tsenov, G., Merigo, F., Di Chio, M., Banciu, D.D., Tognoli, C., Kacer, P., Giorgetti, A., Radu, M., Bertini, G., Fabene, P.F., 2017. All muscarinic acetylcholine receptors (M1-M5) are expressed in murine brain microvascular endothelium. Sci. Rep. 7, 1–15.

Raidoo, D.M., Bhoola, K.D., 1997. Kinin receptors on human neurones. J. Neuroimmunol. 77, 39–44.

Rakkar, K., Bayraktutan, U., 2016. Increases in intracellular calcium perturb blood-brain barrier via protein kinase C-alpha and apoptosis. Biochim. Biophys. Acta 1862, 56–71.

Randolph, S.A., 2016. Ischemic stroke. Workplace Health Saf. 64, 444.

Rapoport, S.I., 2000. Osmotic opening of the blood–brain barrier: principles, mechanism, and therapeutic applications. Cell. Mol. Neurobiol. 20, 217–230.

Rapoport, S.I., 2001. Advances in osmotic opening of the blood-brain barrier to enhance CNS chemotherapy. Expert Opin. Investig. Drugs 10, 1809–1818.

Rapoport, S.I., Robinson, P.J., 1986. Tight-junctional modification as the basis of osmotic opening of the blood-brain barrier. Ann. N. Y. Acad. Sci. 481, 250–267.

Relton, J.K., Beckey, V.E., Hanson, W.L., Whalley, E.T., 1997. CP-0597, a selective bradykinin B2 receptor antagonist, inhibits brain injury in a rat model of reversible middle cerebral artery occlusion. Stroke 28, 1430–1436.

Rempe, R.G., Hartz, A.M.S., Soldner, E.L.B., Sokola, B.S., Alluri, S.R., Abner, E.L., Kryscio, R.J., Pekcec, A., Schlichtiger, J., Bauer, B., 2018. Matrix metalloproteinase-mediated blood-brain barrier dysfunction in epilepsy. J. Neurosci. 38, 4301 LP–4315.

Revest, P.A., Abbott, N.J., Gillespie, J.I., 1991. Receptor-mediated changes in intracellular [Ca^{2+}] in cultured rat brain capillary endothelial cells. Brain Res. 549, 159–161.

Rosado, J.A., Sage, S.O., 2000. The actin cytoskeleton in store-mediated calcium entry. J. Physiol. 526, 221–229.

Rosenkranz, S.C., Shaposhnykov, A., Schnapauff, O., Epping, L., Vieira, V., Heidermann, K., Schattling, B., Tsvilovskyy, V., Liedtke, W., Meuth, S.G., Freichel, M., Gelderblom, M., Friese, M.A., 2020. TRPV4-mediated regulation of the blood brain barrier is abolished during inflammation. Front. Cell Dev. Biol. 8, 849.

Rougerie, P., Miskolci, V., Cox, D., 2008. Generation of membrane structures during phagocytosis and chemotaxis of macrophages: role and regulation of the actin cytoskeleton. Bone 23, 1–7.

Rubin, L.L., Staddon, J.M., 1999. The cell biology of the blood-brain barrier. Annu. Rev. Neurosci. 22, 11–28.

Ryu, H.J., Kim, J.-E., Kim, Y.-J., Kim, J.-Y., Kim, W.I.L., Choi, S.-Y., Kim, M.-J., Kang, T.-C., 2013. Endothelial transient receptor potential conical channel (TRPC)-3 activation induces vasogenic edema formation in the rat piriform cortex following status epilepticus. Cell. Mol. Neurobiol. 33, 575–585.

Saberi, S., Stauffer, J.E., Schulte, D.J., Ravits, J., 2015. Neuropathology of amyotrophic lateral sclerosis and its variants. Neurol. Clin. 33, 855–876.

Sampieri, A., Zepeda, A., Asanov, A., Vaca, L., 2009. Visualizing the store-operated channel complex assembly in real time: identification of SERCA2 as a new member. Cell Calcium 45, 439–446.

Sandoval, K.E., Witt, K.A., 2008. Blood-brain barrier tight junction permeability and ischemic stroke. Neurobiol. Dis. 32, 200–219.

Sandow, S.L., Grayson, T.H., 2009. Limits of isolation and culture: intact vascular endothelium and BKCa. Am. J. Physiol. Heart Circ. Physiol. 297, H1–H7.

Scarpellino, G., Genova, T., Avanzato, D., Bernardini, M., Bianco, S., Petrillo, S., Tolosano, E., de Almeida Vieira, J.R., Bussolati, B., Fiorio Pla, A., Munaron, L., 2019. Purinergic calcium signals in tumor-derived endothelium. Cancers (Basel) 11, 766.

Scheitlin, C.G., Julian, J.A., Shanmughapriya, S., Madesh, M., Tsoukias, N.M., Alevriadou, B.R., 2016. Endothelial mitochondria regulate the intracellular Ca^{2+} response to fluid shear stress. Am. J. Physiol. Cell Physiol. 310, C479–C490.

Schliwa, M., Van Blerkom, J., 1981. Structural interaction of cytoskeletal components. J. Cell Biol. 90, 222–235.

Shasby, D.M., Stevens, T., Ries, D., Moy, A.B., Kamath, J.M., Kamath, A.M., Shabby, S.S., 1997. Thrombin inhibits myosin light chain dephosphorylation in endothelial cells. Am. J. Physiol. Lung Cell. Mol. Physiol. 272, L311–L319.

Shi, Y., Zhang, L., Pu, H., Mao, L., Hu, X., Jiang, X., Xu, N., Stetler, R.A., Zhang, F., Liu, X., Leak, R.K., Keep, R.F., Ji, X., Chen, J., 2016. Rapid endothelial cytoskeletal reorganization enables early blood-brain barrier disruption and long-term ischaemic reperfusion brain injury. Nat. Commun. 7, 10523.

Shinde, A.V., Motiani, R.K., Zhang, X., Abdullaev, I.F., Adam, A.P., González-Cobos, J.C., Zhang, W., Matrougui, K., Vincent, P.A., Trebak, M., 2013. STIM1 controls endothelial barrier function independently of Orai1 and Ca^{2+} entry. Sci. Signal. 6, ra18.

Siegal, T., Rubinstein, R., Bokstein, F., Schwartz, A., Lossos, A., Shalom, E., Chisin, R., Gomori, J.M., 2000. In vivo assessment of the window of barrier opening after osmotic blood-brain barrier disruption in humans. J. Neurosurg. 92, 599–605.

Simmers, M.B., Pryor, A.W., Blackman, B.R., 2007. Arterial shear stress regulates endothelial cell-directed migration, polarity, and morphology in confluent monolayers. Am. J. Physiol. Heart Circ. Physiol. 293, H1937–H1946.

Sita, G., Hrelia, P., Graziosi, A., Ravegnini, G., Morroni, F., 2018. TRPM2 in the brain: role in health and disease. Cells 7 (7), 82.

Smani, T., Gómez, L.J., Regodon, S., Woodard, G.E., Siegfried, G., Khatib, A.-M., Rosado, J.A., 2018. TRP channels in angiogenesis and other endothelial functions. Front. Physiol. 9, 1731.

Song, J.-G., Kostan, J., Drepper, F., Knapp, B., de Almeida Ribeiro, E.J., Konarev, P.V., Grishkovskaya, I., Wiche, G., Gregor, M., Svergun, D.I., Warscheid, B., Djinović-Carugo, K., 2015. Structural insights into Ca^{2+}-calmodulin regulation of Plectin 1a-integrin β4 interaction in hemidesmosomes. Structure 23, 558–570.

Souza, D.G., Lomez, E.S.L., Pinho, V., Pesquero, J.B., Bader, M., Pesquero, J.L., Teixeira, M.M., 2004. Role of bradykinin b2 receptors in the local, remote, and systemic inflammatory responses that follow intestinal ischemia and reperfusion injury. J. Immunol. 172, 2542 LP–2548.

Stamatovic, S., Keep, R., Andjelkovic, A., 2008. Brain endothelial cell-cell junctions: how to open the blood brain barrier. Curr. Neuropharmacol. 6, 179–192.

Stamatovic, S.M., Johnson, A.M., Keep, R.F., Andjelkovic, A.V., 2016. Junctional proteins of the blood-brain barrier: new insights into function and dysfunction. Tissue Barriers 4 (1), e1154641.

Stanimirovic, D.B., Sandhu, J.K., Costain, W.J., 2018. Emerging technologies for delivery of biotherapeutics and gene therapy across the blood–brain barrier. BioDrugs 32, 547–559.

Stankowski, J.N., Gupta, R., 2011. Therapeutic targets for neuroprotection in acute ischemic stroke: lost in translation? Antioxid. Redox Signal. 14, 1841–1851.

Su, K.-H., Lin, S.-J., Wei, J., Lee, K.-I., Zhao, J.-F., Shyue, S.-K., Lee, T.-S., 2014. The essential role of transient receptor potential vanilloid 1 in simvastatin-induced activation of endothelial nitric oxide synthase and angiogenesis. Acta Physiol. (Oxf.) 212, 191–204.

Suadicani, S.O., Flores, C.E., Urban-Maldonado, M., Beelitz, M., Scemes, E., 2004. Gap junction channels coordinate the propagation of intercellular Ca^{2+} signals generated by P2Y receptor activation. Glia 48, 217–229.

Suetsugu, S., Takenawa, T., 2003. Translocation of N-WASP by nuclear localization and export signals into the nucleus modulates expression of HSP90. J. Biol. Chem. 278, 42515–42523.

Sundivakkam, P.C., Natarajan, V., Malik, A.B., Tiruppathi, C., 2013. Store-operated Ca^{2+} entry (SOCE) induced by protease-activated receptor-1 mediates STIM1 protein phosphorylation to inhibit SOCE in endothelial cells through AMP-activated protein kinase and p38β mitogen-activated protein kinase. J. Biol. Chem. 288, 17030–17041.

Swissa, E., Serlin, Y., Vazana, U., Prager, O., Friedman, A., 2019. Blood–brain barrier dysfunction in status epileptics: mechanisms and role in epileptogenesis. Epilepsy Behav. 101, 106285.

Tadahiro, S., Ryoko, K., Masayo, M., Yukari, Y., Sadao, M., 1973. Studies on kinin-like substances in brain. Biochem. Pharmacol. 22, 567–573.

Tallini, Y.N., Brekke, J.F., Shui, B., Doran, R., Hwang, S., Nakai, J., Salama, G., Segal, S.S., Kotlikoff, M.I., 2007. Propagated endothelial Ca^{2+} waves and arteriolar dilation in vivo: measurements in Cx40BAC GCaMP2 transgenic mice. Circ. Res. 101, 1300–1309.

Tang, D.D., Gerlach, B.D., 2017. The roles and regulation of the actin cytoskeleton, intermediate filaments and microtubules in smooth muscle cell migration. Respir. Res. 18, 1–12.

Thakore, P., Earley, S., 2019. Transient receptor potential channels and endothelial cell calcium signaling. Compr. Physiol. 9, 1249–1277.

Thijs, R.D., Surges, R., O'Brien, T.J., Sander, J.W., 2019. Epilepsy in adults. Lancet 393, 689–701.

Thillaiappan, N.B., Chakraborty, P., Hasan, G., Taylor, C.W., 2019. IP3 receptors and Ca^{2+} entry. Biochim. Biophys. Acta, Mol. Cell Res. 1866, 1092–1100.

Thoppil, R.J., Cappelli, H.C., Adapala, R.K., Kanugula, A.K., Paruchuri, S., Thodeti, C.K., 2016. TRPV4 channels regulate tumor angiogenesis via modulation of Rho/Rho kinase pathway. Oncotarget 7, 25849–25861.

Tojkander, S., Gateva, G., Lappalainen, P., 2012. Actin stress fibers—assembly, dynamics and biological roles. J. Cell Sci. 125, 1855–1864.

Tong, Q., Zhang, W., Conrad, K., Mostoller, K., Cheung, J.Y., Peterson, B.Z., Miller, B.A., 2006. Regulation of the transient receptor potential channel TRPM2 by the Ca^{2+} sensor calmodulin. J. Biol. Chem. 281, 9076–9085.

Tran, Q.K., Ohashi, K., Watanabe, H., 2000. Calcium signalling in endothelial cells. Cardiovasc. Res. 48, 13–22.

Tsai, F.C., Meyer, T., 2012. Ca^{2+} pulses control local cycles of lamellipodia retraction and adhesion along the front of migrating cells. Curr. Biol. 22, 837–842.

Tsai, F.C., Kuo, G.H., Chang, S.W., Tsai, P.J., 2015. Ca^{2+} signaling in cytoskeletal reorganization, cell migration, and cancer metastasis. Biomed. Res. Int. 2015, 409245.

Tsai, T.-Y., Leong, I.-L., Cheng, K.-S., Shiao, L.-R., Su, T.-H., Wong, K.-L., Chan, P., Leung, Y.-M., 2019. Lysophosphatidylcholine-induced cytotoxicity and protection by heparin in mouse brain bEND.3 endothelial cells. Fundam. Clin. Pharmacol. 33, 52–62.

Upadhyay, R.K., 2014. Drug delivery systems, CNS protection, and the blood brain barrier. Biomed. Res. Int. 2014, 869269.

Vandamme, W., Braet, K., Cabooter, L., Leybaert, L., 2004. Tumour necrosis factor alpha inhibits purinergic calcium signalling in blood-brain barrier endothelial cells. J. Neurochem. 88, 411–421.

Verin, A.D., Birukova, A., Wang, P., Feng, L., Becker, P., Birukov, K., Garcia, J.G.N., 2001. Microtubule disassembly increases endothelial cell barrier dysfunction: role of MLC phosphorylation. Am. J. Physiol. Lung Cell. Mol. Physiol. 281, 565–574.

Verri, M., Pastoris, O., Dossena, M., Aquilani, R., Guerriero, F., Cuzzoni, G., Venturini, L., Ricevuti, G., Bongiorno, A.I., 2012. Mitochondrial alterations, oxidative stress and neuroinflammation in Alzheimer's disease. Int. J. Immunopathol. Pharmacol. 25, 345–353.

Wade, R.H., 2009. On and around microtubules: an overview. Mol. Biotechnol. 43, 177–191.

Wallez, Y., Huber, P., 2008. Endothelial adherens and tight junctions in vascular homeostasis, inflammation and angiogenesis. Biochim. Biophys. Acta, Biomembr. 1778, 794–809.

Wang, Y., Shibasaki, F., Mizuno, K., 2005. Calcium signal-induced cofilin dephosphorylation is mediated by slingshot via calcineurin. J. Biol. Chem. 280, 12683–12689.

Wang, X., Sykes, D.B., Miller, D.S., 2010. Constitutive androstane receptor-mediated up-regulation of ATP-driven xenobiotic efflux transporters at the blood-brain barrier. Mol. Pharmacol. 78, 376–383.

Wang, N., De Bock, M., Decrock, E., Bol, M., Gadicherla, A., Vinken, M., Rogiers, V., Bukauskas, F.F., Bultynck, G., Leybaert, L., 2013. Paracrine signaling through plasma membrane hemichannels. Biochim. Biophys. Acta 1828, 35–50.

Wang, F., Liu, D., Xu, H., Li, Y., Wang, W., Liu, B., Zhang, L., 2014a. Thapsigargin induces apoptosis by impairing cytoskeleton dynamics in human lung adenocarcinoma cells. ScientificWorldJournal 2014, 619050.

Wang, F., Liu, D.Z., Xu, H., Li, Y., Wang, W., Liu, B.L., Zhang, L.Y., 2014b. Thapsigargin induces apoptosis by impairing cytoskeleton dynamics in human lung adenocarcinoma cells. Sci. World J. 2014, 1–8.

Wang, L., Wei, L.Y., Ding, R., Feng, Y., Li, D., Li, C., Malko, P., Syed Mortadza, S.A., Wu, W., Yin, Y., Jiang, L.-H., 2020. Predisposition to Alzheimer's and age-related brain pathologies by PM2.5 exposure: perspective on the roles of oxidative stress and TRPM2 channel. Front. Physiol. 11, 155.

Warach, S., Latour, L.L., 2004. Evidence of reperfusion injury, exacerbated by thrombolytic therapy, in human focal brain ischemia using a novel imaging marker of early blood-brain barrier disruption. Stroke 35, 2659–2661.

Woodruff, T.M., Thundyil, J., Tang, S.-C., Sobey, C.G., Taylor, S.M., Arumugam, T.V., 2011. Pathophysiology, treatment, and animal and cellular models of human ischemic stroke. Mol. Neurodegener. 6, 11.

Wu, K.-C., Cheng, K.-S., Wong, K.-L., Shiao, L.-R., Leung, Y.-M., Chang, L.-Y., 2019. ARC 118925XX stimulates cation influx in bEND.3 endothelial cells. Fundam. Clin. Pharmacol. 33, 604–611.

Xu, J.-H., Tang, F.-R., 2018. Voltage-dependent calcium channels, calcium binding proteins, and their interaction in the pathological process of epilepsy. Int. J. Mol. Sci. 19, 2735.

Xu, J., Schwarz, W.H., Käs, J.A., Stossel, T.P., Janmey, P.A., Pollard, T.D., 1998. Mechanical properties of actin filament networks depend on preparation, polymerization conditions, and storage of actin monomers. Biophys. J. 74, 2731–2740.

Yamazaki, Y., Kanekiyo, T., 2017. Blood-brain barrier dysfunction and the pathogenesis of Alzheimer's disease. Int. J. Mol. Sci. 18 (9), 1965.

Yang, W., Barth, R.F., Carpenter, D.E., Moeschberger, M.L., Goodman, J.H., 1996. Enhanced delivery of boronophenylalanine for neutron capture therapy by means of intracarotid injection and blood-brain barrier disruption. Neurosurgery 38, 985–992.

Yang, D., Jing, Y., Liu, Y., Xu, Z., Yuan, F., Wang, M., Geng, Z., Tian, H., 2019. Inhibition of transient receptor potential vanilloid 1 attenuates blood–brain barrier disruption after traumatic brain injury in mice. J. Neurotrauma 36 (8), 1279–1290.

Yang, Q., Zhang, X.F., Van Goor, D., Dunn, A.P., Hyland, C., Medeiros, N., Forscher, P., 2013. Protein kinase C activation decreases peripheral actin network density and increases central nonmuscle myosin II contractility in neuronal growth cones. Mol. Biol. Cell 24, 3097–3114.

Yang, F., Zhao, K., Zhang, X., Zhang, J., Xu, B., 2016. ATP induces disruption of tight junction proteins via IL-1 beta-dependent MMP-9 activation of human blood-brain barrier in vitro. Neural Plast. 2016, 8928530.

Yarlagadda, A., Kaushik, S., Clayton, A.H., 2007. Blood brain barrier: the role of calcium homeostasis. Psychiatry (Edgmont) 4, 55–59.

Yuen, A.W.C., Keezer, M.R., Sander, J.W., 2018. Epilepsy is a neurological and a systemic disorder. Epilepsy Behav. 78, 57–61.

Zhao, H., Zhang, K., Tang, R., Meng, H., Zou, Y., Wu, P., Hu, R., Liu, X., Feng, H., Chen, Y., 2018. TRPV4 blockade preserves the blood-brain barrier by inhibiting stress fiber formation in a rat model of intracerebral hemorrhage. Front. Mol. Neurosci. 11, 97.

Zlokovic, B.V., 2008. The blood-brain barrier in health and chronic neurodegenerative disorders. Neuron 57, 178–201.

Zubakova, R., Gille, A., Faussner, A., Hilgenfeldt, U., 2008. Ca^{2+} signalling of kinins in cells expressing rat, mouse and human B1/B2-receptor. Int. Immunopharmacol. 8, 276–281.

Zuccolo, E., Laforenza, U., Negri, S., Botta, L., Berra-Romani, R., Faris, P., Scarpellino, G., Forcaia, G., Pellavio, G., Sancini, G., Moccia, F., 2019. Muscarinic M5 receptors trigger acetylcholine-induced Ca(2+) signals and nitric oxide release in human brain microvascular endothelial cells. J. Cell. Physiol. 234, 4540–4562.

CHAPTER THREE

Mitochondrial calcium homeostasis in hematopoietic stem cell: Molecular regulation of quiescence, function, and differentiation

Massimo Bonora*, Asrat Kahsay, and Paolo Pinton*

Department of Medical Sciences, Section of Experimental Medicine, Laboratory for Technologies of Advanced Therapies (LTTA), University of Ferrara, Ferrara, Italy
*Corresponding authors: e-mail address: bnrmsm1@unife.it; pnp@unife.it

Contents

Abstract

Hematopoiesis is based on the existence of hematopoietic stem cells (HSC) with the capacity to self-proliferate and self-renew or to differentiate into specialized cells. The hematopoietic niche is the essential microenvironment where stem cells reside and integrate various stimuli to determine their fate. Recent studies have identified niche containing high level of calcium (Ca^{2+}) suggesting that HSCs are sensitive to Ca^{2+}. This is a highly versatile and ubiquitous second messenger that regulates a wide variety of cellular functions. Advanced methods for measuring its concentrations, genetic experiments, cell fate tracing data, single-cell imaging, and transcriptomics studies provide information into its specific roles to integrate signaling into an array of mechanisms that determine HSC identity, lineage potential, maintenance, and self-renewal. Accumulating and contrasting evidence, are revealing Ca^{2+} as a previously unacknowledged feature of HSC, involved in functional maintenance, by regulating multiple

International Review of Cell and Molecular Biology, Volume 362
ISSN 1937-6448
https://doi.org/10.1016/bs.ircmb.2021.05.003

actors including transcription and epigenetic factors, Ca^{2+}-dependent kinases and mitochondrial physiology. Mitochondria are significant participants in HSC functions and their responsiveness to cellular demands is controlled to a significant extent via Ca^{2+} signals. Recent reports indicate that mitochondrial Ca^{2+} uptake also controls HSC fate. These observations reveal a physiological feature of hematopoietic stem cells that can be harnessed to improve HSC-related disease. In this review, we discuss the current knowledge Ca^{2+} in hematopoietic stem cell focusing on its potential involvement in proliferation, self-renewal and maintenance of HSC and discuss future research directions.

1. Introduction

The blood system contains more than 10 different blood cell types (lineages) with various functions. All blood cell types arise from hematopoietic stem cells (HSCs) that reside mainly in the bone marrow (BM), the major site of adult hematopoiesis. The pioneering findings by Till and McCulloch revealed the regenerative potential of single BM cells, thus establishing the existence of multipotential HSCs. HSCs are the only cells within the hematopoietic system that possess the potential for both multipotency and self-renewal. BM-HSCs are functionally defined by their unique capacity to self-renew and to differentiate to produce all mature blood cell types (Song et al., 2010). Hematopoiesis is then based on the existence of hematopoietic stem cells with the capacity to proliferate and self-renew or to differentiate into specialized cells (Seita and Weissman, 2010). Multipotency is the ability to differentiate into all types of functional blood cells, while self-renewal is the ability to give rise to identical daughter without differentiation. The HSC's choice between self-renewal and differentiation must be tightly regulated to enable both the generation of differentiated cells and the accurate maintenance of the right HSC number (Ito and Ito, 2018). The cells originating from hematopoietic stem cells can commit to two lineages: the myeloid lineage, including granulocytes, erythrocytes, megakaryocytes/platelets and monocytes; and the lymphoid lineage comprising B and T lymphocytes as well as natural killer cells (Moignard et al., 2013). Leukocytes represent many specialized cell types involved in innate and acquired immunity. Erythrocytes provide O_2 and CO_2 transport, whereas megakaryocytes generate platelets for blood clotting and wound healing (Graf, 2002).

Most of the HSCs within the BM are in a quiescent state. Quiescent HSCs have the ability to self-renew indefinitely, mediating the homeostatic and continuous turnover of blood cells that organisms require throughout their life (Arai et al., 2004; Morita et al., 2006; Seita and Weissman, 2010; Wang and Ema, 2016; Yang et al., 2007). The BM is an intricate tissue that encompasses several hematopoietic and non-hematopoietic cell types that are interconnected by a vascular and innervated network within the cavities of long bones and axial bones (Baccin et al., 2020; Coutu et al., 2017; Pinho and Frenette, 2019; Zhang et al., 2021). The stem cell niche is the essential microenvironment where stem cells reside and integrate various stimuli to determine their fate and provides special support for cell viability. Niche-specific cell populations, extracellular matrix components, varied growth factors, and cell adhesion molecules produced by niche cells are integrated together for the common goal of controlling stem cell behavior. When tissue damage occurs, niches are feedback systems for communicating information about the state of a tissue back to the related stem cells (Ellis et al., 2011; Katayama et al., 2006; Petit et al., 2002; Pinho and Frenette, 2019).

Hematopoietic cells share this bone cavity together with other cells, such as osteoblasts, osteoclasts, macrophages, adipocytes, perivascular cells, mesenchymal stem cells, and endothelial cells. The communications among these different cells are mediated through various cytokines, growth factors, matrix proteins, and cell-cell adhesions. All these factors have been demonstrated to be HSC niche component as they are co-localized with HSCs and have the functional effects on HSC self-renewal and differentiation (Morrison and Scadden, 2014).

Considering the complex pattern of signals regulating HSC activity, it is not surprising that in the last years, investigation of Ca^{2+} signaling have started, bringing light to unexpected clues. In addition, in last decade, mitochondrial physiology have been reported as a significant regulator of HSC function, mostly by means of respiration and reactive oxygen species (ROS) production (Chen et al., 2008; Ito et al., 2019; Rimmelé et al., 2015; Tai-Nagara et al., 2014).

The purpose of this review is then to summarize the early and exciting observation reported so far that try to link mitochondrial Ca^{2+} to normal or defective hematopoiesis. Also, we will try to explain the contradictory results presented and offer a perspective on future directions.

2. Intracellular Ca^{2+} homeostasis in HSC maintenance and commitment

To date, studies investigating the role of Ca^{2+} signaling in HSC are relatively few. Still, thanks to some recent insight we can start delineating some pictures. Intriguingly the evidence now available are partially contradicting and could support two different models. This section then groups the evidence according to the models which are most likely to occur in hematopoiesis.

2.1 Ca^{2+} driven commitment of HSC

Direct measurements of cytoplasmic Ca^{2+} concentration, $[Ca^{2+}]_c$, at resting conditions by Indo-1 in flow cytometry, indicates that mouse HSC have extremely low free cytosolic Ca^{2+} (between 20 and 30 nM). The stimulation with stromal cell-derived factor 1 (SDF-1, also known as CXCL12) activates the G protein-coupled protein receptor C-X-C chemokine receptor 4 (CXCR4) inducing an Ip3 mediated release of Ca^{2+} from ER via inositol trisphosphate (Ip3) receptors (Ip3R) with transient elevation of $[Ca^{2+}]_c$ (Cancilla et al., 2020; Hadad et al., 2013; Xiang et al., 2002) (Fig. 1), which still reaches concentration as low as 100 nM (Luchsinger et al., 2019).

Ca^{2+} levels significantly increase along with hematopoietic commitment. Both basal $[Ca^{2+}]_c$ and SDF-1 induced cytosolic transients are augmented in multipotent (MPP) and lineage-restricted progenitors (CP). In agreement with Indo-1 measurement, analysis of mitochondrial motility (considered as a readout of $[Ca^{2+}]_c$ due to the activity of Miro) displayed high movement frequency in the most primitive hematopoietic compartment compared to progenitors. To explain a so low $[Ca^{2+}]_c$ Luchsinger et al. observed that HSC expressed the highest levels of plasma membrane Ca^{2+} ATPase isoform 4 (PMCA4) among all the tested hematopoietic populations and that its activity is inversely proportional to their differentiation status.

These measurements suggest that low $[Ca^{2+}]_c$ is required for HSC commitment. In agreement, agents which induces HSC activation (exit from quiescence status and commitment) are reported to elicit transient $[Ca^{2+}]_c$ elevation. Several cytokines are known to induce HSC proliferation and commitment for the physiological stimulation of blood production. Among many, the most characterized are probably IL-3, IL-6, Il-7, EPO and GM-CSF (Abkowitz et al., 2003; Ogawa, 1993; Zhao et al., 2014). All these cytokines have been shown to stimulated rapid and transient elevation of $[Ca^{2+}]_c$ in HSPC maintained in vitro on stromal cells (Paredes-Gamero et al., 2008). Authors further characterized the signaling induced by IL-3 and GM-CSF and observed that in vitro co-administration of Ca^{2+} chelator, BAPTA (1,2-bis(2-aminophenoxy)ethane-N,N,N',N'-tetraacetic acid) or the Ip3R

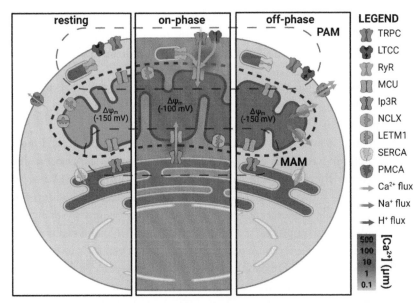

Fig. 1 Mitochondrial calcium homeostasis. At resting conditions (left panel), highest $[Ca^{2+}]$ is found within lumen of Ca^{2+} stores (e.g., ER), guaranteed by the constant activity of SERCA. Its concerted activity with the once of PMCA ensures that $[Ca^{2+}]_c$ is maintained at very low levels (most often within 0.1–0.2 µM range). The strong $\Delta\Psi m$ should be sufficient to induce accumulation of Ca^{2+} within mitochondrial matrix, but the low affinity for Ca^{2+} of the MCU ensures that $[Ca^{2+}]_m$ is kept within the range 0.2–0.5 µM, close to the values of $[Ca^{2+}]_c$. In the on-phase, the opening of Ip3R or RyR (Ip3-operated and voltage-operated channels, respectively) on ER membrane causes the depletion of intracellular stores, that pours its Ca^{2+} in cytoplasm (which reaches 1–2 µM). Due to the strong gradient of $[Ca^{2+}]$ between ER and cytosol, Ca^{2+} is released in the cytoplasm faster than its diffusion generating a microdomain with high $[Ca^{2+}]_c$. At contact sites between mitochondria and ER (MAM) the MCU is submerged into a $[Ca^{2+}]_c$ sufficient to induces its opening therefore the accumulation. A comparable mechanism occurs at mitochondria associated to plasma membrane (PAM) where $[Ca^{2+}]_c$ microdomain are achieved by the opening of ligand-operated or voltage-operated PM channels (e.g., TRPCs and LTCC, respectively). Finally, elevation of $[Ca^{2+}]_m$ activates exchangers on IMM that ensures the extrusion of Ca^{2+} into cytoplasm and engaging the off-phase. These are the Ca^{2+}/Na^+ exchanger, NCLX, and LETM1, the putative Ca^{2+}/H^+ exchanger. Concomitantly, activity of SERCA and PMCA reaches their maximal rates, restoring $[Ca^{2+}]_c$ and $[Ca^{2+}]_{er}$ to the values of resting.

inhibitor, 2APB, significantly impair proliferation in the forming colonies, suggesting the involvement of Ip3–Ca^{2+} signaling in IL-3- and GM-CSF-induced HSC commitment. In the same study authors shown that P2X receptor agonists ATP, ADP, and UTP elicits $[Ca^{2+}]_c$ waves higher than the once induced by cytokine and that their effect on *in vitro* differentiation was partially reverted by BAPTA. Extracellular nucleotides (especially ATP) can be released in the BM by HSC mobilizing factor (e.g., G-CSF) to induce sterile inflammation (Ratajczak et al., 2018). Intraperitoneal injection of ATP in mice, induces mobilization of HSC and isolated HSC exposed to ATP displayed diminished reconstitution capacity (Barbosa et al., 2011). In agreement, mobilization stimulated by C-GSF is strongly inhibited in mice lacking the purinergic receptor P2X7 (Adamiak et al., 2018).

HSC can be induced to exit dormancy also by genotoxic stress. Indeed, agents which induces DNA damage (e.g., 5-FU or γ radiation) mostly affect cycling cells in the BM leading to their apoptotic cell death (Høyer and Nielsen, 1992). After an initial reduction of BM cellularity, some HSCs re-enter cell cycle undergoing commitment rather than self-renewal to replenish the dying populations (Schoedel et al., 2016). Interestingly two independent report displayed increase in $[Ca^{2+}]_c$ in active HSCs isolated from mice exposed to 5-FU (Fukushima et al., 2019; Umemoto et al., 2018).

Also, some molecular effectors downstream Ca^{2+} are involved in HSC differentiation. Elevation of $[Ca^{2+}]_c$ in HSC have been linked to Nuclear factor of activated T-cells, cytoplasmic 2 (NFAT) activation. Sustained increase in $[Ca^{2+}]_c$ activates calcineurin, which dephosphorylates NFAT and promotes its translocation to the nucleus (Beals et al., 1997; Hogan et al., 2003). Noncanonical Wnt signaling, mediated by Fmi-Fz8, restricts NFAT nuclear translocation controlling intracellular Ca^{2+} level, potentially through inhibition of L-type Ca^{2+} channels (LTCC) (Sugimura et al., 2012). 5-FU stimulation induces NFAT nuclear translocation where up regulates the transcription of Interferon gamma (IFNγ) and Cox2 transcription in HSC. INFγ in turn is reported to instruct HSC to exit quiescent status (Baldridge et al., 2010).

In addition to NFAT signaling a Ca^{2+}-Calpain axis was proposed to control HSC function. Calpains are Ca^{2+}-regulated cysteine proteases and, according to GEXC database, HSC express at least 3 of the 14 isoforms. Calpains activity was detectable in cultured HSC that are usually maintained in media with high $[Ca^{2+}]$. On opposite, Ca^{2+} deprivation from media caused significant inhibition of Calpain activity. Interestingly, among the many predicted Calpains target is Ten-eleven translocated (TET) enzymes. These converts 5-methylcytosine (5mC) to 5-hydroxymethyl cytosine (5hmC), an epigenetic marker required for the demethylation of 5mC (Cimmino and Aifantis, 2017; Rasmussen and Helin, 2016; Zhang et al., 2016). TET2 plays a major role in HSCs by suppressing the expression of lineage-specific genes, therefore favoring HSC maintenance (Zhang et al., 2016). Tet2$^{-/-}$ HSCs show progressive myeloid bias *in vivo* that ultimately leads to myeloid malignancy (Ko et al., 2011; Li et al., 2011). Calpains inhibition by small molecules, overexpression of its endogenous inhibitor (calpastatin) or low calcium media, increased TET protein expression and 5hmC deposition in chromosomal DNA. In agreement, $Tet2^{-/-}$ HSCs did not benefit from culturing in low Ca^{2+} media (Luchsinger et al., 2019).

Take together, this evidence indicates that HSCs requires low $[Ca^{2+}]_c$ to maintain their quiescent status and function, on opposite, stimuli which instruct HSC proliferation and commitment, elevate $[Ca^{2+}]_c$ to induce NFAT signaling and Tet2 degradation (Fig. 2). Also, as $[Ca^{2+}]_c$ is a function of intraluminal Ca^{2+} concentration of ER, $[Ca^{2+}]_{er}$, it would be of interest to test if $[Ca^{2+}]_{er}$ have adapted to this mechanism.

2.2 Ca^{2+} control of HSC quiescence

The previous model is nonetheless oversimplified as additional evidence indicates a far more complex participation of Ca^{2+} in HSC function.

One of the earliest evidence calling for Ca^{2+} investigation in HSC was the dependence of HSC function on the calcium-sensing receptor (CasR). This is a G-protein-coupled receptor activated by extracellular Ca^{2+}, $[Ca^{2+}]_o$, mostly known for its participation in parathyroid hormone response to Ca^{2+} levels in tissue milieu (Chavez-Abiega et al., 2020). $CasR^{-/-}$ develop early hypercalcemia and do not survive longer then 7–10 days after birth. $CasR^{-/-}$ mice analyzed within 72h from birth display significant BM hypocellularity and limited amount of hematopoietic stem and progenitor cells (HSPC). Also, fetal liver HSC (FL-HSC, a common source of HSC from non-vital mouse strains) displayed altered maintenance capacity *in vitro* and impaired capacity to reconstitute the hematopoietic system (reconstitution capacity) in competitive bone marrow transplantation (BMT, a gold standard assay for HSC function). Interestingly HSC $CasR^{-/-}$ also displayed aberrant localization in the BM, suggesting alteration in the recognition of their niche (a process usually termed homing) (Adams et al., 2006). This hypothesis was further supported by the observation that isolated HSC $CasR^{-/-}$ have impaired adhesion on collagen coated culturing device. Stimulation of isolated HSPC with the CasR agonist, Cinacalcet, induces elevation of $[Ca^{2+}]_c$, promotes cellular adhesion *in vitro* and favored reconstitution capacity in competitive BMT (Lam et al., 2011). FL-HSC are not quiescent and support hematopoiesis in developing fetus. Only after birth, FL-HSC, migrates to BM, were install in their niche and exit cell cycle in favor of quiescence (Arai et al., 2005; Bowie et al., 2006).

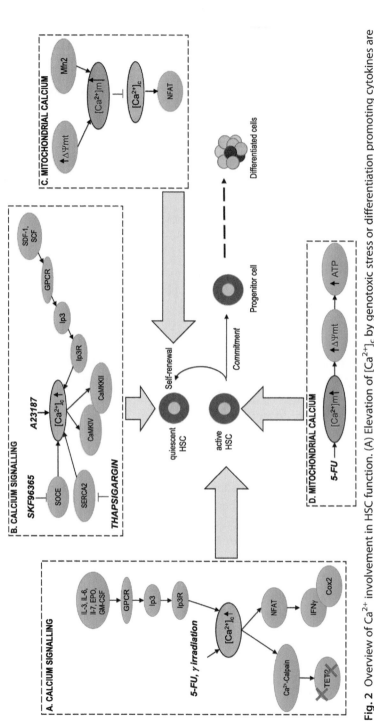

Fig. 2 Overview of Ca^{2+} involvement in HSC function. (A) Elevation of $[Ca^{2+}]_c$ by genotoxic stress or differentiation promoting cytokines are associated to HSC activation and commitment. In these conditions, Calpains induces degradation of the stimulator of self-renewal Tet2, while NFAT translocate to nucleus where can stimulate the transcription of factors promoting commitment. (B) High $[Ca^{2+}]_c$ was also reported in quiescent vs active HSC. Quiescence can be induced by stimulating with SDF-1 or SCF, as well as by the pharmacological elevation of $[Ca^{2+}]_c$ via SERCA-2 or SOCE inhibition and the use of ionophores. (C) HSC have high membrane potential which is expected to maintain high basal $[Ca^{2+}]_m$ in concert with Mfn2 activity. The resulting buffering activity on $[Ca^{2+}]_c$ blocks the NFAT nuclear translocation favoring HSC maintenance. (D) Stimulation with 5-FU induces the elevation of $[Ca^{2+}]_m$ that indirectly potentiates $\Delta\Psi m$ and ATP production, associated with HSC commitment.

This suggest that Ca^{2+} signals in HSC might be important for niche interaction. In agreement, one of the most characterized signaling molecule required for HSC homing is CXCR4. This receptor is highly expressed in HSC where mediates its interaction with multiple SDF-1 expressing cells in the vascular niche (Pinho and Frenette, 2019; Sugiyama et al., 2006). SDF-1 is chemoattractant for HSC and binding of its receptor is required for maintenance of quiescence (Nie et al., 2008). SDF-1 administration to cultured HSC inhibits the spontaneous entry into cell cycle. Interestingly, as previously described, SDF-1 elicit transient $[Ca^{2+}]_c$ elevation (Lam et al., 2011; Luchsinger et al., 2019). *In vivo* antagonization of the SDF-1-CXCR4 interaction through AMD3100 induces HSC mobilization and stimulation of hematopoiesis (Hübel et al., 2004). AMD3100 was demonstrated in fully differentiated hematopoietic cells, to potently inhibits Ca^{2+} mobilization via SDF-1 (Hatse et al., 2002).

Another ligand required for HSC interaction with its niche is the stem cell factor (SCF). This is released in the BM mainly (but not only) by arteriolar endothelial cells where bind the HSC marker c-kit (Asada et al., 2017).

In vivo studies have shown that deletion of SCF using endothelial cell-specific Cre strains, impairs HSC maintenance at steady state, and confirms the role of endothelial cells as niche regulators (Asada et al., 2017; Ding et al., 2012). In cultured HSC also SCF is able to elicit $[Ca^{2+}]_c$ transients (Paredes-Gamero et al., 2008).

If Ca^{2+} signaling is required for HSC installation and maintenance in their niche, there should be evidences that quiescent HSC have elevated $[Ca^{2+}]_c$. In a very recent report Fukushima et al. presented evidence in this favor (Fukushima et al., 2019). They developed a transgenic mouse stably expressing, in the hematopoietic system, a fluorescent marker for quiescent cells (mVenus-p27K$^-$) and compared basal $[Ca^{2+}]_c$ in dormant vs active HSC through the Ca^{2+} sensitive dye CaSiR-1. Interestingly they observed that quiescent HSC displayed higher $[Ca^{2+}]_c$ compared to cycling one, no significant alteration while compared to MPP, while a significant increase in $[Ca^{2+}]_c$ was detectable only in more CP. Also, quiescent cell with high $[Ca^{2+}]_c$ displayed full long term reconstitution capacity on opposite then low $[Ca^{2+}]_c$, in competitive BMT. They addressed the increased $[Ca^{2+}]_c$ to elevated expression of multiple channels involved in Ca^{2+} influx from plasma membrane, including L-type Ca^{2+} channels (LTCC) and transient receptor potential channel (TRPC). Additionally, they cultured HSC for 24 h with the Sarco/endoplasmic reticulum calcium-ATPase (SERCA) inhibitor thapsigargin or the Ca^{2+} ionophore A23187. This treatment induces elevation of $[Ca^{2+}]_c$ (thought unexpected for a long thapsigargin exposure) and slightly promoted reconstitution capacity in BMT.

A partial confirmation comes from investigation of $S100a6^{-/-}$ animals. S100A6 protein is a member of the EF-hands S100 protein family, implicated in the regulation of several cellular functions, such as proliferation, apoptosis, cytoskeleton dynamics, and cellular response to stress factors. S100A6 is highly expressed in HSC and its genetic inactivation leads to major defects in HSC function. $S100a6^{-/-}$ mice have indeed significant BM hypocellularity, including reduction in the HSPC compartment. Also, HSC from $S100a6^{-/-}$ have limited reconstitution capacity in competitive BMT experiments, indicative of impaired HSC function. Interestingly, SCF induced $[Ca^{2+}]_c$ transient was ablated by S100A6 null cells, suggesting a that its regulation of HSC function is mediated by Ca^{2+}, but no overt alteration of quiescence were reported (Grahn et al., 2020).

Another study tested the effect of the store operated calcium entry (SOCE) inhibitor, SKF96365 on HSC function *in vitro* and *in vivo*. Isolated HSPC exposed to SKF96365 displayed reduced basal $[Ca^{2+}]_c$ and exit from quiescence causing expansion of HSC and committed progenitors. Accordingly, SKF96365-treated HSC transplanted into non–irradiated immune compromised mice displayed slightly better reconstitution capacity after 1 month, but not in the long term (4 month), suggesting that SOCE inhibition slightly promoted HSC commitment (Uslu et al., 2020).

In agreement for a role in Ca^{2+} signaling in regulation of HSC maintenance is the observation that Calmodulin-dependent Protein Kinase IV (CaMKIV) was demonstrated fundamental in the regulation of HSC quiescence. $CaMKIV^{-/-}$ animals have significant BM hypocellularity with reduced amount of HSPC. Reconstitution capacity of HSPC $CaMKIV^{-/-}$ was significantly altered, being able to provide hematopoiesis only for the first 6 weeks. Further, HSPC $CaMKIV^{-/-}$ were significantly faster cycling and failed to activated Bcl-2 expression, due to low phosphorylation status of CREB (Kitsos et al., 2005). Likewise, an impairment of the quiescence status was observed in Calcium/calmodulin-dependent kinase kinase 2 $(CaMKK2)^{-/-}$ HSC. Especially, HSCP $CaMKK2^{-/-}$ have increased proliferation and a transcriptional profile consistent with poor quiescence and increased differentiation. Should be noted that transplanted $CaMKK2^{-/-}$ progenitors have only slightly affected reconstitution capacity (at least after 4 months), but apparently are significantly resistant to irradiative stress

(Racioppi et al., 2017). These data indicate that CaMKIV and CaMKK2 can transduce Ca^{2+} signal of some sort to transcriptional control of quiescence (Fig. 2).

Many of the contrasting results described so far could be attributed to the differences on the experimental methods and to the complexity in properly measure Ca^{2+} in HSC (see Box 1). A better comprehension of the phenomenon could be achieved by the description of the Ca^{2+} dynamic rather than solely basal $[Ca^{2+}]_c$ as is clear that Ca^{2+} signals differing in frequency, amplitude or duration can lead to completely different outcome.

BOX 1 Challenges and pitfalls of Ca^{2+} measurements in HSC.

As for all primary cells, investigation of Ca^{2+} signaling in HSC is challenging. HSC investigation requires in first place their isolation from mouse BM in most cases, but also mouse FL as well as human BM. Though isolation procedures vary from the origin of the donor, these all share significant amount of time in Ca^{2+}-free PBS or PBS + 2% FBS (with expected $[Ca^{2+}]_o$ around 0.1 mM). Adult HSC resides in the BM where $[Ca^{2+}]$ have been recently estimated around 0.5 mM, though it could be higher in the arteriolar niche surrounding a significant proportion of quiescent HSC. These aspects represent a first significant limitation to investigation of Ca^{2+} signals, indeed the activity of PMCA usually outcompete the activity of SERCAs, especially when cellular samples are maintained in buffers with low $[Ca^{2+}]_o$. This cause the progressive depletion of Ca^{2+} from intracellular store and cytoplasm (which is poor in endogenous Ca^{2+} buffers). HSC isolation is then expected to significantly alter the intracellular distribution of Ca^{2+}, though the phenomenon should be limited by maintaining samples at 4 °C. This Ca^{2+} depletion artifact might be particularly significant when comparing HSC with committed progenitors, considering the high PMCA activity reported for this cell type, and could explain the dramatic low $[Ca^{2+}]_c$ observed.

As HSC are strongly affected by culturing, then Ca^{2+} recording is usually performed immediately at the end of the isolation procedure. Future investigation should then consider a Ca^{2+} repletion step (e.g., 30 min incubation at 37 °C in a buffer with $[Ca^{2+}]_o$ resembling BM milieu). Also, because of this limitation, Ca^{2+} investigation have been obtained so far only using fluorescent dyes. Fluorescent small molecules should be used carefully in HSCs, because of the high activity of the xenobiotic efflux pumps characterizing these cells and that can result in dye extrusion and uneven staining between HSC and committed progenitors (Morganti et al., 2019b). In the past, the extrusion of fluorescent dyes (e.g., Rhodamine 123 or Hoechst Blue vs Hoechst Red) was used as a technique to identify and isolate HSCs (Zhou et al., 2001) before of the characterization of CD150 as

Continued

BOX 1 Challenges and pitfalls of Ca^{2+} measurements in HSC.—cont'd

a marker of HSC (Oguro et al., 2013). This phenomenon should not impact the recording of steady state [Ca^{2+}]$_c$ using robust radiometric dyes as Indo-1, when a sufficient signal to noise ratio is achieved, but it will jeopardize recording obtained by non ratiometric dyes (e.g., Fluo-3, Fluo-4, or CaSir-1). Therefore, the correlation between dye intensity and quiescence should be considered carefully.

Another strong limitation in fluorescent dyes is poor targetability of fluorescent dyes to cellular district. All the AM dye indeed can virtually localize into all cellular compartments, but their K$_D$ for Ca^{2+} usually allows the recording of Ca^{2+} in regions with low concentrations as cytoplasm. Some fluorescent dyes developed for Mg^{2+} investigation have been proven to report Ca^{2+} in intracellular stores (e.g., ER or Golgi) due to their low K$_D$ (Rossi and Taylor, 2020).

A notable exception is represented by the rhodamine derivative Rhod-2 or X-Rhod-1. Thanks to the presence of one net positive charge, these compounds can easily accumulate the most electronegative intracellular compartments, mitochondria (the most represented) and peroxisomes. Rhod-2 was then considered the best option for [Ca^{2+}]$_m$ analysis in HSCs and provided so far valuable preliminary insights. Apart from being sensitive to xenobiotic fluxes as other dyes, these derivatives are nonetheless extremely sensitive to $\Delta\Psi$mt. Also, Rhod-2 have a high affinity for Ca^{2+} while mitochondria have been reporter that can accumulate [Ca^{2+}]$_m$ to concentration far above 0.1 mM. Therefore, Rhod-2 will easily saturate in cells with mitochondria very active in terms of Ca^{2+} signaling.

The ideal investigation of Ca^{2+} signaling in hematopoiesis should be obtained by the far superiors GECIs. A plethora of fluorescent or luminescent indicators have been developed so far, allowing Ca^{2+} recordings, virtually, in every cellular compartment (Bonora et al., 2013; Horikawa, 2015).

Isolated mouse HSCs can be transduced in vitro to allow GECIs expression then transplanted to investigate their Ca^{2+} homeostasis in a more native environment. Most intriguingly, BM structure and hematopoietic niche have been demonstrated to undergoes significant remodeling after γ-irradiation, a procedure usually required to allow proper HSC reconstitution during transplantation (Batsivari et al., 2020).

It could be then of extreme interest to compared Ca^{2+} homeostasis in HSC that undergoes to transplantation, compared to quiescent HSCs from untreated animals (e.g., from transgenic mice with stable expression of GECI), as it could provide useful insight in the participation of Ca^{2+} in the communications between HSCs and its niche as well as niche remodeling during stress.

3. Mitochondrial calcium in HSC regulation

The impact of mitochondria on HSC function has been intensively investigated in the last decade (Gurumurthy et al., 2010; Ito et al., 2019; Nakamura-Ishizu et al., 2020; Rimmelé et al., 2015). The BM niche which maintains HSC quiescent is hypoxic and allows stabilization of the Hypoxia-inducible factor 1-alpha. This is a transcription factor which configure HSC's metabolism so to sustain glucose consumption while limiting pyruvate entry into mitochondria (Takubo et al., 2010, 2013).

In agreement, HSC have been shown rely mostly on glycolysis rather than mitochondrial respiration for ATP recycling. Further, the activation of mitochondrial respiration is determinant in cell fate decision by means of the production of reactive oxygen species (ROS), which favors commitment against self-renewal (Chen et al., 2008; Ito et al., 2004, 2006).

The deeper investigation of HSC's mitochondria indicates that this organelle is also important in HSC maintenance. Indeed HSCs pharmacological inhibition of fatty acid oxidation (FAO)—the catabolism of fatty acid in the mitochondrial matrix that feeds the tricarboxylic acid cycle (TCA)—as well as deletion of FAO driver genes, strongly dampens self-renewal capacity (Ito et al., 2012). In agreement, genetic inactivation of electron transport chain (ETC) activity by deletion of genes coding for of an essential subunit of respiratory chain complex III (*Uqcrfs1*) or complex II (*SdhD*) were reported to cause severe pancytopenia and reduction in HSCs, respectively, indicating a meaning for ETC in HSC, regardless the low respiration that these cells can carry on (Ansó et al., 2017; Bejarano-García et al., 2016).

The use of common mitochondrial dyes (e.g., TMRM, Mitotrackers, Rhod 123) in combination with blockade of xenobiotic efflux (e.g., by exposure to Verapamil), revealed that mitochondrial content in HSC was largely underestimated and that display, *in vitro*, a higher mitochondrial membrane potential ($\Delta\Psi$m) compared to committed cells (Bonora et al., 2018; Mansell et al., 2021). HSC are characterized by a minimal configuration of ETC. Especially, we observed that HSC display low levels of respiratory complexes I and V, and that electrons enters ETC most entirely through the activity of respiratory complex II (Bonora et al., 2018; Morganti et al., 2019a). The elevated $\Delta\Psi$mt, despite low respiration rate, is most likely due to the ratio between respiratory complexes I–IV to complex V.

Being $\Delta\Psi$m the driving force for mitochondrial Ca^{2+}, $[Ca^{2+}]_m$ (Fig. 1), it should be expected that HSC might have higher basal $[Ca^{2+}]_m$ compared to the differentiating progeny. To date, no studies reported $[Ca^{2+}]_m$ in HSC compared to progenitors or fully differentiated cells. Still, Luchsinger et al. reported that isolated HSC cultured in low Ca^{2+} media significantly diminished the brightness of the Ca^{2+} sensitive rhodamine derivative Rhod-2. Also, they observed that respiration levels correlate with the Ca^{2+} availability in the media (Luchsinger et al., 2019). This was explained by the Ca^{2+} dependence of TCA cycle and F1/FO ATP synthase. The same study reported (as previously discussed) extremely low $[Ca^{2+}]_c$ (below 100 nM), considering that mitochondrial Ca^{2+}-dependent dehydrogenases have relatively low K_D for Ca^{2+}, it is plausible that HSC have a relatively high $[Ca^{2+}]_m$ at resting conditions, thanks to elevated $\Delta\Psi m$. In agreement, the same group reported evidence that mitofusin-2 (mfn2) is required for proper hematopoiesis. Mfn2 is a well-known regulator of mitochondrial dynamics, and the mfn2 KO HSC display a prototypical fragmentation phenotype. Also, loss of mfn2 leads to significant reduction of mitochondrial Ca^{2+} uptake capacity. In HSC, mfn2 KO displayed increased $[Ca^{2+}]_c$ elicited by SDF-1 and an elevation of the nuclear/cytoplasmic ratio of NFAT-C1 (Luchsinger et al., 2016).

Finally, mfn2 KO cells displayed impaired reconstitution capacity in competitive BMT experiments, though apparently biased for the lymphoid potential. In agreement, $S100a6^{-/-}$ HSC (which have defective HSC function) displayed blunted $[Ca^{2+}]_m$ uptake in response to SCF compared to wild-type as well as lower $\Delta\Psi m$ and maximal respiratory capacity (Grahn et al., 2020). This suggests that mitochondrial Ca^{2+} uptake (at least its buffer capacity) is instrumental for the maintenance of HSC function (Fig. 2).

This model is nonetheless not confirmed by investigation of HSC function in Drp1 KO cells. Drp1 is the main executor of the mitochondrial fission. Its regulation of mitochondrial network dynamics have been reported to impact $[Ca^{2+}]_m$ in multiple cellular (Favaro et al., 2019; Szabadkai et al., 2004). Especially, overexpression of Drp1 induces hyper-fragmentation, causing isolation of mitochondrial particles which do not have a direct contact with ER and, therefore, reduces mitochondrial Ca^{2+} uptake. In parallel Drp1 inactivation, generates an hyperfused mitochondrial network, favoring the diffusion of Ca^{2+} within mitochondrial matrix. While $[Ca^{2+}]_m$ in HSC was never reported during Drp1 modulation, it is expected that its genetic inactivation should lead to increased buffering capacity, therefore favoring HSC maintenance. A recent report displayed instead that $Drp1^{-/-}$ HSC have reduced reconstitution capacity, thought it was explained by the inability of mitochondria to properly segregate during HSC division (Hinge et al., 2020).

A more recent report investigated basal $[Ca^{2+}]_m$ in HSCs during stress hematopoiesis. Umemoto et al. reported that HSC activation by intraperitoneal injection of 5-FU elevates basal $[Ca^{2+}]_c$, $\Delta\Psi$m and basal $[Ca^{2+}]_m$. Also, exposure to Nifedipine, a LTCC blocker, inhibited mitochondrial functions and prolonged the cell cycle phase of HSCs which resulted in slower recovery of stress hematopoiesis (Umemoto et al., 2018).

In agreement, a RNA sequencing study from mouse HSC display that some members of the mitochondrial calcium uniporter complex (Micu1, Micu2 and Smdt1), have a tendency in increase their expression from quiescent HSC to active to MPP (Cabezas-Wallscheid et al., 2017).

Finally, mitochondrial Ca^{2+} transfer from ER to mitochondria regulate autophagic flux and mitochondrial degradation (Calì et al., 2013; Cárdenas and Foskett, 2012; Gomez-Suaga et al., 2017). HSC have been shown strong dependency on autophagy and its subroutine mitophagy. Especially, it was demonstrated that genetic inactivation of autophagy is detrimental for HSC maintenance, leading to premature aging (Ho et al., 2017) and that HSC maintain a high rate of mitophagy to ensure mitochondrial clearance (Ito et al., 2016; Jin et al., 2018) which results in favored HSC self-renewal capacity (Fig. 2).

4. Calcium homeostasis deregulation in malignant hematopoiesis

Alterations in the control of HSC quiescence as well as their division balance causes hematological malignancies (Heidel et al., 2011; Park et al., 2019) such as chronic leukemia (CL) or acute myelogenous leukemia (AML). A panel of recurrent mutations have been described in these conditions, offering insights on the molecular mechanism underlining their insurgence and development (Kim et al., 2017; Lindsley et al., 2015; Stevens et al., 2018; Thol et al., 2017). In CL or *de novo* AML the earliest mutation is the trigger event for the development of the disease. On opposite, a subgroup of AML (secondary AML) develop from other hematological conditions, often generalized as pre-leukemia (e.g., clonal hematopoiesis of indeterminate potential, CHIP or myelodysplastic syndromes, MDS). The transition from pre-leukemia to secondary AML is driven by the acquisition of additional mutations, conferring the neoplastic phenotype (Chen et al., 2019; Corces-Zimmerman et al., 2014; Koeffler and Leong, 2017; Makishima et al., 2017; Nagata and Maciejewski, 2019; Saeed et al., 2021; Shlush et al., 2014).

To date, no direct measurement reports $[Ca^{2+}]_m$ handling in pre-leukemic HSC or malignant stem cell (SC) compared to their normal counterparts. Still, many observations could help us depicting model for $[Ca^{2+}]_m$ in malignant hematopoiesis. As for each malignant condition also preleukemic-HSC or malignant-SC have been reported as resistant to apoptosis (Jilg et al., 2016; Pandolfi et al., 2013; Xu et al., 2012).

Mitochondria participates to regulated cell death (RCD) via at least two pathways—the mitochondrial outer membrane permeabilization (MOMP) and the mitochondrial permeability transition (MPT)—both of which can be triggered by transfer of Ca^{2+} from ER to mitochondria characterized by long duration but low amplitude (Adachi et al., 1997; Bonora et al., 2015; Marchetti et al., 1996). Not surprisingly, multiple reports indicate that this kind of Ca^{2+} signals can be elicited in leukemic cells by a variety of compounds, often favoring RCD engagement in the malignant cells compared to HSPCs (Bouchet et al., 2016; Ge et al., 2019; Metts et al., 2017; Pinton et al., 2011).

Most significantly, several genes involved in the emergence of malignant hematopoiesis have been related to the control of ER to mitochondria Ca^{2+} transfer. Bcl-2 is one of the first oncogene ever discovered and it was isolated from a B-cell lymphoma. Bcl-2 is well characterized as a master suppressor of RCD in almost all its sub-routines. It is also capable of suppressing mitochondrial Ca^{2+} uptake by either lowering $[Ca^{2+}]_{er}$ at steady state and inhibiting the activity of the Ip3Rs (Bittremieux et al., 2016; Pinton et al., 2001). A comparable phenotype is also attributed to another anti-apoptotic member of the Bcl-2 family, Mcl-1 (Bittremieux et al., 2016; Eckenrode et al., 2010). Bcl-2 is expressed in hematopoietic system, though is not enriched in HSC, but rather in lymphoid progenitors. Interestingly, its overexpression in HSC do not cause spontaneous development of a malignant phenotype but suppresses spontaneous apoptosis and support the quiescent state (Domen et al., 2000; Kollek et al., 2016). In contrast, *Mcl-1* is highly expressed in the stem/progenitor's compartment and critically required for the maintenance of HSCs. *Mcl-1* conditional deletion induces depletion of HSPC due to increased apoptotic rates, leading to bone marrow failure (Opferman et al., 2005). Also its pharmacological inhibition support a role for Mcl-1 in the control of HSC function (Campbell et al., 2010). Same line of evidence is regarding the promyelocytic leukemia (Pml) and phosphatase and tensin homolog (Pten) (Morotti et al., 2015; Testa and Lo-Coco, 2016). Both genes are affected by loss of function mutations frequently associated to myeloid malignancies (Ito et al., 2008;

Zhang et al., 2006). Interestingly, both proteins can localize at the contact sites between ER and mitochondria, where sustains the activity of Ip3Rs and therefore mitochondrial Ca^{2+} uptake to control both metabolic demands and RCD engagement (Bononi et al., 2013; Giorgi et al., 2010; Missiroli et al., 2016). In terms of HSC function both genes are required for stem cell maintenance. Indeed, genetic inactivation of both Pten and Pml lead to excess of HSC proliferation which can easily repopulate BM during competitive transplantation only in the short term, while failing in the long term, indicative of exhaustion of HSC capacity (Ito et al., 2008; Zhang et al., 2006). Also, for Pten was reported spontaneous development of myeloid malignancy.

The tumor suppressor p53 is non-surprisingly another fundamental regulator of hematopoiesis. Data from tp53$^{-/-}$ mice indicates that in steady state conditions, p53 is required to positively regulate HSC quiescence, apparently via a GFI-1/NECDIN pathway rather than via p21 (Liu et al., 2009; Yamashita et al., 2016). At the mitochondrial level, p53 is reported to have multiple functions. It transcriptionally modulate the expression of assembly factors that favors respiratory chain (Matoba et al., 2006). During stress conditions, it can translocate to outer mitochondrial membrane to directly activated MOMP and MPT (Talos et al., 2005; Vaseva et al., 2012). Most interestingly, at steady state p53 localize to ER and favor SERCA2 activity via direct binding (Giorgi et al., 2015a). This elevates $[Ca^{2+}]_{er}$ then Ca^{2+} transfer to mitochondria (Giorgi et al., 2015b).

These pieces of evidence suggest that deregulation of $[Ca^{2+}]_m$ could be involved in the development of HSC-related leukemic condition. Surprisingly, the described conditions, though all sharing alterations in $[Ca^{2+}]_m$, only in a minor of cases results in the spontaneous development of a malignancy, while the most affected HSC property appears to be quiescence. This suggest that the relevance for $[Ca^{2+}]_m$ in this group of malignancies might require the collaboration of additional mutations.

Of the many genes identified so far that are target of preleukemic mutation, none was related to $[Ca^{2+}]_m$ homeostasis, with exception of enhancer of zeste homolog 2 (EZH2) (Bowman et al., 2018; Saeed et al., 2021; Thol et al., 2017). This is a histone methyltransferase part of the polycomb repressive complex 2. Its frequently altered in hematological disorders, spanning from CHIP to MDS and sAML. Its activity in HSC function is still controversial though most evidence indicates that is required for the control of HSC commitment (Herviou et al., 2016). Its genetic manipulation appears to be significantly more efficient in FL-HSC rather than BM-HSC, most likely because of a compensation by EZH1. In cancer cells of epithelial origin, pharmacological inhibition, or shRNA inactivation of EZH2 significantly impairs the expression of mitochondrial calcium uptake protein 1 (MICU1) (Zhou et al., 2015). This is the gatekeeper of the mitochondrial calcium uniporter (MCU), that confers to the channel low affinity for Ca^{2+} (Marchi et al., 2020). MICU1 down-regulation, increases channel permeability, favoring elevation of basal $[Ca^{2+}]_m$. This also often associated with increased ROS production and predisposed cells to succumb RCD.

5. Conclusion

HSC function is of vital importance for a physiological hematopoiesis throughout all life and is therefore the results of multiple, balanced signals that regulates quiescence and self/renewal vs commitment. Considering the emerged meaning of mitochondrial physiology in the regulation of these function and the well-established role of intracellular Ca^{2+} as signaling molecules, it was just expectable that investigation of mitochondrial Ca^{2+} homeostasis would also reach the hematopoietic field.

To date seminal investigation are starting to depict a model of this topic. We now know that (i) there is a remodeling of all intracellular (including mitochondrial) Ca^{2+} homeostasis in HSC exiting quiescence and undergoing commitment, (ii) HSC can sense $[Ca^{2+}]_o$, (iii) pharmacological or genetic regulation of Ca^{2+} homeostasis can alter HSC quiescence status as well as in vitro and in vivo function, (iv) there are Ca^{2+} sensitive molecular participant involved in direct regulation of HSC quiescence or commitment programs.

These are all important information that prove the participation in Ca^{2+} signaling in hematopoiesis. Still, information is currently somewhat fragmentary and sometimes contradictory. We can currently propose at least two models, in the first Ca^{2+} signals are resting in the quiescent HSC, while activating when commitment is needed (e.g., during 5-FU or adenine nucleotide stimulation). Activation of a "pro-commitment" Ca^{2+} signal favors nuclear translocation of the commitment promoter NFAT and degradation of the self-renewal promoter Tet2. On the mitochondrial side, quiescent cells mitochondria actively participate to clear $[Ca^{2+}]_c$ via buffering while in active HSC of $[Ca^{2+}]_m$ activates respiration.

On the other side, a pro-quiescence model indicates that quiescent cells have higher Ca^{2+} and that niche factors favor $[Ca^{2+}]_c$ elevation to sustain quiescence (possibly via Calmodulin/CaMKs pathway).

As we still do not know amplitude, frequency, and duration of these Ca^{2+} signals or if these are different from apparent steady state conditions, the two model might be coexisting.

Two major lines of evidence are now required: (1) direct measurements of Ca^{2+} dynamics via the use of high performance genetically encoded calcium indicators (GECI) in cells from different stages of hematopoiesis and (2) investigate HSC function in models that specifically ablate fundamental members of the Ca^{2+} toolkit.

Indeed, many evidence are collected by investigating molecular actors which participates in multiple signaling pathways other than Ca^{2+}. Of extreme interest will be to understand the HSC function in transgenic animals which inactivates MCU, the mitochondrial Ca^{2+} exchanger NCLX (involved in the off-phase of mitochondrial Ca^{2+} signals, Fig. 1) or the ER channels Ip3Rs and ryanodine receptors (Fig. 1).

This information will allow to better define the mechanism of HSC regulation, especially in the definition of therapeutic regiment for hematological disease. Indeed, we have now available multiple compounds that can selectively inactivate the MCU or NCLX which should reduce or elevate $[Ca^{2+}]_m$, respectively. Of interest, Mitoxantrone, an anthracenedione structurally related to the anthracycline antibiotics, is recognized as a useful drug in first line therapy against AML (Patzke and Emadi, 2020) and have been recently reported as a potent inhibitor of MCU (Arduino et al., 2017).

Future investigation then might expose unexpected therapeutic strategies, based on the manipulation of mitochondrial Ca^{2+} signaling, for the treatment of hematological malignancies as well as improving the success of BM or cord blood transplantation.

Author contributions

M.B. and P.P. conceived the article; M.B. and A.K. wrote the first version of the manuscript with constructive input from P.P.; A.K. prepared display items under the supervision of M.B. and P.P. Figures are original and have not been published before. All authors reviewed and edited the manuscript before submission. All authors have read and agreed to the published version of the manuscript.

Funding

P.P. is grateful to Camilla degli Scrovegni for continuous support. The Signal Transduction Laboratory is supported by the Italian Association for Cancer Research Grant IG-23670 (to P.P.) and Grant IG-19803 (to C.G.), A-ROSE, Telethon Grant GGP11139B (to P.P.); Progetti di Rilevante Interesse Nazionale Grants PRIN2017E5L5P3 (to P.P.) and PRIN20177E9EPY (to C.G.); Italian Ministry of Health Grant GR-2013-02356747 (to C.G.); local funds from the University of Ferrara (to P.P. and M.B.).

Conflicts of interest

The authors declare no conflict of interest.

References

Abkowitz, J.L., Robinson, A.E., Kale, S., Long, M.W., Chen, J., 2003. Mobilization of hematopoietic stem cells during homeostasis and after cytokine exposure. Blood 102, 1249–1253. https://doi.org/10.1182/blood-2003-01-0318.

Adachi, S., Cross, A.R., Babior, B.M., Gottlieb, R.A., 1997. Bcl-2 and the outer mitochondrial membrane in the inactivation of cytochrome c during Fas-mediated apoptosis. J. Biol. Chem. 272, 21878–21882. https://doi.org/10.1074/jbc.272.35.21878.

Adamiak, M., Bujko, K., Cymer, M., Plonka, M., Glaser, T., Kucia, M., Ratajczak, J., Ulrich, H., Abdel-Latif, A., Ratajczak, M.Z., 2018. Novel evidence that extracellular nucleotides and purinergic signaling induce innate immunity-mediated mobilization of hematopoietic stem/progenitor cells. Leukemia 32, 1920–1931. https://doi.org/10.1038/s41375-018-0122-0.

Adams, G.B., Chabner, K.T., Alley, I.R., Olson, D.P., Szczepiorkowski, Z.M., Poznansky, M.C., Kos, C.H., Pollak, M.R., Brown, E.M., Scadden, D.T., 2006. Stem cell engraftment at the endosteal niche is specified by the calcium-sensing receptor. Nature 439, 599–603. https://doi.org/10.1038/nature04247.

Ansó, E., Weinberg, S.E., Diebold, L.P., Thompson, B.J., Malinge, S., Schumacker, P.T., Liu, X., Zhang, Y., Shao, Z., Steadman, M., Marsh, K.M., Xu, J., Crispino, J.D., Chandel, N.S., 2017. The mitochondrial respiratory chain is essential for haematopoietic stem cell function. Nat. Cell Biol. 19, 614–625. https://doi.org/10.1038/ncb3529.

Arai, F., Hirao, A., Ohmura, M., Sato, H., Matsuoka, S., Takubo, K., Ito, K., Koh, G.Y., Suda, T., 2004. Tie2/angiopoietin-1 signaling regulates hematopoietic stem cell quiescence in the bone marrow niche. Cell 118, 149–161. https://doi.org/10.1016/j.cell.2004.07.004.

Arai, F., Hirao, A., Suda, T., 2005. Regulation of hematopoietic stem cells by the niche. Trends Cardiovasc. Med. 15, 75–79. https://doi.org/10.1016/j.tcm.2005.03.002.

Arduino, D.M., Wettmarshausen, J., Vais, H., Navas-Navarro, P., Cheng, Y., Leimpek, A., Ma, Z., Delrio-Lorenzo, A., Giordano, A., Garcia-Perez, C., Médard, G., Kuster, B., García-Sancho, J., Mokranjac, D., Foskett, J.K., Alonso, M.T., Perocchi, F., 2017. Systematic identification of MCU modulators by orthogonal interspecies chemical screening. Mol. Cell 67, 711–723.e7. https://doi.org/10.1016/j.molcel.2017.07.019.

Asada, N., Kunisaki, Y., Pierce, H., Wang, Z., Fernandez, N.F., Birbrair, A., Ma'ayan, A., Frenette, P.S., 2017. Differential cytokine contributions of perivascular haematopoietic stem cell niches. Nat. Cell Biol. 19, 214–223. https://doi.org/10.1038/ncb3475.

Baccin, C., Al-Sabah, J., Velten, L., Helbling, P.M., Grünschläger, F., Hernández-Malmierca, P., Nombela-Arrieta, C., Steinmetz, L.M., Trumpp, A., Haas, S., 2020. Combined single-cell and spatial transcriptomics reveal the molecular, cellular and spatial bone marrow niche organization. Nat. Cell Biol. 22, 38–48. https://doi.org/10.1038/s41556-019-0439-6.

Baldridge, M.T., King, K.Y., Boles, N.C., Weksberg, D.C., Goodell, M.A., 2010. Quiescent haematopoietic stem cells are activated by IFN-γ in response to chronic infection. Nature 465, 793–797. https://doi.org/10.1038/nature09135.

Barbosa, C.M., Leon, C.M., Nogueira-Pedro, A., Wasinsk, F., Araujo, R.C., Miranda, A., Ferreira, A.T., Paredes-Gamero, E.J., 2011. Differentiation of hematopoietic stem cell and myeloid populations by ATP is modulated by cytokines. Cell Death Dis. 2, e165. https://doi.org/10.1038/cddis.2011.49.

Batsivari, A., Haltalli, M.L.R., Passaro, D., Pospori, C., Lo Celso, C., Bonnet, D., 2020. Dynamic responses of the haematopoietic stem cell niche to diverse stresses. Nat. Cell Biol. 22, 7–17. https://doi.org/10.1038/s41556-019-0444-9.

Beals, C.R., Clipstone, N.A., Ho, S.N., Crabtree, G.R., 1997. Nuclear localization of NF-ATc by a calcineurin-dependent, cyclosporin-sensitive intramolecular interaction. Genes Dev. 11, 824–834. https://doi.org/10.1101/gad.11.7.824.

Bejarano-García, J.A., Millán-Uclés, Á., Rosado, I.V., Sánchez-Abarca, L.I., Caballero-Velázquez, T., Durán-Galván, M.J., Pérez-Simón, J.A., Piruat, J.I., 2016. Sensitivity of hematopoietic stem cells to mitochondrial dysfunction by SdhD gene deletion. Cell Death Dis. 7, e2516. https://doi.org/10.1038/cddis.2016.411.

Bittremieux, M., Parys, J.B., Pinton, P., Bultynck, G., 2016. ER functions of oncogenes and tumor suppressors: Modulators of intracellular Ca2+ signaling. Biochim. Biophys. Acta 1863, 1364–1378. https://doi.org/10.1016/j.bbamcr.2016.01.002.

Bononi, A., Bonora, M., Marchi, S., Missiroli, S., Poletti, F., Giorgi, C., Pandolfi, P.P.P., Pinton, P., 2013. Identification of PTEN at the ER and MAMs and its regulation of Ca(2+) signaling and apoptosis in a protein phosphatase-dependent manner. Cell Death Differ. 20, 1631–1643. https://doi.org/10.1038/cdd.2013.77.

Bonora, M., Giorgi, C., Bononi, A., Marchi, S., Patergnani, S., Rimessi, A., Rizzuto, R., Pinton, P., 2013. Subcellular calcium measurements in mammalian cells using jellyfish photoprotein aequorin-based probes. Nat. Protoc. 8, 2105–2118. https://doi.org/10.1038/nprot.2013.127.

Bonora, M., Wieckowski, M.R.R., Chinopoulos, C., Kepp, O., Kroemer, G., Galluzzi, L., Pinton, P., 2015. Molecular mechanisms of cell death: central implication of ATP synthase in mitochondrial permeability transition. Oncogene 34, 1475–1486. https://doi.org/10.1038/onc.2014.96.

Bonora, M., Ito, K.K., Morganti, C., Pinton, P., Ito, K.K., 2018. Membrane-potential compensation reveals mitochondrial volume expansion during HSC commitment. Exp. Hematol. 68, 30–37.e1. https://doi.org/10.1016/j.exphem.2018.10.012.

Bouchet, S., Tang, R., Fava, F., Legrand, O., Bauvois, B., 2016. The CNGRC-GG-D (KLAKLAK)2 peptide induces a caspase-independent, Ca2+-dependent death in human leukemic myeloid cells by targeting surface aminopeptidase N/CD13. Oncotarget 7, 19445–19467. https://doi.org/10.18632/oncotarget.6523.

Bowie, M.B., McKnight, K.D., Kent, D.G., McCaffrey, L., Hoodless, P.A., Eaves, C.J., 2006. Hematopoietic stem cells proliferate until after birth and show a reversible phase-specific engraftment defect. J. Clin. Invest. 116, 2808–2816. https://doi.org/10.1172/JCI28310.

Bowman, R.L., Busque, L., Levine, R.L., 2018. Clonal hematopoiesis and evolution to hematopoietic malignancies. Cell Stem Cell 22, 157–170. https://doi.org/10.1016/j.stem.2018.01.011.

Cabezas-Wallscheid, N., Buettner, F., Sommerkamp, P., Klimmeck, D., Ladel, L., Thalheimer, F.B., Pastor-Flores, D., Roma, L.P., Renders, S., Zeisberger, P., Przybylla, A., Schönberger, K., Scognamiglio, R., Altamura, S., Florian, C.M., Fawaz, M., Vonficht, D., Tesio, M., Collier, P., Pavlinic, D., Geiger, H., Schroeder, T., Benes, V., Dick, T.P., Rieger, M.A., Stegle, O., Trumpp, A., 2017. Vitamin A-retinoic acid signaling regulates hematopoietic stem cell dormancy. Cell 169, 807–823.e19. https://doi.org/10.1016/j.cell.2017.04.018.

Calì, T., Ottolini, D., Negro, A., Brini, M., 2013. Enhanced parkin levels favor ER-mitochondria crosstalk and guarantee Ca2+ transfer to sustain cell bioenergetics. Biochim. Biophys. Acta, Mol. Basis Dis. 1832, 495–508. https://doi.org/10.1016/j.bbadis.2013.01.004.

Campbell, C.J.V., Lee, J.B., Levadoux-Martin, M., Wynder, T., Xenocostas, A., Leber, B., Bhatia, M., 2010. The human stem cell hierarchy is defined by a functional dependence on Mcl-1 for self-renewal capacity. Blood 116, 1433–1442. https://doi.org/10.1182/blood-2009-12-258095.

Cancilla, D., Rettig, M.P., DiPersio, J.F., 2020. Targeting CXCR4 in AML and ALL. Front. Oncol. 10, 1672. https://doi.org/10.3389/fonc.2020.01672.

Cárdenas, C., Foskett, J.K., 2012. Mitochondrial Ca(2+) signals in autophagy. Cell Calcium 52, 44–51. https://doi.org/10.1016/j.ceca.2012.03.001.

Chavez-Abiega, S., Mos, I., Centeno, P.P., Elajnaf, T., Schlattl, W., Ward, D.T., Goedhart, J., Kallay, E., 2020. Sensing extracellular calcium—an insight into the structure and function of the calcium-sensing receptor (CaSR). In: Advances in Experimental Medicine and Biology. Springer, New York LLC, pp. 1031–1063. https://doi.org/10.1007/978-3-030-12457-1_41.

Chen, C., Liu, Y., Liu, R., Ikenoue, T., Guan, K.-L., Liu, Y., Zheng, P., 2008. TSC-mTOR maintains quiescence and function of hematopoietic stem cells by repressing mitochondrial biogenesis and reactive oxygen species. J. Exp. Med. 205, 2397–2408. https://doi.org/10.1084/jem.20081297.

Chen, J., Kao, Y.-R., Sun, D., Todorova, T.I., Reynolds, D., Narayanagari, S.-R., Montagna, C., Will, B., Verma, A., Steidl, U., 2019. Myelodysplastic syndrome progression to acute myeloid leukemia at the stem cell level. Nat. Med. 25, 103–110. https://doi.org/10.1038/s41591-018-0267-4.

Cimmino, L., Aifantis, I., 2017. Alternative roles for oxidized mCs and TETs. Curr. Opin. Genet. Dev. 42, 1–7. https://doi.org/10.1016/j.gde.2016.11.003.

Corces-Zimmerman, M.R., Hong, W.J., Weissman, I.L., Medeiros, B.C., Majeti, R., 2014. Preleukemic mutations in human acute myeloid leukemia affect epigenetic regulators and persist in remission. Proc. Natl. Acad. Sci. U. S. A. 111, 2548–2553. https://doi.org/10.1073/pnas.1324297111.

Coutu, D.L., Kokkaliaris, K.D., Kunz, L., Schroeder, T., 2017. Three-dimensional map of nonhematopoietic bone and bone-marrow cells and molecules. Nat. Biotechnol. 35, 1202–1210. https://doi.org/10.1038/nbt.4006.

Ding, L., Saunders, T.L., Enikolopov, G., Morrison, S.J., 2012. Endothelial and perivascular cells maintain haematopoietic stem cells. Nature 481, 457–462. https://doi.org/10.1038/nature10783.

Domen, J., Cheshier, S.H., Weissman, I.L., 2000. The role of apoptosis in the regulation of hematopoietic stem cells: overexpression of Bcl-2 increases both their number and repopulation potential. J. Exp. Med. 191, 253–264. https://doi.org/10.1084/jem.191.2.253.

Eckenrode, E.F., Yang, J., Velmurugan, G.V., Kevin Foskett, J., White, C., 2010. Apoptosis protection by Mcl-1 and Bcl-2 modulation of inositol 1,4,5-trisphosphate receptor-dependent Ca2+ signaling. J. Biol. Chem. 285, 13678–13684. https://doi.org/10.1074/jbc.M109.096040.

Ellis, S.L., Grassinger, J., Jones, A., Borg, J., Camenisch, T., Haylock, D., Bertoncello, I., Nilsson, S.K., 2011. The relationship between bone, hemopoietic stem cells, and vasculature. Blood 118, 1516–1524. https://doi.org/10.1182/blood-2010-08-303800.

Favaro, G., Romanello, V., Varanita, T., Andrea Desbats, M., Morbidoni, V., Tezze, C., Albiero, M., Canato, M., Gherardi, G., De Stefani, D., Mammucari, C., Blaauw, B., Boncompagni, S., Protasi, F., Reggiani, C., Scorrano, L., Salviati, L., Sandri, M., 2019. DRP1-mediated mitochondrial shape controls calcium homeostasis and muscle mass. Nat. Commun. 10, 2576. https://doi.org/10.1038/s41467-019-10226-9.

Fukushima, T., Tanaka, Y., Hamey, F.K., Chang, C.H., Oki, T., Asada, S., Hayashi, Y., Fujino, T., Yonezawa, T., Takeda, R., Kawabata, K.C., Fukuyama, T., Umemoto, T., Takubo, K., Takizawa, H., Goyama, S., Ishihama, Y., Honda, H., Göttgens, B., Kitamura, T., 2019. Discrimination of dormant and active hematopoietic stem cells by G0 marker reveals dormancy regulation by cytoplasmic calcium. Cell Rep. 29, 4144–4158.e7. https://doi.org/10.1016/j.celrep.2019.11.061.

Ge, C., Huang, H., Huang, F., Yang, T., Zhang, T., Wu, H., Zhou, H., Chen, Q., Shi, Y., Sun, Y., Liu, L., Wang, X., Pearson, R.B., Cao, Y., Kang, J., Fu, C., 2019.

Neurokinin-1 receptor is an effective target for treating leukemia by inducing oxidative stress through mitochondrial calcium overload. Proc. Natl. Acad. Sci. U. S. A. 116, 19635–19645. https://doi.org/10.1073/pnas.1908998116.

Giorgi, C., Ito, K., Lin, H.K., Santangelo, C., Wieckowski, M.R., Lebiedzinska, M., Bononi, A., Bonora, M., Duszynski, J., Bernardi, R., Rizzuto, R., Tacchetti, C., Pinton, P., Pandolfi, P.P., 2010. PML regulates apoptosis at endoplasmic reticulum by modulating calcium release. Science 330, 1247–1251. https://doi.org/10.1126/science.1189157.

Giorgi, C., Bonora, M., Sorrentino, G., Missiroli, S., Poletti, F., Suski, J.M., Ramirez, F.G., Rizzuto, R., Di Virgilio, F., Zito, E., Pandolfi, P.P., Wieckowski, M.R., Mammano, F., Del Sal, G., Pinton, P., 2015a. P53 at the endoplasmic reticulum regulates apoptosis in a Ca2+-dependent manner. Proc. Natl. Acad. Sci. U. S. A. 112, 1779–1784. https://doi.org/10.1073/pnas.1410723112.

Giorgi, C., Bonora, M., Missiroli, S., Poletti, F., Ramirez, F.G.F.G., Morciano, G., Morganti, C., Pandolfi, P.P.P.P., Mammano, F., Pinton, P., 2015b. Intravital imaging reveals p53-dependent cancer cell death induced by phototherapy via calcium signaling. Oncotarget 6, 1435–1445. https://doi.org/10.18632/oncotarget.2935.

Gomez-Suaga, P., Paillusson, S., Stoica, R., Noble, W., Hanger, D.P., Miller, C.C.J., 2017. The ER-mitochondria tethering complex VAPB-PTPIP51 regulates autophagy. Curr. Biol. 27, 371–385. https://doi.org/10.1016/j.cub.2016.12.038.

Graf, T., 2002. Differentiation plasticity of hematopoietic cells. Blood 99, 3089–3101. https://doi.org/10.1182/blood.V99.9.3089.

Grahn, T.H.M., Niroula, A., Végvári, Á., Oburoglu, L., Pertesi, M., Warsi, S., Safi, F., Miharada, N., Garcia, S.C., Siva, K., Liu, Y., Rörby, E., Nilsson, B., Zubarev, R.A., Karlsson, S., 2020. S100A6 is a critical regulator of hematopoietic stem cells. Leukemia 34, 3323–3337. https://doi.org/10.1038/s41375-020-0901-2.

Gurumurthy, S., Xie, S.Z., Alagesan, B., Kim, J., Yusuf, R.Z., Saez, B., Tzatsos, A., Ozsolak, F., Milos, P., Ferrari, F., Park, P.J., Shirihai, O.S., Scadden, D.T., Bardeesy, N., 2010. The Lkb1 metabolic sensor maintains haematopoietic stem cell survival. Nature 468, 659–663. https://doi.org/10.1038/nature09572.

Hadad, I., Veithen, A., Springael, J.Y., Sotiropoulou, P.A., Mendes Da Costa, A., Miot, F., Naeije, R., De Deken, X., Entee, K.M., 2013. Stroma cell-derived factor-1α signaling enhances calcium transients and beating frequency in rat neonatal cardiomyocytes. PLoS One 8, e56007. https://doi.org/10.1371/journal.pone.0056007.

Hatse, S., Princen, K., Bridger, G., De Clercq, E., Schols, D., 2002. Chemokine receptor inhibition by AMD3100 is strictly confined to CXCR4. FEBS Lett. 527, 255–262. https://doi.org/10.1016/S0014-5793(02)03143-5.

Heidel, F.H., Mar, B.G., Armstrong, S.A., 2011. Self-renewal related signaling in myeloid leukemia stem cells. Int. J. Hematol. 94, 109–117. https://doi.org/10.1007/s12185-011-0901-0.

Herviou, L., Cavalli, G., Cartron, G., Klein, B., Moreaux, J., 2016. EZH2 in normal hematopoiesis and hematological malignancies. Oncotarget 7, 2284–2296. https://doi.org/10.18632/oncotarget.6198.

Hinge, A., He, J., Bartram, J., Javier, J., Xu, J., Fjellman, E., Sesaki, H., Li, T., Yu, J., Wunderlich, M., Mulloy, J., Kofron, M., Salomonis, N., Grimes, H.L., Filippi, M.D., 2020. Asymmetrically segregated mitochondria provide cellular memory of hematopoietic stem cell replicative history and drive HSC attrition. Cell Stem Cell 26, 420–430.e6. https://doi.org/10.1016/j.stem.2020.01.016.

Ho, T.T., Warr, M.R., Adelman, E.R., Lansinger, O.M., Flach, J., Verovskaya, E.V., Figueroa, M.E., Passegué, E., 2017. Autophagy maintains the metabolism and function of young and old stem cells. Nature 543, 205–210. https://doi.org/10.1038/nature21388.

Hogan, P.G., Chen, L., Nardone, J., Rao, A., 2003. Transcriptional regulation by calcium, calcineurin, and NFAT. Genes Dev. 17, 2205–2232. https://doi.org/10.1101/gad.1102703.

Horikawa, K., 2015. Recent progress in the development of genetically encoded Ca2+ indicators. J. Med. Invest. 62, 24–28. https://doi.org/10.2152/jmi.62.24.

Høyer, M., Nielsen, O.S., 1992. Influence of dose on regeneration of murine hematopoietic stem cells after total body irradiation and 5-fluorouracil. Oncologia 49, 166–172. https://doi.org/10.1159/000227033.

Hübel, K., Liles, W.C., Broxmeyer, H.E., Rodger, E., Wood, B., Cooper, S., Hangoc, G., MacFarland, R., Bridger, G.J., Henson, G.W., Calandra, G., Dale, D.C., 2004. Leukocytosis and mobilization of CD34+ hematopoietic progenitor cells by AMD3100, a CXCR4 antagonist. Support. Cancer Ther. 1, 165–172. https://doi.org/10.3816/sct.2004.n.008.

Ito, K., Ito, K., 2018. Hematopoietic stem cell fate through metabolic control. Exp. Hematol. 64, 1–11. https://doi.org/10.1016/j.exphem.2018.05.005.

Ito, K., Hirao, A., Arai, F., Matsuoka, S., Takubo, K., Hamaguchi, I., Nomiyama, K., Hosokawa, K., Sakurada, K., Nakagata, N., Ikeda, Y., Mak, T.W., Suda, T., 2004. Regulation of oxidative stress by ATM is required for self-renewal of haematopoietic stem cells. Nature 431, 997–1002. https://doi.org/10.1038/nature02989.

Ito, K., Hirao, A., Arai, F., Takubo, K., Matsuoka, S., Miyamoto, K., Ohmura, M., Naka, K., Hosokawa, K., Ikeda, Y., Suda, T., 2006. Reactive oxygen species act through p38 MAPK to limit the lifespan of hematopoietic stem cells. Nat. Med. 12, 446–451. https://doi.org/10.1038/nm1388.

Ito, K., Bernardi, R., Morotti, A., Matsuoka, S., Saglio, G., Ikeda, Y., Rosenblatt, J., Avigan, D.E., Teruya-Feldstein, J., Pandolfi, P.P., 2008. PML targeting eradicates quiescent leukaemia-initiating cells. Nature 453, 1072–1078. https://doi.org/10.1038/nature07016.

Ito, K., Carracedo, A., Weiss, D., Arai, F., Ala, U., Avigan, D.E., Schafer, Z.T., Evans, R.M., Suda, T., Lee, C.-H., Pandolfi, P.P., 2012. A PML–PPAR-δ pathway for fatty acid oxidation regulates hematopoietic stem cell maintenance. Nat. Med. 18, 1350–1358. https://doi.org/10.1038/nm.2882.

Ito, K., Turcotte, R., Cui, J., Zimmerman, S.E., Pinho, S., Mizoguchi, T., Arai, F., Runnels, J.M., Alt, C., Teruya-Feldstein, J., Mar, J.C., Singh, R., Suda, T., Lin, C.P., Frenette, P.S., Ito, K., 2016. Self-renewal of a purified Tie2+ hematopoietic stem cell population relies on mitochondrial clearance. Science 354, 1156–1160. https://doi.org/10.1126/science.aaf5530.

Ito, K., Bonora, M., Ito, K., 2019. Metabolism as master of hematopoietic stem cell fate. Int. J. Hematol. 109, 18–27. https://doi.org/10.1007/s12185-018-2534-z.

Jilg, S., Reidel, V., Müller-Thomas, C., König, J., Schauwecker, J., Höckendorf, U., Huberle, C., Gorka, O., Schmidt, B., Burgkart, R., Ruland, J., Kolb, H.J., Peschel, C., Oostendorp, R.A.J., Götze, K.S., Jost, P.J., 2016. Blockade of BCL-2 proteins efficiently induces apoptosis in progenitor cells of high-risk myelodysplastic syndromes patients. Leukemia 30, 112–123. https://doi.org/10.1038/leu.2015.179.

Jin, G., Xu, C., Zhang, X., Long, J., Rezaeian, A.H., Liu, C., Furth, M.E., Kridel, S., Pasche, B., Bian, X.W., Lin, H.K., 2018. Atad3a suppresses Pink1-dependent mitophagy to maintain homeostasis of hematopoietic progenitor cells article. Nat. Immunol. 19, 29–40. https://doi.org/10.1038/s41590-017-0002-1.

Katayama, Y., Battista, M., Kao, W.M., Hidalgo, A., Peired, A.J., Thomas, S.A., Frenette, P.S., 2006. Signals from the sympathetic nervous system regulate hematopoietic stem cell egress from bone marrow. Cell 124, 407–421. https://doi.org/10.1016/j.cell.2005.10.041.

Kim, T., Tyndel, M.S., Kim, H.J., Ahn, J.S., Choi, S.H., Park, H.J., Kim, Y.K., Yang, D.H., Lee, J.J., Jung, S.H., Kim, S.Y., Min, Y.H., Cheong, J.W., Sohn, S.K., Moon, J.H., Choi, M., Lee, M., Zhang, Z., Kim, D.D.H., 2017. The clonal origins of leukemic progression of myelodysplasia. Leukemia 31, 1928–1935. https://doi.org/10.1038/leu.2017.17.

Kitsos, C.M., Sankar, U., Illario, M., Colomer-Font, J.M., Duncan, A.W., Ribar, T.J., Reya, T., Means, A.R., 2005. Calmodulin-dependent protein kinase IV regulates hematopoietic stem cell maintenance. J. Biol. Chem. 280, 33101–33108. https://doi.org/10.1074/jbc.M505208200.

Ko, M., Bandukwala, H.S., An, J., Lamperti, E.D., Thompson, E.C., Hastie, R., Tsangaratou, A., Rajewsky, K., Koralov, S.B., Rao, A., 2011. Ten-eleven-translocation 2 (TET2) negatively regulates homeostasis and differentiation of hematopoietic stem cells in mice. Proc. Natl. Acad. Sci. U. S. A. 108, 14566–14571. https://doi.org/10.1073/pnas.1112317108.

Koeffler, H.P., Leong, G., 2017. Preleukemia: one name, many meanings. Leukemia 31, 534–542. https://doi.org/10.1038/leu.2016.364.

Kollek, M., Müller, A., Egle, A., Erlacher, M., 2016. Bcl-2 proteins in development, health, and disease of the hematopoietic system. FEBS J. 283, 2779–2810. https://doi.org/10.1111/febs.13683.

Lam, B.S., Cunningham, C., Adams, G.B., 2011. Pharmacologic modulation of the calcium-sensing receptor enhances hematopoietic stem cell lodgment in the adult bone marrow. Blood 117, 1167–1175. https://doi.org/10.1182/blood-2010-05-286294.

Li, Z., Wang, J., Xu, M., Cai, X., Cai, C.-L., Zhang, W., Petersen, B.E., Yang, F.-C., 2011. Deletion of Tet2 in mice leads to dysregulated hematopoietic stem cells and subsequent development of myeloid malignancies. Blood 118, 4509–4518. https://doi.org/10.1182/blood-2010-12-325241.

Lindsley, R.C., Mar, B.G., Mazzola, E., Grauman, P.V., Shareef, S., Allen, S.L., Pigneux, A., Wetzler, M., Stuart, R.K., Erba, H.P., Damon, L.E., Powell, B.L., Lindeman, N., Steensma, D.P., Wadleigh, M., DeAngelo, D.J., Neuberg, D., Stone, R.M., Ebert, B.L., 2015. Acute myeloid leukemia ontogeny is defined by distinct somatic mutations. Blood 125, 1367–1376. https://doi.org/10.1182/blood-2014-11-610543.

Liu, Y., Elf, S.E., Miyata, Y., Sashida, G., Liu, Y., Huang, G., Di Giandomenico, S., Lee, J.M., Deblasio, A., Menendez, S., Antipin, J., Reva, B., Koff, A., Nimer, S.D., 2009. p53 regulates hematopoietic stem cell quiescence. Cell Stem Cell 4, 37–48. https://doi.org/10.1016/j.stem.2008.11.006.

Luchsinger, L.L., de Almeida, M.J., Corrigan, D.J., Mumau, M., Snoeck, H.-W., 2016. Mitofusin 2 maintains haematopoietic stem cells with extensive lymphoid potential. Nature 529, 528–531. https://doi.org/10.1038/nature16500.

Luchsinger, L.L., Strikoudis, A., Danzl, N.M., Bush, E.C., Finlayson, M.O., Satwani, P., Sykes, M., Yazawa, M., Snoeck, H.W., 2019. Harnessing hematopoietic stem cell low intracellular calcium improves their maintenance in vitro. Cell Stem Cell 25, 225–240.e7. https://doi.org/10.1016/j.stem.2019.05.002.

Makishima, H., Yoshizato, T., Yoshida, K., Sekeres, M.A., Radivoyevitch, T., Suzuki, H., Przychodzen, B.J., Nagata, Y., Meggendorfer, M., Sanada, M., Okuno, Y., Hirsch, C., Kuzmanovic, T., Sato, Y., Sato-Otsubo, A., Laframboise, T., Hosono, N., Shiraishi, Y., Chiba, K., Haferlach, C., Kern, W., Tanaka, H., Shiozawa, Y., Gómez-Seguí, I., Husseinzadeh, H.D., Thota, S., Guinta, K.M., Dienes, B., Nakamaki, T., Miyawaki, S., Saunthararajah, Y., Chiba, S., Miyano, S., Shih, L.Y., Haferlach, T., Ogawa, S., MacIejewski, J.P., 2017. Dynamics of clonal evolution in myelodysplastic syndromes. Nat. Genet. 49, 204–212. https://doi.org/10.1038/ng.3742.

Mansell, E., Sigurdsson, V., Deltcheva, E., Brown, J., James, C., Miharada, K., Soneji, S., Larsson, J., Enver, T., 2021. Mitochondrial potentiation ameliorates age-related heterogeneity in hematopoietic stem cell function. Cell Stem Cell 28, 241–256.e6. https://doi.org/10.1016/j.stem.2020.09.018.

Marchetti, P., Castedo, M., Susin, S.A., Zamzami, N., Hirsch, T., Macho, A., Haeffner, A., Hirsch, F., Geuskens, M., Kroemer, G., 1996. Mitochondrial permeability transition is a central coordinating event of apoptosis. J. Exp. Med. 184, 1155–1160. https://doi.org/10.1084/jem.184.3.1155.

Marchi, S., Giorgi, C., Galluzzi, L., Pinton, P., 2020. Ca2+ fluxes and cancer. Mol. Cell 78, 1055–1069. https://doi.org/10.1016/j.molcel.2020.04.017.

Matoba, S., Kang, J.-G.G., Patino, W.D., Wragg, A., Boehm, M., Gavrilova, O., Hurley, P.J., Bunz, F., Hwang, P.M., 2006. p53 regulates mitochondrial respiration. Science 312, 1650–1653. https://doi.org/10.1126/science.1126863.

Metts, J., Bradley, H.L., Wang, Z., Shah, N.P., Kapur, R., Arbiser, J.L., Bunting, K.D., 2017. Imipramine blue sensitively and selectively targets FLT3-ITD positive acute myeloid leukemia cells. Sci. Rep. 7, 1–10. https://doi.org/10.1038/s41598-017-04796-1.

Missiroli, S., Bonora, M., Patergnani, S., Poletti, F., Perrone, M., Gafà, R., Magri, E., Raimondi, A., Lanza, G., Tacchetti, C., Kroemer, G., Pandolfi, P.P., Pinton, P., Giorgi, C., 2016. PML at mitochondria-associated membranes is critical for the repression of autophagy and cancer development. Cell Rep. 16, 2415–2427. https://doi.org/10.1016/j.celrep.2016.07.082.

Moignard, V., Macaulay, I.C., Swiers, G., Buettner, F., Schütte, J., Calero-Nieto, F.J., Kinston, S., Joshi, A., Hannah, R., Theis, F.J., Jacobsen, S.E., de Bruijn, M.F., Göttgens, B., 2013. Characterization of transcriptional networks in blood stem and progenitor cells using high-throughput single-cell gene expression analysis. Nat. Cell Biol. 15, 363–372. https://doi.org/10.1038/ncb2709.

Morganti, C., Bonora, M., Ito, K., Ito, K., 2019a. Electron transport chain complex II sustains high mitochondrial membrane potential in hematopoietic stem and progenitor cells. Stem Cell Res. 40, 101573. https://doi.org/10.1016/j.scr.2019.101573.

Morganti, C., Bonora, M., Ito, K., 2019b. Improving the accuracy of flow cytometric assessment of mitochondrial membrane potential in hematopoietic stem and progenitor cells through the inhibition of efflux pumps. J. Vis. Exp. 149. https://doi.org/10.3791/60057.

Morita, Y., Ema, H., Yamazaki, S., Nakauch, H., 2006. Non-side-population hematopoietic stem cells in mouse bone marrow. Blood 108, 2850–2856. https://doi.org/10.1182/blood-2006-03-010207.

Morotti, A., Panuzzo, C., Crivellaro, S., Carrà, G., Torti, D., Guerrasio, A., Saglio, G., 2015. The role of PTEN in myeloid malignancies. Hematol. Rep. 7, 5844. https://doi.org/10.4081/hr.2015.6027.

Morrison, S.J., Scadden, D.T., 2014. The bone marrow niche for haematopoietic stem cells. Nature 505, 327–334. https://doi.org/10.1038/nature12984.

Nagata, Y., Maciejewski, J.P., 2019. The functional mechanisms of mutations in myelodysplastic syndrome. Leukemia 33, 2779–2794. https://doi.org/10.1038/s41375-019-0617-3.

Nakamura-Ishizu, A., Ito, K., Suda, T., 2020. Hematopoietic stem cell metabolism during development and aging. Dev. Cell 54, 239–255. https://doi.org/10.1016/j.devcel.2020.06.029.

Nie, Y., Han, Y.C., Zou, Y.R., 2008. CXCR4 is required for the quiescence of primitive hematopoietic cells. J. Exp. Med. 205, 777–783. https://doi.org/10.1084/jem.20072513.

Ogawa, M., 1993. Differentiation and proliferation of hematopoietic stem cells. Blood 81, 2844–2853. https://doi.org/10.1182/blood.v81.11.2844.2844.

Oguro, H., Ding, L., Morrison, S.J., 2013. SLAM family markers resolve functionally distinct subpopulations of hematopoietic stem cells and multipotent progenitors. Cell Stem Cell 13, 102–116. https://doi.org/10.1016/j.stem.2013.05.014.

Opferman, J.T., Iwasaki, H., Ong, C.C., Suh, H., Mizuno, S.I., Akashi, K., Korsmeyer, S.J., 2005. Obligate role of anti-apoptotic MCL-1 in the survival of hematopoietic stem cells. Science 307, 1101–1104. https://doi.org/10.1126/science.1106114.

Pandolfi, A., Barreyro, L., Steidl, U., 2013. Concise review: preleukemic stem cells: molecular biology and clinical implications of the precursors to Leukemia stem cells. Stem Cells Transl. Med. 2, 143–150. https://doi.org/10.5966/sctm.2012-0109.

Paredes-Gamero, E.J., Leon, C.M.M.P., Borojevic, R., Oshiro, M.E.M., Ferreira, A.T., 2008. Changes in intracellular Ca2+ levels induced by cytokines and P2 agonists differentially modulate proliferation or commitment with macrophage differentiation in murine hematopoietic cells. J. Biol. Chem. 283, 31909–31919. https://doi.org/10.1074/jbc.M801990200.

Park, C.S., Lewis, A., Chen, T., Lacorazza, D., 2019. Concise review: regulation of self-renewal in normal and malignant hematopoietic stem cells by Krüppel-like factor 4. Stem Cells Transl. Med. 8, 568–574. https://doi.org/10.1002/sctm.18-0249.

Patzke, C.L., Emadi, A., 2020. High dose cytarabine, mitoxantrone, pegasapargase (HAM-pegA) in combination with dasatinib for the first-line treatment of Philadelphia chromosome positive mixed phenotype acute leukemia. Am. J. Leuk. Res. 4.

Petit, I., Ponomaryov, T., Zipori, D., Tsvee, L., 2002. G-CSF induces stem cell mobilization by decreasing bone marrow SDF-1 and up-regulating CXCR4. Nat. Immunol. 3, 687–694. https://doi.org/10.1038/ni813.

Pinho, S., Frenette, P.S., 2019. Haematopoietic stem cell activity and interactions with the niche. Nat. Rev. Mol. Cell Biol. 20, 303–320. https://doi.org/10.1038/s41580-019-0103-9.

Pinton, P., Ferrari, D., Rapizzi, E., Di Virgilio, F., Pozzan, T., Rizzuto, R., 2001. The Ca2+ concentration of the endoplasmic reticulum is a key determinant of ceramide-induced apoptosis: significance for the molecular mechanism of Bcl-2 action. EMBO J. 20, 2690–2701. https://doi.org/10.1093/emboj/20.11.2690.

Pinton, P., Giorgi, C., Pandolfi, P.P., 2011. The role of PML in the control of apoptotic cell fate: a new key player at ER-mitochondria sites. Cell Death Differ. 18, 1450–1456. https://doi.org/10.1038/cdd.2011.31.

Racioppi, L., Lento, W., Huang, W., Arvai, S., Doan, P.L., Harris, J.R., Marcon, F., Nakaya, H.I., Liu, Y., Chao, N., 2017. Calcium/calmodulin-dependent kinase kinase 2 regulates hematopoietic stem and progenitor cell regeneration. Cell Death Dis. 8, e3076. https://doi.org/10.1038/cddis.2017.474.

Rasmussen, K.D., Helin, K., 2016. Role of TET enzymes in DNA methylation, development, and cancer. Genes Dev. 30, 733–750. https://doi.org/10.1101/gad.276568.115.

Ratajczak, M.Z., Adamiak, M., Plonka, M., Abdel-Latif, A., Ratajczak, J., 2018. Mobilization of hematopoietic stem cells as a result of innate immunity-mediated sterile inflammation in the bone marrow microenvironment—the involvement of extracellular nucleotides and purinergic signaling. Leukemia 32, 1116–1123. https://doi.org/10.1038/s41375-018-0087-z.

Rimmelé, P., Liang, R., Bigarella, C.L., Kocabas, F., Xie, J., Serasinghe, M.N., Chipuk, J., Sadek, H., Zhang, C.C., Ghaffari, S., 2015. Mitochondrial metabolism in hematopoietic stem cells requires functional FOXO3. EMBO Rep. 16, 1164–1176. https://doi.org/10.15252/embr.201439704.

Rossi, A.M., Taylor, C.W., 2020. Reliable measurement of free Ca2+ concentrations in the ER lumen using Mag-Fluo-4. Cell Calcium 87, 102188. https://doi.org/10.1016/j.ceca.2020.102188.

Saeed, B.R., Manta, L., Raffel, S., Pyl, P.T., Buss, E.C., Wang, W., Eckstein, V., Jauch, A., Trumpp, A., Huber, W., Ho, A.D., Lutz, C., 2021. Analysis of nonleukemic cellular subcompartments reconstructs clonal evolution of acute myeloid leukemia and identifies therapy-resistant preleukemic clones. Int. J. Cancer 148, 2825–2838. https://doi.org/10.1002/ijc.33461.

Schoedel, K.B., Morcos, M.N.F., Zerjatke, T., Roeder, I., Grinenko, T., Voehringer, D., Göthert, J.R., Waskow, C., Roers, A., Gerbaulet, A., 2016. The bulk of the hematopoietic stem cell population is dispensable for murine steady-state and stress hematopoiesis. Blood 128, 2285–2296. https://doi.org/10.1182/blood-2016-03-706010.

Seita, J., Weissman, I.L., 2010. Hematopoietic stem cell: self-renewal versus differentiation. Wiley Interdiscip. Rev. Syst. Biol. Med. 2, 640–653. https://doi.org/10.1002/wsbm.86.

Shlush, L.I., Zandi, S., Mitchell, A., Chen, W.C., Brandwein, J.M., Gupta, V., Kennedy, J.A., Schimmer, A.D., Schuh, A.C., Yee, K.W., McLeod, J.L., Doedens, M., Medeiros, J.J.F., Marke, R., Kim, H.J., Lee, K., McPherson, J.D., Hudson, T.J., Pan-Leukemia Gene Panel Consortium, H.A.L.T., Brown, A.M.K., Yousif, F., Trinh, Q.M., Stein, L.D., Minden, M.D., Wang, J.C.Y., Dick, J.E., 2014. Identification of pre-leukaemic haematopoietic stem cells in acute leukaemia. Nature 506, 328–333. https://doi.org/10.1038/nature13038.

Song, J., Kiel, M.J., Wang, Z., Wang, J., Taichman, R.S., Morrison, S.J., Krebsbach, P.H., 2010. An in vivo model to study and manipulate the hematopoietic stem cell niche. Blood 115, 2592–2600. https://doi.org/10.1182/blood-2009-01-200071.

Stevens, B.M., Khan, N., D'Alessandro, A., Nemkov, T., Winters, A., Jones, C.L., Zhang, W., Pollyea, D.A., Jordan, C.T., 2018. Characterization and targeting of malignant stem cells in patients with advanced myelodysplastic syndromes. Nat. Commun. 9, 1–14. https://doi.org/10.1038/s41467-018-05984-x.

Sugimura, R., He, X.C., Venkatraman, A., Arai, F., Box, A., Semerad, C., Haug, J.S., Peng, L., Zhong, X.B., Suda, T., Li, L., 2012. Noncanonical Wnt signaling maintains hematopoietic stem cells in the niche. Cell 150, 351–365. https://doi.org/10.1016/j.cell.2012.05.041.

Sugiyama, T., Kohara, H., Noda, M., Nagasawa, T., 2006. Maintenance of the hematopoietic stem cell Pool by CXCL12-CXCR4 chemokine signaling in bone marrow stromal cell niches. Immunity 25, 977–988. https://doi.org/10.1016/j.immuni.2006.10.016.

Szabadkai, G., Simoni, A.M., Chami, M., Wieckowski, M.R., Youle, R.J., Rizzuto, R., 2004. Drp-1-dependent division of the mitochondrial network blocks intraorganellar Ca2+ waves and protects against Ca2+-mediated apoptosis. Mol. Cell 16, 59–68. https://doi.org/10.1016/j.molcel.2004.09.026.

Tai-Nagara, I., Matsuoka, S., Ariga, H., Suda, T., 2014. Mortalin and DJ-1 coordinately regulate hematopoietic stem cell function through the control of oxidative stress. Blood 123, 41–50. https://doi.org/10.1182/blood-2013-06-508333.

Takubo, K., Goda, N., Yamada, W., Iriuchishima, H., Ikeda, E., Kubota, Y., Shima, H., Johnson, R.S., Hirao, A., Suematsu, M., Suda, T., 2010. Regulation of the HIF-1alpha level is essential for hematopoietic stem cells. Cell Stem Cell 7, 391–402. https://doi.org/10.1016/j.stem.2010.06.020.

Takubo, K., Nagamatsu, G., Kobayashi, C.I., Nakamura-Ishizu, A., Kobayashi, H., Ikeda, E., Goda, N., Rahimi, Y., Johnson, R.S., Soga, T., Hirao, A., Suematsu, M., Suda, T., 2013. Regulation of glycolysis by Pdk functions as a metabolic checkpoint for cell cycle quiescence in hematopoietic stem cells. Cell Stem Cell 12, 49–61. https://doi.org/10.1016/j.stem.2012.10.011.

Talos, F., Petrenko, O., Mena, P., Moll, U.M., 2005. Mitochondrially targeted p53 has tumor suppressor activities in vivo. Cancer Res. 65, 9971–9981. https://doi.org/10.1158/0008-5472.CAN-05-1084.

Testa, U., Lo-Coco, F., 2016. Prognostic factors in acute promyelocytic leukemia: strategies to define high-risk patients. Ann. Hematol. 95, 673–680. https://doi.org/10.1007/s00277-016-2622-1.

Thol, F., Klesse, S., Köhler, L., Gabdoulline, R., Kloos, A., Liebich, A., Wichmann, M., Chaturvedi, A., Fabisch, J., Gaidzik, V.I., Paschka, P., Bullinger, L., Bug, G., Serve, H., Göhring, G., Schlegelberger, B., Lübbert, M., Kirchner, H., Wattad, M., Kraemer, D., Hertenstein, B., Heil, G., Fiedler, W., Krauter, J., Schlenk, R.F., Döhner, K., Döhner, H., Ganser, A., Heuser, M., 2017. Acute myeloid leukemia derived from lympho-myeloid clonal hematopoiesis. Leukemia 31, 1286–1295. https://doi.org/10.1038/leu.2016.345.

Umemoto, T., Hashimoto, M., Matsumura, T., Nakamura-Ishizu, A., Suda, T., 2018. Ca2+-mitochondria axis drives cell division in hematopoietic stem cells. J. Exp. Med. 215, 2097–2113. https://doi.org/10.1084/jem.20180421.

Uslu, M., Albayrak, E., Kocabaş, F., 2020. Temporal modulation of calcium sensing in hematopoietic stem cells is crucial for proper stem cell expansion and engraftment. J. Cell. Physiol. 235, 9644–9666. https://doi.org/10.1002/jcp.29777.

Vaseva, A.V., Marchenko, N.D., Ji, K., Tsirka, S.E., Holzmann, S., Moll, U.M., 2012. p53 opens the mitochondrial permeability transition pore to trigger necrosis. Cell 149, 1536–1548. https://doi.org/10.1016/j.cell.2012.05.014.

Wang, Z., Ema, H., 2016. Mechanisms of self-renewal in hematopoietic stem cells. Int. J. Hematol. 103, 498–509. https://doi.org/10.1007/s12185-015-1919-5.

Xiang, Y., Li, Y., Zhang, Z., Cui, K., Wang, S., Yuan, X.B., Wu, C.P., Poo, M.M., Duan, S., 2002. Nerve growth cone guidance mediated by G protein-coupled receptors. Nat. Neurosci. 5, 843–848. https://doi.org/10.1038/nn899.

Xu, F., Yang, R., Wu, L., He, Q., Zhang, Z., Zhang, Q., Yang, Y., Guo, J., Chang, C., Li, X., 2012. Overexpression of BMI1 confers clonal cells resistance to apoptosis and contributes to adverse prognosis in myelodysplastic syndrome. Cancer Lett. 317, 33–40. https://doi.org/10.1016/j.canlet.2011.11.012.

Yamashita, M., Nitta, E., Suda, T., 2016. Regulation of hematopoietic stem cell integrity through p53 and its related factors. Ann. N. Y. Acad. Sci. 1370, 45–54. https://doi.org/10.1111/nyas.12986.

Yang, L., Wang, L., Geiger, H., Cancelas, J.A., Mo, J., Zheng, Y., 2007. Rho GTPase Cdc42 coordinates hematopoietic stem cell quiescence and niche interaction in the bone marrow. Proc. Natl. Acad. Sci. U. S. A. 104, 5091–5096. https://doi.org/10.1073/pnas.0610819104.

Zhang, J., Grindley, J.C., Yin, T., Jayasinghe, S., He, X.C., Ross, J.T., Haug, J.S., Rupp, D., Porter-Westpfahl, K.S., Wiedemann, L.M., Wu, H., Li, L., 2006. PTEN maintains haematopoietic stem cells and acts in lineage choice and leukaemia prevention. Nature 441, 518–522. https://doi.org/10.1038/nature04747.

Zhang, X., Su, J., Jeong, M., Ko, M., Huang, Y., Park, H.J., Guzman, A., Lei, Y., Huang, Y.H., Rao, A., Li, W., Goodell, M.A., 2016. DNMT3A and TET2 compete and cooperate to repress lineage-specific transcription factors in hematopoietic stem cells. Nat. Genet. 48, 1014–1023. https://doi.org/10.1038/ng.3610.

Zhang, J., Wu, Q., Johnson, C.B., Pham, G., Kinder, J.M., Olsson, A., Slaughter, A., May, M., Weinhaus, B., D'Alessandro, A., Engel, J.D., Jiang, J.X., Kofron, J.M., Huang, L.F., Prasath, V.B.S., Way, S.S., Salomonis, N., Grimes, H.L., Lucas, D., 2021. In situ mapping identifies distinct vascular niches for myelopoiesis. Nature 590, 457–462. https://doi.org/10.1038/s41586-021-03201-2.

Zhao, J.L., Ma, C., O'Connell, R.M., Mehta, A., Diloreto, R., Heath, J.R., Baltimore, D., 2014. Conversion of danger signals into cytokine signals by hematopoietic stem and progenitor cells for regulation of stress-induced hematopoiesis. Cell Stem Cell 14, 445–459. https://doi.org/10.1016/j.stem.2014.01.007.

Zhou, S., Schuetz, J.D., Bunting, K.D., Colapietro, A.M., Sampath, J., Morris, J.J., Lagutina, I., Grosveld, G.C., Osawa, M., Nakauchi, H., Sorrentino, B.P., 2001. The ABC transporter Bcrp1/ABCG2 is expressed in a wide variety of stem cells and is a molecular determinant of the side-population phenotype. Nat. Med. 7, 1028–1034.

Zhou, X., Ren, Y., Kong, L., Cai, G., Sun, S., Song, W., Wang, Y., Jin, R., Qi, L., Mei, M., Wang, X., Kang, C., Li, M., Zhang, L., 2015. Targeting EZH2 regulates tumor growth and apoptosis through modulating mitochondria dependent cell-death pathway in HNSCC. Oncotarget 6, 33720–33732. https://doi.org/10.18632/oncotarget.5606.

CHAPTER FOUR

Lysosomal calcium and autophagy

Diego L. Medina[a,b,*]

[a]Telethon Institute of Genetics and Medicine (TIGEM), Pozzuoli, Naples, Italy
[b]Medical Genetics Unit, Department of Medical and Translational Science, Federico II University, Naples, Italy
*Corresponding author: e-mail address: medina@tigem.it

Contents

Abstract

Lysosomal calcium is emerging as a modulator of autophagy and lysosomal compartment, an obligatory partner to complete the autophagic pathway. A variety of specific signals such as nutrient deprivation or oxidative stress can trigger lysosomal calcium-mediated nuclear translocation of the transcription factor EB (TFEB), a master regulator of global lysosomal function. Also, lysosomal calcium can promote the formation of autophagosome vesicles (AVs) by a mechanism that requires the production of the phosphoinositide PI3P by the VPS34 autophagic complex and the activation of the energy-sensing kinase AMPK. Additionally, lysosomal calcium plays a role in membrane fusion and fission events involved in cellular processes such as endocytic maturation, autophagosome-lysosome fusion, lysosomal exocytosis, and lysosomal reformation upon autophagy completion. Lysosomal calcium-dependent functions are defective in cellular and animal models of the non-selective cation channel TRPML1, whose mutations in humans cause the neurodegenerative lysosomal storage disease mucolipidosis type IV (MLIV). Lysosomal calcium is not only acting as a positive regulator of autophagy, but it is also responsible for turning-off this process through the reactivation of the mTOR kinase during prolonged starvation. More recently, it has been described the role of lysosomal calcium on an elegant sequence of intracellular signaling events such as

International Review of Cell and Molecular Biology, Volume 362
ISSN 1937-6448
https://doi.org/10.1016/bs.ircmb.2021.03.002

membrane repair, lysophagy, and lysosomal biogenesis upon the induction of different grades of lysosomal membrane damage. Here, we will discuss these novel findings that re-define the importance of the lysosome and lysosomal calcium signaling at regulating cellular metabolism.

1. Introduction

Macroautophagy, hereafter referred to as autophagy, is a well-conserved multi-step intracellular catabolic process that mediates the sequestration of cytoplasmic material such as damaged organelles and macromolecules, into specialized double-membrane vesicles, called autophagosomes. Upon maturation, autophagosomes must fuse to lysosomes for cargo delivery and degradation, and therefore lysosomes are a required step to complete this process. Autophagy exists at basal levels and can be activated in response to stimuli including nutrient deprivation. In mammalian cells, the initiation and formation of the autophagosome after amino acid depletion inhibits one of the main regulators of macroautophagy, the mTOR complex 1 (mTORC1), which in normal nutrient conditions suppresses autophagosome formation by phosphorylating ATG13 and ULK1 (Jung et al., 2009; Nazio et al., 2013; Nicklin et al., 2009). More recently, it has been shown that mTORC1-mediated phosphorylation of the transcription factor EB (TFEB, a master transcriptional regulator of lysosomal biogenesis and autophagy) inhibits autophagy by preventing its nuclear translocation (Medina et al., 2015; Settembre et al., 2011, 2012). Conversely, TFEB nuclear translocation is induced by its de-phosphorylation in a process that requires calcium released by the lysosomal calcium channel TRPML1, that subsequently activates the calmodulin and calcium-dependent serine/threonine phosphatase calcineurin (CaN) (Medina et al., 2015).

In the formation of autophagosomes can participate several organelles, the most well-established process occurs on the membrane of the endoplasmic reticulum (ER) through the formation of the omegasome (Axe et al., 2008). As in many steps of the autophagic pathway, this process requires the assembly and activation of multiprotein complex. Thus, omegasome formation involves the complex formed by ATG13, ATG101, FIP200 and ULK1 (Joachim et al., 2015; Karanasios et al., 2016; Orsi et al., 2012; Papinski et al., 2014; Stanley et al., 2014). This event initiates the elongation of pre-autophagosomal membranes that subsequently permits the recruitment of the multiprotein complex with Class III phosphatidylinositol 3-kinase (PI3K) activity, which contains Beclin 1, VPS34, VPS15

(Kihara et al., 2001a, 2001b), and ATG14 (Fan et al., 2011; Itakura et al., 2008; Matsunaga et al., 2009; Sun et al., 2008; Zhong et al., 2009). VPS34 complex produces phosphatidylinositol 3-phosphate (PI3P), which favors the expansion of autophagosomal membranes by recruiting PI3P-binding ATG WIPI proteins (Proikas-Cezanne et al., 2015) such as WIPI2B, which recruits the E3-like complex ATG12–ATG5-ATG16L1 involved in the lipididation of LC3 and GABARAPs, involved in the elongation of the autophagosome (Dooley et al., 2014). The expanding phagophore must eventually mature and close to form the autophagosome, which traffics to and fuses with an endosome and/or lysosome, becoming an autolysosome. In order to fuse with lysosomes the autophagosomes move through microtubule railways (Monastyrska et al., 2009), while proteins involved in the fusion are the VTIIB protein, for the AV-endosome fusion (Atlashkin et al., 2003), the GTPase RAB7, which promotes fusion with lysosomes (Jäger et al., 2004; Liang et al., 2008), and the components of the SNARE machinery, VAMP7, VAMP9, and more recently syntaxin 17 which localizes to mature autophagosomes and is required for fusion with the endosome/lysosome through an interaction with SNAP29 and the endosomal/lysosomal SNARE VAMP8 (Fader et al., 2009; Furuta et al., 2010; Itakura et al., 2012). Finally, autophagy is completed when the toolkit of acidic hydrolases and other catabolic enzymes degrade autophagic cargo.

Calcium is a universal second messenger involved in many biological processes including autophagic signaling. The simplest experiment demonstrating the involvement of calcium in autophagy came from the use of calcium chelators such as BAPTA-AM, to prevent cytosolic Ca2+ signals in response to many different stimuli. Thus the treatment with BAPTA-AM during the induction of autophagy by nutrient starvation blocks this process at different levels (Decuypere et al., 2011; Medina et al., 2015). In agreement with a positive role of calcium at inducing autophagy, the nutrient sensing AMPK can be activated by calmodulin kinases and phosphorylates ULK1 on S555, leading to activation and autophagy (Gómez-Suaga et al., 2012; Kim et al., 2011; Scotto Rosato et al., 2019). However, the effects of calcium levels on autophagy are more complex than expected, and in some context calcium signals can exert anti-autophagic actions too (Bootman et al., 2018).

Cytosolic calcium levels are finely regulated and kept at extremely low concentrations by its storage in intracellular organelles such as the endoplasmic reticulum and the mitochondria. The storage of calcium allows a fast and transient response by the elevation of intracellular calcium levels upon specific stimulus, and also its re-uptake to the protection from the deleterious effects of sustained cytosolic calcium levels (Scotto Rosato et al., 2019). Cumulative experimental evidence establishes that the lysosomal compartment can store calcium at a concentration which is similar to the ER, $\sim 500\,\mu M$. However, it is important to note that the cellular volume occupied by the lysosomal compartment is much smaller compared to the ER, rendering lysosomes better suited for generating calcium micro-domains for signaling (Patel and Kilpatrick, 2018; Xu and Ren, 2015). In accordance with this signaling role, lysosomal calcium can modulate the autophagic pathway at different levels (Medina et al., 2015; Scotto Rosato et al., 2019; Wang et al., 2015; Zhang et al., 2016a). In this review, we will focus on the role of the lysosome as source of calcium and the effects that lysosomal calcium homeostasis has in the autophagic pathway.

2. Endolysosomal calcium channels

Ca^{2+} efflux from the lysosomal compartment has been involved in intracellular processes such as signal transduction, organelle homeostasis, lysosomal biogenesis, organelle exocytosis, organelle acidification, and autophagy (Di Paola et al., 2018; Luzio et al., 2007a, 2007b; Medina et al., 2011, 2015; Morgan et al., 2011; Scotto Rosato et al., 2019; Xu and Ren, 2015). In order to modulate these processes, the gradient of calcium between the lysosomal and the cytosol need to be established by an uni-identified Ca^{2+}/H^+ exchanger or Ca^{2+} transporter which probably requires ATP (Morgan et al., 2011). The compartmentalization of calcium within the lysosome allows its rapid mobilization upon specific stimuli through the activation of non-selective cation channels. Members of two different families of channels have been described within the lysosomal compartment, the TPCs and the TRPMLs (Di Paola et al., 2018; Xu and Ren, 2015). Additionally, multiple calcium sensors allow lysosomal calcium release to regulate distinct steps of lysosomal trafficking such as synaptotagmin for lysosomal exocytosis (Czibener et al., 2006), calmodulin and ALG-2 for LE-lysosome fusion (Pryor et al., 2000; Vergarajauregui et al., 2009).

3. Two-pore channels (TPCs)

Two-pore channels (TPCs) belong to the voltage-gated ion channel superfamily. Both, TPC1 and TPC2, are voltage- and ligand gated cation channels that are permeable to Na^+, Ca^{2+}, and H^+. TPCs are found in membranes of endosomes and lysosomes (Calcraft et al., 2009; Morgan et al., 2011). Originally TPCs were described as nicotinic acid adenine dinucleotide phosphate (NAADP)-sensitive Ca^{2+} channels (Calcraft et al., 2009; Rosato et al., 2020). Other studies, however, reported that TPCs are Na^+ channels activated by $PI(3,5)P_2$ or voltage (Boccaccio et al., 2014; Cang et al., 2013, 2014; Lagostena et al., 2017; Marchant and Patel, 2015; Patel and Kilpatrick, 2018). These contradictory observations have been solved in a recent study by Christian Grimm' laboratory. Thus, by using novel small molecule agonists of TPC2 they propose two possible activation modes of the channel that confer specific permeation properties that may result in very different physiological outcome such as lysosomal exocytosis versus pH changes (Gerndt et al., 2020a, 2020b). Thus, TPCs are involved in the release of calcium from the vesicles into the cytoplasm, vesicular pH control, cytokine release from immune cells, membrane trafficking and fusion events (Clementi et al., 2020; Filippini et al., 2020; Gerndt et al., 2020a; Plesch et al., 2018; Stokłosa et al., 2020). Although mutations on TPCs have not been associated with human disease, several studies suggest a role of these channels in Parkinson (Gómez-Suaga et al., 2012), non-alcoholic fatty liver disease (Grimm et al., 2014), infectious diseases such as Ebola (Sakurai et al., 2015), cardiac dysfunction (García-Rúa et al., 2016), defects in pigmentation (Chao et al., 2017), and diabetes (Arredouani et al., 2015; Patel and Kilpatrick, 2018). These experimental evidence makes the TPC channels potential therapeutic targets in human disease.

4. Two-pore channels (TPCs) and autophagy pathway

The first functional link between TPC channels and a crucial regulator of autophagy was described by Cang et al. although in this manuscript the TPCs were defined as lysosomal sodium channels (Cang et al., 2013). Thus, they report an ATP-sensitive Na^+ channel (lysoNa$_{ATP}$) on endolysosomal membranes and formed by TPC1 and TPC2 that associates with mTORC1 and senses cellular nutrient status. TPCs are constitutively open when nutrients are depleted and when mTOR translocates away from the complex at the lysosomal membrane. LysoNa$_{ATP}$ determines the sensitivity of endolysosome's resting membrane potential to Na^+ and cytosolic ATP, controls lysosomal pH stability and regulates whole-body amino acid homeostasis. In accordance with these observations, they show that mutant mice lacking *tpc1* and *tpc2* have severely reduced endurance after fasting (Cang et al., 2013). Ogunbayo et al. has recently demonstrated that mTORC1 is not only regulating lysosomal Na^+ release, but also TPC2/NAADP-mediated Ca^{2+} release both in pulmonary arterial smooth muscle cells (PASMCs) and stably expressing HEK293 cells (Ogunbayo et al., 2018). Indeed, the mTOR inhibitor rapamycin elicited similar Ca^{2+} signals as NAADP in wild-type PASMCs through lysosomal Ca^{2+} release, while neither NAADP nor rapamycin evoked similar signals in $Tpc2^{-/-}$ PAMSCs (Ogunbayo et al., 2018). In agreement with the mTORC1-mediated inhibition of TPCs channel activities, Chao et al. recently observed that mTOR kinase inhibitors such as torin1 and rapamycin can enhance TPC2 channel activity (Chao et al., 2017). In contrast, a more recent work shows that in some cell types, rapamycin can act as an agonist of the lysosomal calcium channel TRPML1 in a mTOR-independent manner, whereas in the same report whole-endolysosome TPC2 were not affected by rapamycin (Zhang et al., 2019). Additionally, although TPC2 sensitivity to ATP seems to require mTORC1 kinase activity, the mTOR target site on TPC2 remains un-characterized. Further studies by independent laboratories would be necessary to exclude or conciliate these contrasting observations.

Functionally, TPCs have been involved in several cellular processes such as endo-lysosomal morphology and endosome-lysosome fusion (Hockey et al., 2015; Lin-Moshier et al., 2014), retrograde transport from endosomes to the Golgi (Ruas et al., 2010), membrane contact site formation between late endosomes and the ER (Kilpatrick et al., 2017), and in regulating autophagy (Pereira et al., 2011, 2017). Pereira et al. showed that the treatment with the TPC agonist NAADP can induce an elevation of autophagic markers such as LC3 and Beclin1 (Pereira et al., 2011). Similar results were obtained by the same group using neural cells stimulated with glutamate (Pereira et al., 2017). In another study, the stimulation with photoactivatable sphingosine leads to an TPC1-mediated elevation of cytosolic calcium that promotes the nuclear translocation of TFEB (Höglinger et al., 2015). In rat cardiomyocytes, nutrient starvation and stimulated ischemia elevate the levels of TPCs while the depletion of TPC2 inhibits autophagic flux and decreases cardiomyocyte viability, thus concluding that TPC1 and TPC2 are essential for basal and induced autophagic flux in cardiomyocytes (García-Rúa et al., 2016). On the contrary, others have shown that TPC2 overexpression increases LC3 and p62 levels, and inhibits autophagosome-lysosome fusion in two cancer cell lines (Sun and Yue, 2018).

5. TRPMLs

The Transient Receptor Potential Mucolipin (TRPML) channel group belongs to the Transient Receptor Potential (TRP) multigene super-family, and is composed of three members, TRPML1, TRPML2 and

TRPML3, which share about 75% similarity in the amino acid (Di Paola et al., 2018; Xu and Ren, 2015). TRPML1 is widely expressed in many tissues, whereas TRPML2 is mainly present in immune cells, particularly in thymus, spleen, and lymphatic nodes. TRPML3 is also expressed in a limited number of tissues, including skin, lung, intestine, and colon (Cuajungco et al., 2016). All mammalian TRPML channels are stimulated on the cytoplasmic side by phosphatidylinositol 3,5-bisphosphate $(PI(3,5)P_2)$, a phosphoinositide enriched in intracellular membranes, (Dong et al., 2010) and, at least for TRPML1, antagonized by phosphatidylinositol 4,5-bisphosphate $(PI(4,5)P_2)$ (Zhang et al., 2012) enriched in the PM, suggesting elevated activity of TRPML channels when localized at lysosomal membranes. So far, phosphoinositides species are the only endogenous modulators of these channels. TRPML1 and TRPML3 have been involved in autophagy while the role of TRPML2 in modulating this process is not clear (Cuajungco et al., 2016; Di Paola et al., 2018; Medina et al., 2015; Scotto Rosato et al., 2019).

6. TRPML1 and autophagy

TRPML1 acts as a non-selective channel permeable to various cations such as Ca^{2+}, Zn^{2+}, Fe^{2+}, Na^+ and K^+ (Dong et al., 2008; Eichelsdoerfer et al., 2010; LaPlante et al., 2002; Xu and Ren, 2015). The late endolysosome (LEL) compartment is the primary site of TRPML1 localization in mammalian cells (Manzoni et al., 2004; Pryor et al., 2006). TRPML1 can also reach the plasma membrane, through the biosynthetic pathway from the Golgi apparatus or by lysosomal exocytosis, a process responsible for the repair/resealing of plasma membrane injuries, secretion of lysosomal enzymes or clearance of lysosomal content (Medina et al., 2011; Pryor et al., 2006; Rodríguez et al., 1997; Vergarajauregui and Puertollano, 2006). Mutations in TRPML1 lead to a very rare lysosomal storage disease (LSD) called Mucolipidosis type 4 (MLIV) (OMIM 252650). At the cellular level, lack of TRPML1 causes impairment in lysosomal functions, with an abnormal accumulation of heterogeneous material within lysosomes. As mentioned above, TRPML1 can mobilize heavy metals such as Fe^{2+} and Zn^{2+} from the lumen of the lysosome (Dong et al., 2008; Eichelsdoerfer et al., 2010). Therefore, the accumulation of Fe^{2+} and Zn^{2+} might be involved in the neurodegenerative phenotype observed in MLIV disease. More studies are necessary to evaluate the real impact of such accumulation in the progression of MLIV pathology.

One of the first evidence linking TRPML1 function to autophagy came from the study of cellular and animal models of MLIV. Thus, impaired autophagy and mitophagy are present in various models of MLIV with an accumulation of both LC3 and SQSTM1/P62 proteins (Jennings et al., 2006; Vergarajauregui et al., 2008). The analysis by immunofluorescence of these autophagic markers has shown marked increase in the co-localization of LC3-puncta with SQSTM1/P62 and delayed fusion of AV with lysosomes during nutrient deprivation (Vergarajauregui et al., 2008). Similar findings were also described in neuronal cells derived from a MLIV mouse model (Curcio-Morelli et al., 2010; Micsenyi et al., 2009; Venugopal et al., 2009), and Drosophila trpml1 mutants (Venkatachalam et al., 2008). In mutant flies a blocking in the fusion of amphisomes (vesicles derived from the fusion of autophagosomes and late endosomes) with lysosomes have been observed (Wong et al., 2012). A role of TRPML1 in autophagy has also been reported in cup-5 null mutant in *C. elegans* resulting in impairment of degradation of autophagy substrates with accumulation of T12G3.1 and LGG-1 (SQSTM1/P62 and LC3 homologs) in enlarged LEL structures indicating that dysfunctional cup-5 does not affect AV-lysosomal fusion, but interferes with autolysosome degradation (Sun et al., 2011). Also, cup-5 deficiency is responsible for maternal-lethal effect (Hersh et al., 2002) due to the onset of a starvation-like phenotype caused by a decrease in nutrient availability (Schaheen et al., 2006). Indeed, the inhibition of autophagy by addition of nutrients can partially relieves mutant embryo lethality (Schaheen et al., 2006; Sun et al., 2011). These observations link the cup-5 to the modulation of mTOR activity.

More recently we have shown that TRPML1 modulates TFEB, a master gene of lysosomal and autophagic functions (Medina et al., 2015). Thus, while seeking for the phosphatase involved in the required activating de-phosphorylation of TFEB, we found that starvation induces lysosomal calcium release through TRPML1 (Medina et al., 2015; Wang et al., 2015), and the subsequent activation of the Ca^{2+} and CaM-dependent phosphatase Calcineurin (CaN). CaN binds and de-phosphorylates two key serines (S142 and S211) on TFEB, and de-phosphorylated TFEB shuttles into the nucleus to activate the transcription of lysosomal and autophagic genes (Medina et al., 2015). Since TRPML1 is a transcriptional target of TFEB, there is a positive feedback loop that boosts TRPML1-TFEB response during starvation conditions (Medina et al., 2015; Wang et al., 2015). Also, we observed that the overexpression of TRPML1 can induce AV formation whereas its depletion during starvation reduces the recruitment of the nascent AV marker WIPI2, raising the possibility that TRPML1 might be involved in AV biogenesis (Medina et al., 2015). This hypothesis has been recently confirmed (Scotto Rosato et al., 2019). Thus, TRPML1 activation induces autophagic vesicle (AV) biogenesis through the generation of phosphatidylinositol 3-phosphate (PI3P) and the recruitment of essential PI3P-binding proteins, such DFCP1 and WIPI2, to the nascent phagophore in a TFEB-independent manner (Scotto Rosato et al., 2019). TRPML1 modulates phagophore formation through the activation of the calcium-dependent kinase CaMKKβ and AMPK, which increase the activation of ULK1 and VPS34 autophagic protein complexes. Consistently, cells from MLIV patients show a reduced recruitment of PI3P-binding proteins to the phagophore during autophagy induction, suggesting that altered AV biogenesis is part of the pathological features of this disease. Together, these observations suggest that TRPML1 impacts the autophagic pathway at several levels (Medina et al., 2015; Scotto Rosato et al., 2019). The unresolved question yet is how nutrient deprivation modulates TRPML1 channel activity (Medina et al., 2015; Wang et al., 2015; Zhang et al., 2012). One hypothesis might be that the changes in cytosolic pH during starvation modulates TRPML1 activity (Li et al., 2016a). Recently, it has been described that TRPML1 can act as a sensor for reactive oxygen species (ROS) and becomes activated or sensitized upon high ROS condition, such as mitochondrial damage. Downstream ROS-mediated TRPML1 activation, CaN binds and de-phosphorylates TFEB which induce autophagic and lysosomal biogenesis related genes (Zhang et al., 2016a). The latter observation together with impaired mitophagy in MLIV clearly links the lysosomes with mitochondrial dysfunction, raising the question of how these organelles communicate to keep cellular homeostasis during stress conditions such as environmental cues (i.e., nutrients) or mitochondria derived ROS. In this context, TRPML1 depleted cells present lysosomal accumulation of Fe^{2+} that may contribute to ROS production, impairment of mitochondrial membrane potential and accumulation of damaged mitochondria which cannot be recycled by the impairment of mitophagy (Coblentz et al., 2014; Zhang et al., 2016a). As we mentioned before, the mTOR complex 1 (mTORC1) is recruited to the lysosomal surface and can sense lysosomal amino acid content to adapt cell metabolism. In normal nutrient rich conditions mTORC1 is localized on lysosomes, is active and inhibits autophagy. In drosophila trpml1 mutants, the lower catabolic activity of the autophagic/lysosomal pathway reduce the amino acid levels. This condition leads to an inhibition of TORC1 that mimics nutrient depletion conditions, and due to the impaired AV-lysosome fusion, causes aberrant accumulation of autophagic substrates and cell death (Wong et al., 2012). Thus, it has been suggested that Ca^{2+} release via TRPML channels is required for activation of mTORC1, which, in turn, inhibits autophagy (Fliniaux et al., 2018; Wong et al., 2012). Also, it has been recently reported that lysosomal calcium release through the calcium channel TRPML1 is required for mTORC1 activation. TRPML1 depletion inhibits mTORC1 activity, while overexpression or pharmacologic activation of TRPML1 has the opposite effect by inducing association of calmodulin (CaM) with mTOR (Li et al., 2016b). On the other hand, mTOR can phosphorylate and inactivate the TRPML1 (Onyenwoke et al., 2015). A later report suggested that the mechanism described by Li et al. might be relevant during starvation. Thus, starvation activates TRPML1 by relieving mTORC1's inhibition of the channel. Subsequently, TRPML1 and the calcium sensor CaM are necessary for mTORC1 reactivation during prolonged starvation. This negative feedback regulation of mTORC1 through TRPML1 may prevent excessive loss of mTORC1 function during starvation (Sun et al., 2018). The complex regulation involving TRPML1 and mTORC1 may explain the role of TRPML1 in both vesicular fusion and fission events. Thus, during starvation, a condition that induces autophagic flux, TRPML1 promotes the centripetal movement of the lysosomes toward the perinuclear region, where AVs accumulates (Li et al., 2016a). Therefore, the coexistence of AVs and lysosomes in the same intracellular region may favor AV-lysosomal fusion. Also, synthetic agonist of TRPML1 increases AV-lysosome fusion, a process that can be inhibited by vinblastine, a microtubule inhibitor, and by silencing the autophagic soluble N-ethylmaleimide-sensitive factor attachment protein receptor (SNARE) factor, syntaxin 17 (Scotto Rosato et al., 2019). Conversely, cellular and animal models of MLIV present a partial block of AV-lysosomal fusion that promotes the accumulation of autophagic markers (Curcio-Morelli et al., 2010; Micsenyi et al., 2009; Venkatachalam et al., 2008; Vergarajauregui et al., 2008; Wong et al., 2012). In contrast, a role for TRPML1 in fission events during lysosomal reformation has been hypothesized (Cao et al., 2017). This function may be linked to its role in the reactivation of mTORC1 during prolonged starvation (Sun et al., 2018). Reactivation of mTOR attenuates autophagy and generates protolysosomal tubules and vesicles that extrude from autolysosomes and ultimately mature into functional lysosomes, an important cellular adaptation mechanism critical for lysosomal homeostasis, especially during prolonged starvation (Yu et al., 2010). Interestingly, a recent report shows that autophagic lysosomal reformation depends on mTOR reactivation also in H_2O_2-induced autophagy, a condition that increases intracellular ROS, an endogenous activator of TRPML1 (Zhang et al., 2016b). Further studies are required to understand the intriguing relationship of the TRPML1-mTORC1 axis in the different steps of the autophagic pathway.

7. TRPML1 and chaperone-mediated autophagy (CMA)

Chaperone-mediated autophagy is other type of autophagy that it does not require formation of intermediate vesicular compartments (autophagosomes or microvesicles) for the import of cargo into lysosomes (Kaushik et al., 2011). Instead, the proposed mechanism is that the CMA substrates translocate across the lysosomal membrane through the action of a protein complex that includes HSC70 (located in the cytosol and lysosome lumen), and the lysosome membrane protein LAMP2A (Klionsky et al., 2016). Both yeast two-hybrid and co-immunoprecipitation (coIP) experiments identified interactions between TRPML1 and two member proteins, Hsc70 and Hsp40, of the molecular chaperone complex required for protein transport into the lysosome during CMA (Venugopal et al., 2009). Interestingly, MLIV cells present an impairment in CMA and have increased levels of target oxidized proteins compared to control fibroblasts (Venugopal et al., 2009), suggesting that defects in CMA might be part of the pathologic hallmarks of MLIV.

8. TRPML3 and autophagy

TRPML3 is expressed in many intracellular compartments, including endosomes, lysosomes, and APs, and it has been involved in autophagy (Di Paola et al., 2018). Thus, TRPML3 overexpression increases LC3 levels (Kim et al., 2009). Upon starvation, TRPML3 overexpression stimulates the formation of autophagosomes. Interestingly, it has been speculated that TRPML3 is recruited to autophagosomes to provide the calcium that is required for fusion or fission events during autophagy (Kim et al., 2009, 2018). More recently, it has been observed that TRPML3 specifically binds to GATE16, a mammalian ATG8 homolog which is important for autophagosome maturation (Choi and Kim, 2014). However, the role of TRPML3 during autophagy is not yet fully addressed. In particular, the link between TRPML3, GATE16, and autophagosome maturation remains to be fully clarified.

9. Lysosomal calcium and selective autophagy

Selective autophagy recognizes specific targets, including damaged mitochondria (mitophagy), aggregated proteins (aggrephagy), and invading bacteria (xenophagy) to engulf by isolation membrane, and degrades toxic materials within lysosomes. Thus, the inhibition of host macroautophagy/autophagy is one of the strategies used by several intracellular pathogens, including *H. pylori*, to escape killing. A very recent report revealed that vacuolating cytotoxin A (VacA) produced by this pathogen inhibits lysosomal and autophagic killing by impairing the activity of TRPML1 (Capurro et al., 2020).

Recently, it has been revealed that membrane-damaged lysosomes also become targets of autophagy, a process called lysophagy. By eliminating ruptured lysosomes, lysophagy prevents the subsequent activation of the inflammasome complex and innate immune response (Hung et al., 2013; Maejima et al., 2013). Also, it has been shown that TFEB is activated during the lysosomal damage by a mechanism that requires LC3 recruitment on lysosomes, where lipidated LC3 interacts with TRPML1, facilitating calcium efflux for TFEB activation. This mechanism might be important in some pathological conditions, such as in the kidney of a mouse model for oxalate nephropathy (Nakamura et al., 2020). Future studies using the MLIV mouse model are needed to determine whether in vivo the lack of TRPML1 impairs lysosomal membrane repair on specific organs or tissues, as well as to determine whether these repairing defects are related to the pathology of MLIV.

10. Concluding remarks

Despite the emerging experimental evidence supporting the role of lysosomal calcium on autophagy, there are still many open questions about the precise molecular mechanisms governing this regulation. In this context, most of these studies are focused on the release of lysosomal calcium through TPCs and TRPMLs channels (Fig. 1) but very few have been focused in the regulation of vesicular calcium uptake (Garrity et al., 2016; Wang et al., 2017; Yang et al., 2019). Moreover, the transporter involved in the refill of lysosomal calcium has not been identified yet. A more technical concern is that most of the studies in mammalian cells are limited to MEFs, human fibroblasts or heterologous cell lines that have been depleted of specific

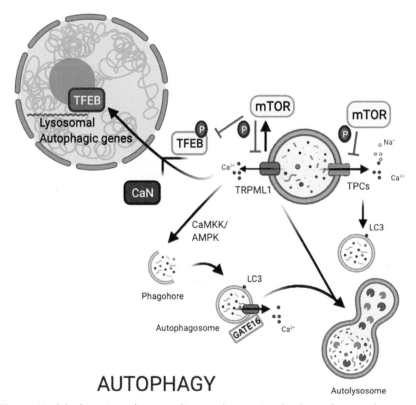

Fig. 1 Model depicting the signaling pathways involved in the regulation of macroautophagy by lysosomal calcium.

proteins involved in lysosomal calcium signaling such as TPCs or TRPMLs using transient siRNA-silencing. The additional use of new approaches such as CRISPR/Cas9 technology and MLIV human fibroblasts reprogramming (such as induced pluripotent stem cell, iPSCs) might contribute to the generation of better models to study lysosomal signaling in specific cell types that are relevant in health and disease (i.e., lysosomal storage diseases).

Finally, a few lysosomal Ca^{2+} effectors involved in specific TRPML1 functions have been identified. Some examples are CaN, and ALG-2 (Li et al., 2016a; Medina et al., 2015; Vergarajauregui et al., 2009, p. 2). Recently, Krogsaeter et al. reviewed data on interactomes for TPC1, TPC2, TRPML1, and TRPML3 (Krogsaeter et al., 2018). Hopefully future data using classical proteomics and novel approaches such as immuno-purification of lysosomes (Lyso-IP) (Abu-Remaileh et al., 2017, p.) will allow to better identify novel players in the regulation of lysosomal calcium signaling and autophagy.

Acknowledgments

We acknowledge financial support from Italian Telethon Foundation (TMDMHCSIITT) (D.L.M.), Mucolipidosis Type IV Foundation (D.L.M.), Horizon 2020 grant BATCure (666918) (D.L.M.). We acknowledge Alessio Reggio for his help in the figure scheme. The figure has been created with Biorender.com

References

Abu-Remaileh, M., Wyant, G.A., Kim, C., Laqtom, N.N., Abbasi, M., Chan, S.H., Freinkman, E., Sabatini, D.M., 2017. Lysosomal metabolomics reveals V-ATPase- and mTOR-dependent regulation of amino acid efflux from lysosomes. Science 358, 807–813. https://doi.org/10.1126/science.aan6298.

Arredouani, A., Ruas, M., Collins, S.C., Parkesh, R., Clough, F., Pillinger, T., Coltart, G., Rietdorf, K., Royle, A., Johnson, P., Braun, M., Zhang, Q., Sones, W., Shimomura, K., Morgan, A.J., Lewis, A.M., Chuang, K.-T., Tunn, R., Gadea, J., Teboul, L., Heister, P.M., Tynan, P.W., Bellomo, E.A., Rutter, G.A., Rorsman, P., Churchill, G.C., Parrington, J., Galione, A., 2015. Nicotinic acid adenine dinucleotide phosphate (NAADP) and endolysosomal two-pore channels modulate membrane excitability and stimulus-secretion coupling in mouse pancreatic β cells. J. Biol. Chem. 290, 21376–21392. https://doi.org/10.1074/jbc.M115.671248.

Atlashkin, V., Kreykenbohm, V., Eskelinen, E.-L., Wenzel, D., Fayyazi, A., Fischer von Mollard, G., 2003. Deletion of the SNARE vti1b in mice results in the loss of a single SNARE partner, syntaxin 8. Mol. Cell. Biol. 23, 5198–5207. https://doi.org/10.1128/MCB.23.15.5198-5207.2003.

Axe, E.L., Walker, S.A., Manifava, M., Chandra, P., Roderick, H.L., Habermann, A., Griffiths, G., Ktistakis, N.T., 2008. Autophagosome formation from membrane compartments enriched in phosphatidylinositol 3-phosphate and dynamically connected to the endoplasmic reticulum. J. Cell Biol. 182, 685–701. https://doi.org/10.1083/jcb.200803137.

Boccaccio, A., Scholz-Starke, J., Hamamoto, S., Larisch, N., Festa, M., Gutla, P.V.K., Costa, A., Dietrich, P., Uozumi, N., Carpaneto, A., 2014. The phosphoinositide PI(3,5)P$_2$ mediates activation of mammalian but not plant TPC proteins: functional expression of endolysosomal channels in yeast and plant cells. Cell. Mol. Life Sci. 71, 4275–4283. https://doi.org/10.1007/s00018-014-1623-2.

Bootman, M.D., Chehab, T., Bultynck, G., Parys, J.B., Rietdorf, K., 2018. The regulation of autophagy by calcium signals: do we have a consensus? Cell Calcium 70, 32–46. https://doi.org/10.1016/j.ceca.2017.08.005.

Calcraft, P.J., Ruas, M., Pan, Z., Cheng, X., Arredouani, A., Hao, X., Tang, J., Rietdorf, K., Teboul, L., Chuang, K.-T., Lin, P., Xiao, R., Wang, C., Zhu, Y., Lin, Y., Wyatt, C.N., Parrington, J., Ma, J., Evans, A.M., Galione, A., Zhu, M.X., 2009. NAADP mobilizes calcium from acidic organelles through two-pore channels. Nature 459, 596–600. https://doi.org/10.1038/nature08030.

Cang, C., Bekele, B., Ren, D., 2014. The voltage-gated sodium channel TPC1 confers endolysosomal excitability. Nat. Chem. Biol. 10, 463–469. https://doi.org/10.1038/nchembio.1522.

Cang, C., Zhou, Y., Navarro, B., Seo, Y.-J., Aranda, K., Shi, L., Battaglia-Hsu, S., Nissim, I., Clapham, D.E., Ren, D., 2013. mTOR regulates lysosomal ATP-sensitive two-pore Na(+) channels to adapt to metabolic state. Cell 152, 778–790. https://doi.org/10.1016/j.cell.2013.01.023.

Cao, Q., Yang, Y., Zhong, X.Z., Dong, X.-P., 2017. The lysosomal Ca2+ release channel TRPML1 regulates lysosome size by activating calmodulin. J. Biol. Chem. 292, 8424–8435. https://doi.org/10.1074/jbc.M116.772160.

Capurro, M.I., Prashar, A., Jones, N.L., 2020. MCOLN1/TRPML1 inhibition—a novel strategy used by helicobacter pylori to escape autophagic killing and antibiotic eradication therapy in vivo. Autophagy 16, 169. https://doi.org/10.1080/15548627.2019.1677322.

Chao, Y.-K., Schludi, V., Chen, C.-C., Butz, E., Nguyen, O.N.P., Müller, M., Krüger, J., Kammerbauer, C., Ben-Johny, M., Vollmar, A.M., Berking, C., Biel, M., Wahl-Schott, C.A., Grimm, C., 2017. TPC2 polymorphisms associated with a hair pigmentation phenotype in humans result in gain of channel function by independent mechanisms. Proc. Natl. Acad. Sci. U. S. A. 114, E8595–E8602. https://doi.org/10.1073/pnas.1705739114.

Choi, S., Kim, H.J., 2014. The Ca2+ channel TRPML3 specifically interacts with the mammalian ATG8 homologue GATE16 to regulate autophagy. Biochem. Biophys. Res. Commun. 443, 56–61. https://doi.org/10.1016/j.bbrc.2013.11.044.

Clementi, N., Scagnolari, C., D'Amore, A., Palombi, F., Criscuolo, E., Frasca, F., Pierangeli, A., Mancini, N., Antonelli, G., Clementi, M., Carpaneto, A., Filippini, A., 2020. Naringenin is a powerful inhibitor of SARS-CoV-2 infection in vitro. Pharmacol. Res. 163, 105255. https://doi.org/10.1016/j.phrs.2020.105255.

Coblentz, J., St Croix, C., Kiselyov, K., 2014. Loss of TRPML1 promotes production of reactive oxygen species: is oxidative damage a factor in mucolipidosis type IV? Biochem. J. 457, 361–368. https://doi.org/10.1042/BJ20130647.

Cuajungco, M.P., Silva, J., Habibi, A., Valadez, J.A., 2016. The mucolipin-2 (TRPML2) ion channel: a tissue-specific protein crucial to normal cell function. Pflugers Arch. 468, 177–192. https://doi.org/10.1007/s00424-015-1732-2.

Curcio-Morelli, C., Charles, F.A., Micsenyi, M.C., Cao, Y., Venugopal, B., Browning, M.F., Dobrenis, K., Cotman, S.L., Walkley, S.U., Slaugenhaupt, S.A., 2010. Macroautophagy is defective in mucolipin-1-deficient mouse neurons. Neurobiol. Dis. 40, 370–377. https://doi.org/10.1016/j.nbd.2010.06.010.

Czibener, C., Sherer, N.M., Becker, S.M., Pypaert, M., Hui, E., Chapman, E.R., Mothes, W., Andrews, N.W., 2006. Ca2+ and synaptotagmin VII-dependent delivery of lysosomal membrane to nascent phagosomes. J. Cell Biol. 174, 997–1007. https://doi.org/10.1083/jcb.200605004.

Decuypere, J.-P., Welkenhuyzen, K., Luyten, T., Ponsaerts, R., Dewaele, M., Molgó, J., Agostinis, P., Missiaen, L., De Smedt, H., Parys, J.B., Bultynck, G., 2011. Ins(1,4,5) P3 receptor-mediated Ca2+ signaling and autophagy induction are interrelated. Autophagy 7, 1472–1489. https://doi.org/10.4161/auto.7.12.17909.

Di Paola, S., Scotto-Rosato, A., Medina, D.L., 2018. TRPML1: the Ca(2+)retaker of the lysosome. Cell Calcium 69, 112–121. https://doi.org/10.1016/j.ceca.2017.06.006.

Dong, X.-P., Cheng, X., Mills, E., Delling, M., Wang, F., Kurz, T., Xu, H., 2008. The type IV mucolipidosis-associated protein TRPML1 is an endolysosomal iron release channel. Nature 455, 992–996. https://doi.org/10.1038/nature07311.

Dong, X., Shen, D., Wang, X., Dawson, T., Li, X., Zhang, Q., Cheng, X., Zhang, Y., Weisman, L.S., Delling, M., Xu, H., 2010. PI(3,5)P(2) controls membrane trafficking by direct activation of mucolipin Ca(2+) release channels in the endolysosome. Nat. Commun. 1, 38. https://doi.org/10.1038/ncomms1037.

Dooley, H.C., Razi, M., Polson, H.E.J., Girardin, S.E., Wilson, M.I., Tooze, S.A., 2014. WIPI2 links LC3 conjugation with PI3P, autophagosome formation, and pathogen clearance by recruiting Atg12-5-16L1. Mol. Cell 55, 238–252. https://doi.org/10.1016/j.molcel.2014.05.021.

Eichelsdoerfer, J.L., Evans, J.A., Slaugenhaupt, S.A., Cuajungco, M.P., 2010. Zinc dyshomeostasis is linked with the loss of mucolipidosis IV-associated TRPML1 ion channel. J. Biol. Chem. 285, 34304–34308. https://doi.org/10.1074/jbc.C110.165480.

Fader, C.M., Sánchez, D.G., Mestre, M.B., Colombo, M.I., 2009. TI-VAMP/VAMP7 and VAMP3/cellubrevin: two v-SNARE proteins involved in specific steps of the autophagy/multivesicular body pathways. Biochim. Biophys. Acta 1793, 1901–1916. https://doi.org/10.1016/j.bbamcr.2009.09.011.

Fan, W., Nassiri, A., Zhong, Q., 2011. Autophagosome targeting and membrane curvature sensing by Barkor/Atg14(L). Proc. Natl. Acad. Sci. U. S. A. 108, 7769–7774. https://doi.org/10.1073/pnas.1016472108.

Filippini, A., D'Amore, A., Palombi, F., Carpaneto, A., 2020. Could the inhibition of endo-lysosomal two-pore channels (TPCs) by the natural flavonoid naringenin represent an option to fight SARS-CoV-2 infection? Front. Microbiol. 11, 970. https://doi.org/10.3389/fmicb.2020.00970.

Fliniaux, I., Germain, E., Farfariello, V., Prevarskaya, N., 2018. TRPs and Ca2 + in cell death and survival. Cell Calcium 69, 4–18. https://doi.org/10.1016/j.ceca.2017.07.002.

Furuta, N., Fujita, N., Noda, T., Yoshimori, T., Amano, A., 2010. Combinational soluble N-Ethylmaleimide-sensitive factor attachment protein receptor proteins VAMP8 and Vti1b mediate fusion of antimicrobial and canonical Autophagosomes with lysosomes. Mol. Biol. Cell 21, 1001–1010. https://doi.org/10.1091/mbc.E09-08-0693.

García-Rúa, V., Feijóo-Bandín, S., Rodríguez-Penas, D., Mosquera-Leal, A., Abu-Assi, E., Beiras, A., María Seoane, L., Lear, P., Parrington, J., Portolés, M., Roselló-Lletí, E., Rivera, M., Gualillo, O., Parra, V., Hill, J.A., Rothermel, B., González-Juanatey, J.R., Lago, F., 2016. Endolysosomal two-pore channels regulate autophagy in cardiomyocytes. J. Physiol. 594, 3061–3077. https://doi.org/10.1113/JP271332.

Garrity, A.G., Wang, W., Collier, C.M., Levey, S.A., Gao, Q., Xu, H., 2016. The endoplasmic reticulum, not the pH gradient, drives calcium refilling of lysosomes. eLife 5, e15887. https://doi.org/10.7554/eLife.15887.

Gerndt, S., Chen, C.-C., Chao, Y.-K., Yuan, Y., Burgstaller, S., Scotto Rosato, A., Krogsaeter, E., Urban, N., Jacob, K., Nguyen, O.N.P., Miller, M.T., Keller, M., Vollmar, A.M., Gudermann, T., Zierler, S., Schredelseker, J., Schaefer, M., Biel, M., Malli, R., Wahl-Schott, C., Bracher, F., Patel, S., Grimm, C., 2020a. Agonist-mediated switching of ion selectivity in TPC2 differentially promotes lysosomal function. eLife 9, e54712. https://doi.org/10.7554/eLife.54712.

Gerndt, S., Krogsaeter, E., Patel, S., Bracher, F., Grimm, C., 2020b. Discovery of lipophilic two-pore channel agonists. FEBS J. 287, 5284–5293. https://doi.org/10.1111/febs.15432.

Gómez-Suaga, P., Luzón-Toro, B., Churamani, D., Zhang, L., Bloor-Young, D., Patel, S., Woodman, P.G., Churchill, G.C., Hilfiker, S., 2012. Leucine-rich repeat kinase 2 regulates autophagy through a calcium-dependent pathway involving NAADP. Hum. Mol. Genet. 21, 511–525. https://doi.org/10.1093/hmg/ddr481.

Grimm, C., Holdt, L.M., Chen, C.-C., Hassan, S., Müller, C., Jörs, S., Cuny, H., Kissing, S., Schröder, B., Butz, E., Northoff, B., Castonguay, J., Luber, C.A., Moser, M., Spahn, S., Lüllmann-Rauch, R., Fendel, C., Klugbauer, N., Griesbeck, O., Haas, A., Mann, M., Bracher, F., Teupser, D., Saftig, P., Biel, M., Wahl-Schott, C., 2014. High susceptibility to fatty liver disease in two-pore channel 2-deficient mice. Nat. Commun. 5, 4699. https://doi.org/10.1038/ncomms5699.

Hersh, B.M., Hartwieg, E., Horvitz, H.R., 2002. The Caenorhabditis elegans mucolipin-like gene cup-5 is essential for viability and regulates lysosomes in multiple cell types. Proc. Natl. Acad. Sci. U. S. A. 99, 4355–4360. https://doi.org/10.1073/pnas.062065399.

Hockey, L.N., Kilpatrick, B.S., Eden, E.R., Lin-Moshier, Y., Brailoiu, G.C., Brailoiu, E., Futter, C.E., Schapira, A.H., Marchant, J.S., Patel, S., 2015. Dysregulation of lysosomal morphology by pathogenic LRRK2 is corrected by TPC2 inhibition. J. Cell Sci. 128, 232–238. https://doi.org/10.1242/jcs.164152.

Höglinger, D., Haberkant, P., Aguilera-Romero, A., Riezman, H., Porter, F.D., Platt, F.M., Galione, A., Schultz, C., 2015. Intracellular sphingosine releases calcium from lysosomes. eLife 4, e10616. https://doi.org/10.7554/eLife.10616.

Hung, Y.-H., Chen, L.M.-W., Yang, J.-Y., Yuan Yang, W., 2013. Spatiotemporally controlled induction of autophagy-mediated lysosome turnover. Nat. Commun. 4, 2111. https://doi.org/10.1038/ncomms3111.

Itakura, E., Kishi, C., Inoue, K., Mizushima, N., 2008. Beclin 1 forms two distinct phosphatidylinositol 3-kinase complexes with mammalian Atg14 and UVRAG. Mol. Biol. Cell 19, 5360–5372. https://doi.org/10.1091/mbc.e08-01-0080.

Itakura, E., Kishi-Itakura, C., Mizushima, N., 2012. The hairpin-type tail-anchored SNARE syntaxin 17 targets to Autophagosomes for fusion with endosomes/lysosomes. Cell 151, 1256–1269. https://doi.org/10.1016/j.cell.2012.11.001.

Jäger, S., Bucci, C., Tanida, I., Ueno, T., Kominami, E., Saftig, P., Eskelinen, E.-L., 2004. Role for Rab7 in maturation of late autophagic vacuoles. J. Cell Sci. 117, 4837–4848. https://doi.org/10.1242/jcs.01370.

Jennings, J.J., Zhu, J.-H., Rbaibi, Y., Luo, X., Chu, C.T., Kiselyov, K., 2006. Mitochondrial aberrations in mucolipidosis type IV. J. Biol. Chem. 281, 39041–39050. https://doi.org/10.1074/jbc.M607982200.

Joachim, J., Jefferies, H.B.J., Razi, M., Frith, D., Snijders, A.P., Chakravarty, P., Judith, D., Tooze, S.A., 2015. Activation of ULK kinase and autophagy by GABARAP trafficking from the centrosome is regulated by WAC and GM130. Mol. Cell 60, 899–913. https://doi.org/10.1016/j.molcel.2015.11.018.

Jung, C.H., Jun, C.B., Ro, S.-H., Kim, Y.-M., Otto, N.M., Cao, J., Kundu, M., Kim, D.-H., 2009. ULK-Atg13-FIP200 complexes mediate mTOR signaling to the autophagy machinery. Mol. Biol. Cell 20, 1992–2003. https://doi.org/10.1091/mbc.e08-12-1249.

Karanasios, E., Walker, S.A., Okkenhaug, H., Manifava, M., Hummel, E., Zimmermann, H., Ahmed, Q., Domart, M.-C., Collinson, L., Ktistakis, N.T., 2016. Autophagy initiation by ULK complex assembly on ER tubulovesicular regions marked by ATG9 vesicles. Nat. Commun. 7, 12420. https://doi.org/10.1038/ncomms12420.

Kaushik, S., Bandyopadhyay, U., Sridhar, S., Kiffin, R., Martinez-Vicente, M., Kon, M., Orenstein, S.J., Wong, E., Cuervo, A.M., 2011. Chaperone-mediated autophagy at a glance. J. Cell Sci. 124, 495–499. https://doi.org/10.1242/jcs.073874.

Kihara, A., Kabeya, Y., Ohsumi, Y., Yoshimori, T., 2001a. Beclin-phosphatidylinositol 3-kinase complex functions at the trans-Golgi network. EMBO Rep. 2, 330–335. https://doi.org/10.1093/embo-reports/kve061.

Kihara, A., Noda, T., Ishihara, N., Ohsumi, Y., 2001b. Two distinct Vps34 phosphatidylinositol 3-kinase complexes function in autophagy and carboxypeptidase Y sorting in Saccharomyces cerevisiae. J. Cell Biol. 152, 519–530. https://doi.org/10.1083/jcb.152.3.519.

Kilpatrick, B.S., Eden, E.R., Hockey, L.N., Yates, E., Futter, C.E., Patel, S., 2017. An endosomal NAADP-sensitive two-pore Ca2+ channel regulates ER-endosome membrane contact sites to control growth factor Signaling. Cell Rep. 18, 1636–1645. https://doi.org/10.1016/j.celrep.2017.01.052.

Kim, S.W., Kim, D.H., Park, K.S., Kim, M.K., Park, Y.M., Muallem, S., So, I., Kim, H.J., 2018. Palmitoylation controls trafficking of the intracellular Ca2+ channel MCOLN3/TRPML3 to regulate autophagy. Autophagy 15, 327–340. https://doi.org/10.1080/15548627.2018.1518671.

Kim, J., Kundu, M., Viollet, B., Guan, K.-L., 2011. AMPK and mTOR regulate autophagy through direct phosphorylation of Ulk1. Nat. Cell Biol. 13, 132–141. https://doi.org/ 10.1038/ncb2152.

Kim, H.J., Soyombo, A.A., Tjon-Kon-Sang, S., So, I., Muallem, S., 2009. The Ca(2+) channel TRPML3 regulates membrane trafficking and autophagy. Traffic 10, 1157–1167. https://doi.org/10.1111/j.1600-0854.2009.00924.x.

Klionsky, D.J., Abdelmohsen, K., Abe, A., Abedin, M.J., Abeliovich, H., Acevedo Arozena, A., Adachi, H., Adams, C.M., Adams, P.D., Adeli, K., Adhihetty, P.J., Adler, S.G., Agam, G., Agarwal, R., Aghi, M.K., Agnello, M., Agostinis, P., Aguilar, P.V., Aguirre-Ghiso, J., Airoldi, E.M., Ait-Si-Ali, S., Akematsu, T., Akporiaye, E.T., Al-Rubeai, M., Albaiceta, G.M., Albanese, C., Albani, D., Albert, M.L., Aldudo, J., Algül, H., Alirezaei, M., Alloza, I., Almasan, A., Almonte-Beceril, M., Alnemri, E.S., Alonso, C., Altan-Bonnet, N., Altieri, D.C., Alvarez, S., Alvarez-Erviti, L., Alves, S., Amadoro, G., Amano, A., Amantini, C., Ambrosio, S., Amelio, I., Amer, A.O., Amessou, M., Amon, A., An, Z., Anania, F.A., Andersen, S.U., Andley, U.P., Andreadi, C.K., Andrieu-Abadie, N., Anel, A., Ann, D.K., Anoopkumar-Dukie, S., Antonioli, M., Aoki, H., Apostolova, N., Aquila, S., Aquilano, K., Araki, K., Arama, E., Aranda, A., Araya, J., Arcaro, A., Arias, E., Arimoto, H., Ariosa, A.R., Armstrong, J.L., Arnould, T., Arsov, I., Asanuma, K., Askanas, V., Asselin, E., Atarashi, R., Atherton, S.S., Atkin, J.D., Attardi, L.D., Auberger, P., Auburger, G., Aurelian, L., Autelli, R., Avagliano, L., Avantaggiati, M.L., Avrahami, L., Awale, S., Azad, N., Bachetti, T., Backer, J.M., Bae, D.-H., Bae, J.-S., Bae, O.-N., Bae, S.H., Baehrecke, E.H., Baek, S.-H., Baghdiguian, S., Bagniewska-Zadworna, A., Bai, H., Bai, J., Bai, X.-Y., Bailly, Y., Balaji, K.N., Balduini, W., Ballabio, A., Balzan, R., Banerjee, R., Bánhegyi, G., Bao, H., Barbeau, B., Barrachina, M.D., Barreiro, E., Bartel, B., Bartolomé, A., Bassham, D.C., Bassi, M.T., Bast, R.C., Basu, A., Batista, M.T., Batoko, H., Battino, M., Bauckman, K., Baumgarner, B.L., Bayer, K.U., Beale, R., Beaulieu, J.-F., Beck, G.R., Becker, C., Beckham, J.D., Bédard, P.-A., Bednarski, P.J., Begley, T.J., Behl, C., Behrends, C., Behrens, G.M., Behrns, K.E., Bejarano, E., Belaid, A., Belleudi, F., Bénard, G., Berchem, G., Bergamaschi, D., Bergami, M., Berkhout, B., Berliocchi, L., Bernard, A., Bernard, M., Bernassola, F., Bertolotti, A., Bess, A.S., Besteiro, S., Bettuzzi, S., Bhalla, S., Bhattacharyya, S., Bhutia, S.K., Biagosch, C., Bianchi, M.W., Biard-Piechaczyk, M., Billes, V., Bincoletto, C., Bingol, B., Bird, S.W., Bitoun, M., Bjedov, I., Blackstone, C., Blanc, L., Blanco, G.A., Blomhoff, H.K., Boada-Romero, E., Böckler, S., Boes, M., Boesze-Battaglia, K., Boise, L.H., Bolino, A., Boman, A., Bonaldo, P., Bordi, M., Bosch, J., Botana, L.M., Botti, J., Bou, G., Bouché, M., Bouchecareilh, M., Boucher, M.-J., Boulton, M.E., Bouret, S.G., Boya, P., Boyer-Guittaut, M., Bozhkov, P.V., Brady, N., Braga, V.M., Brancolini, C., Braus, G.H., Bravo-San Pedro, J.M., Brennan, L.A., Bresnick, E.H., Brest, P., Bridges, D., Bringer, M.-A., Brini, M., Brito, G.C., Brodin, B., Brookes, P.S., Brown, E.J., Brown, K., Broxmeyer, H.E., Bruhat, A., Brum, P.C., Brumell, J.H., Brunetti-Pierri, N., Bryson-Richardson, R.J., Buch, S., Buchan, A.M., Budak, H., Bulavin, D.V., Bultman, S.J., Bultynck, G., Bumbasirevic, V., Burelle, Y., Burke, R.E., Burmeister, M., Bütikofer, P., Caberlotto, L., Cadwell, K., Cahova, M., Cai, D., Cai, J., Cai, Q., Calatayud, S., Camougrand, N., Campanella, M., Campbell, G.R., Campbell, M., Campello, S., Candau, R., Caniggia, I., Cantoni, L., Cao, L., Caplan, A.B., Caraglia, M., Cardinali, C., Cardoso, S.M., Carew, J.S., Carleton, L.A., Carlin, C.R., Carloni, S., Carlsson, S.R., Carmona-Gutierrez, D., Carneiro, L.A., Carnevali, O., Carra, S., Carrier, A., Carroll, B., Casas, C., Casas, J., Cassinelli, G., Castets, P., Castro-Obregon, S., Cavallini, G., Ceccherini, I.,

Cecconi, F., Cederbaum, A.I., Ceña, V., Cenci, S., Cerella, C., Cervia, D., Cetrullo, S., Chaachouay, H., Chae, H.-J., Chagin, A.S., Chai, C.-Y., Chakrabarti, G., Chamilos, G., Chan, E.Y., Chan, M.T., Chandra, D., Chandra, P., Chang, C.-P., Chang, R.C.-C., Chang, T.Y., Chatham, J.C., Chatterjee, S., Chauhan, S., Che, Y., Cheetham, M.E., Cheluvappa, R., Chen, C.-J., Chen, G., Chen, G.-C., Chen, G., Chen, H., Chen, J.W., Chen, J.-K., Chen, M., Chen, M., Chen, P., Chen, Q., Chen, Q., Chen, S.-D., Chen, S., Chen, S.S.-L., Chen, W., Chen, W.-J., Chen, W.Q., Chen, W., Chen, X., Chen, Y.-H., Chen, Y.-G., Chen, Y., Chen, Y., Chen, Y., Chen, Y.-J., Chen, Y.-Q., Chen, Y., Chen, Z., Chen, Z., Cheng, A., Cheng, C.H., Cheng, H., Cheong, H., Cherry, S., Chesney, J., Cheung, C.H.A., Chevet, E., Chi, H.C., Chi, S.-G., Chiacchiera, F., Chiang, H.-L., Chiarelli, R., Chiariello, M., Chieppa, M., Chin, L.-S., Chiong, M., Chiu, G.N., Cho, D.-H., Cho, S.-G., Cho, W.C., Cho, Y.-Y., Cho, Y.-S., Choi, A.M., Choi, E.-J., Choi, E.-K., Choi, J., Choi, M.E., Choi, S.-I., Chou, T.-F., Chouaib, S., Choubey, D., Choubey, V., Chow, K.-C., Chowdhury, K., Chu, C.T., Chuang, T.-H., Chun, T., Chung, H., Chung, T., Chung, Y.-L., Chwae, Y.-J., Cianfanelli, V., Ciarcia, R., Ciechomska, I.A., Ciriolo, M.R., Cirone, M., Claerhout, S., Clague, M.J., Clària, J., Clarke, P.G., Clarke, R., Clementi, E., Cleyrat, C., Cnop, M., Coccia, E.M., Cocco, T., Codogno, P., Coers, J., Cohen, E.E., Colecchia, D., Coletto, L., Coll, N.S., Colucci-Guyon, E., Comincini, S., Condello, M., Cook, K.L., Coombs, G.H., Cooper, C.D., Cooper, J.M., Coppens, I., Corasaniti, M.T., Corazzari, M., Corbalan, R., Corcelle-Termeau, E., Cordero, M.D., Corral-Ramos, C., Corti, O., Cossarizza, A., Costelli, P., Costes, S., Cotman, S.L., Coto-Montes, A., Cottet, S., Couve, E., Covey, L.R., Cowart, L.A., Cox, J.S., Coxon, F.P., Coyne, C.B., Cragg, M.S., Craven, R.J., Crepaldi, T., Crespo, J.L., Criollo, A., Crippa, V., Cruz, M.T., Cuervo, A.M., Cuezva, J.M., Cui, T., Cutillas, P.R., Czaja, M.J., Czyzyk-Krzeska, M.F., Dagda, R.K., Dahmen, U., Dai, C., Dai, W., Dai, Y., Dalby, K.N., Dalla Valle, L., Dalmasso, G., D'Amelio, M., Damme, M., Darfeuille-Michaud, A., Dargemont, C., Darley-Usmar, V.M., Dasarathy, S., Dasgupta, B., Dash, S., Dass, C.R., Davey, H.M., Davids, L.M., Dávila, D., Davis, R.J., Dawson, T.M., Dawson, V.L., Daza, P., de Belleroche, J., de Figueiredo, P., de Figueiredo, R.C.B.Q., de la Fuente, J., De Martino, L., De Matteis, A., De Meyer, G.R., De Milito, A., De Santi, M., de Souza, W., De Tata, V., De Zio, D., Debnath, J., Dechant, R., Decuypere, J.-P., Deegan, S., Dehay, B., Del Bello, B., Del Re, D.P., Delage-Mourroux, R., Delbridge, L.M., Deldicque, L., Delorme-Axford, E., Deng, Y., Dengjel, J., Denizot, M., Dent, P., Der, C.J., Deretic, V., Derrien, B., Deutsch, E., Devarenne, T.P., Devenish, R.J., Di Bartolomeo, S., Di Daniele, N., Di Domenico, F., Di Nardo, A., Di Paola, S., Di Pietro, A., Di Renzo, L., DiAntonio, A., Díaz-Araya, G., Díaz-Laviada, I., Diaz-Meco, M.T., Diaz-Nido, J., Dickey, C.A., Dickson, R.C., Diederich, M., Digard, P., Dikic, I., Dinesh-Kumar, S.P., Ding, C., Ding, W.-X., Ding, Z., Dini, L., Distler, J.H., Diwan, A., Djavaheri-Mergny, M., Dmytruk, K., Dobson, R.C., Doetsch, V., Dokladny, K., Dokudovskaya, S., Donadelli, M., Dong, X.C., Dong, X., Dong, Z., Donohue, T.M., Doran, K.S., D'Orazi, G., Dorn, G.W., Dosenko, V., Dridi, S., Drucker, L., Du, J., Du, L.-L., Du, L., du Toit, A., Dua, P., Duan, L., Duann, P., Dubey, V.K., Duchen, M.R., Duchosal, M.A., Duez, H., Dugail, I., Dumit, V.I., Duncan, M.C., Dunlop, E.A., Dunn, W.A., Dupont, N., Dupuis, L., Durán, R.V., Durcan, T.M., Duvezin-Caubet, S., Duvvuri, U., Eapen, V., Ebrahimi-Fakhari, D., Echard, A., Eckhart, L., Edelstein, C.L., Edinger, A.L., Eichinger, L., Eisenberg, T., Eisenberg-Lerner, A., Eissa, N.T., El-Deiry, W.S., El-Khoury, V., Elazar, Z.,

Eldar-Finkelman, H., Elliott, C.J., Emanuele, E., Emmenegger, U., Engedal, N., Engelbrecht, A.-M., Engelender, S., Enserink, J.M., Erdmann, R., Erenpreisa, J., Eri, R., Eriksen, J.L., Erman, A., Escalante, R., Eskelinen, E.-L., Espert, L., Esteban-Martínez, L., Evans, T.J., Fabri, M., Fabrias, G., Fabrizi, C., Facchiano, A., Færgeman, N.J., Faggioni, A., Fairlie, W.D., Fan, C., Fan, D., Fan, J., Fang, S., Fanto, M., Fanzani, A., Farkas, T., Faure, M., Favier, F.B., Fearnhead, H., Federici, M., Fei, E., Felizardo, T.C., Feng, H., Feng, Y., Feng, Y., Ferguson, T.A., Fernández, Á.F., Fernandez-Barrena, M.G., Fernandez-Checa, J.C., Fernández-López, A., Fernandez-Zapico, M.E., Feron, O., Ferraro, E., Ferreira-Halder, C.V., Fesus, L., Feuer, R., Fiesel, F.C., Filippi-Chiela, E.C., Filomeni, G., Fimia, G.M., Fingert, J.H., Finkbeiner, S., Finkel, T., Fiorito, F., Fisher, P.B., Flajolet, M., Flamigni, F., Florey, O., Florio, S., Floto, R.A., Folini, M., Follo, C., Fon, E.A., Fornai, F., Fortunato, F., Fraldi, A., Franco, R., Francois, A., François, A., Frankel, L.B., Fraser, I.D., Frey, N., Freyssenet, D.G., Frezza, C., Friedman, S.L., Frigo, D.E., Fu, D., Fuentes, J.M., Fueyo, J., Fujitani, Y., Fujiwara, Y., Fujiya, M., Fukuda, M., Fulda, S., Fusco, C., Gabryel, B., Gaestel, M., Gailly, P., Gajewska, M., Galadari, S., Galili, G., Galindo, I., Galindo, M.F., Galliciotti, G., Galluzzi, L., Galluzzi, L., Galy, V., Gammoh, N., Gandy, S., Ganesan, A.K., Ganesan, S., Ganley, I.G., Gannagé, M., Gao, F.-B., Gao, F., Gao, J.-X., García Nannig, L., García Véscovi, E., Garcia-Macía, M., Garcia-Ruiz, C., Garg, A.D., Garg, P.K., Gargini, R., Gassen, N.C., Gatica, D., Gatti, E., Gavard, J., Gavathiotis, E., Ge, L., Ge, P., Ge, S., Gean, P.-W., Gelmetti, V., Genazzani, A.A., Geng, J., Genschik, P., Gerner, L., Gestwicki, J.E., Gewirtz, D.A., Ghavami, S., Ghigo, E., Ghosh, D., Giammarioli, A.M., Giampieri, F., Giampietri, C., Giatromanolaki, A., Gibbings, D.J., Gibellini, L., Gibson, S.B., Ginet, V., Giordano, A., Giorgini, F., Giovannetti, E., Girardin, S.E., Gispert, S., Giuliano, S., Gladson, C.L., Glavic, A., Gleave, M., Godefroy, N., Gogal, R.M., Gokulan, K., Goldman, G.H., Goletti, D., Goligorsky, M.S., Gomes, A.V., Gomes, L.C., Gomez, H., Gomez-Manzano, C., Gómez-Sánchez, R., Gonçalves, D.A., Goncu, E., Gong, Q., Gongora, C., Gonzalez, C.B., Gonzalez-Alegre, P., Gonzalez-Cabo, P., González-Polo, R.A., Goping, I.S., Gorbea, C., Gorbunov, N.V., Goring, D.R., Gorman, A.M., Gorski, S.M., Goruppi, S., Goto-Yamada, S., Gotor, C., Gottlieb, R.A., Gozes, I., Gozuacik, D., Graba, Y., Graef, M., Granato, G.E., Grant, G.D., Grant, S., Gravina, G.L., Green, D.R., Greenhough, A., Greenwood, M.T., Grimaldi, B., Gros, F., Grose, C., Groulx, J.-F., Gruber, F., Grumati, P., Grune, T., Guan, J.-L., Guan, K.-L., Guerra, B., Guillen, C., Gulshan, K., Gunst, J., Guo, C., Guo, L., Guo, M., Guo, W., Guo, X.-G., Gust, A.A., Gustafsson, Å.B., Gutierrez, E., Gutierrez, M.G., Gwak, H.-S., Haas, A., Haber, J.E., Hadano, S., Hagedorn, M., Hahn, D.R., Halayko, A.J., Hamacher-Brady, A., Hamada, K., Hamai, A., Hamann, A., Hamasaki, M., Hamer, I., Hamid, Q., Hammond, E.M., Han, F., Han, W., Handa, J.T., Hanover, J.A., Hansen, M., Harada, M., Harhaji-Trajkovic, L., Harper, J.W., Harrath, A.H., Harris, A.L., Harris, J., Hasler, U., Hasselblatt, P., Hasui, K., Hawley, R.G., Hawley, T.S., He, C., He, C.Y., He, F., He, G., He, R.-R., He, X.-H., He, Y.-W., He, Y.-Y., Heath, J.K., Hébert, M.-J., Heinzen, R.A., Helgason, G.V., Hensel, M., Henske, E.P., Her, C., Herman, P.K., Hernández, A., Hernandez, C., Hernández-Tiedra, S., Hetz, C., Hiesinger, P.R., Higaki, K., Hilfiker, S., Hill, B.G., Hill, J.A., Hill, W.D., Hino, K., Hofius, D., Hofman, P., Höglinger, G.U., Höhfeld, J., Holz, M.K., Hong, Y., Hood, D.A., Hoozemans, J.J., Hoppe, T., Hsu, C., Hsu, C.-Y., Hsu, L.-C., Hu, D., Hu, G., Hu, H.-M., Hu, H., Hu, M.C., Hu, Y.-C., Hu, Z.-W., Hua, F., Hua, Y., Huang, C., Huang, H.-L., Huang, K.-H., Huang, K.-Y., Huang, S., Huang, S., Huang, W.-P., Huang, Y.-R., Huang, Y., Huang, Y., Huber, T.B., Huebbe, P.,

Huh, W.-K., Hulmi, J.J., Hur, G.M., Hurley, J.H., Husak, Z., Hussain, S.N., Hussain, S., Hwang, J.J., Hwang, S., Hwang, T.I., Ichihara, A., Imai, Y., Imbriano, C., Inomata, M., Into, T., Iovane, V., Iovanna, J.L., Iozzo, R.V., Ip, N.Y., Irazoqui, J.E., Iribarren, P., Isaka, Y., Isakovic, A.J., Ischiropoulos, H., Isenberg, J.S., Ishaq, M., Ishida, H., Ishii, I., Ishmael, J.E., Isidoro, C., Isobe, K.-I., Isono, E., Issazadeh-Navikas, S., Itahana, K., Itakura, E., Ivanov, A.I., Iyer, A.K.V., Izquierdo, J.M., Izumi, Y., Izzo, V., Jäättelä, M., Jaber, N., Jackson, D.J., Jackson, W.T., Jacob, T.G., Jacques, T.S., Jagannath, C., Jain, A., Jana, N.R., Jang, B.K., Jani, A., Janji, B., Jannig, P.R., Jansson, P.J., Jean, S., Jendrach, M., Jeon, J.-H., Jessen, N., Jeung, E.-B., Jia, K., Jia, L., Jiang, H., Jiang, H., Jiang, L., Jiang, T., Jiang, X., Jiang, X., Jiang, X., Jiang, Y., Jiang, Y., Jiménez, A., Jin, C., Jin, H., Jin, L., Jin, M., Jin, S., Jinwal, U.K., Jo, E.-K., Johansen, T., Johnson, D.E., Johnson, G.V., Johnson, J.D., Jonasch, E., Jones, C., Joosten, L.A., Jordan, J., Joseph, A.-M., Joseph, B., Joubert, A.M., Ju, D., Ju, J., Juan, H.-F., Juenemann, K., Juhász, G., Jung, H.S., Jung, J.U., Jung, Y.-K., Jungbluth, H., Justice, M.J., Jutten, B., Kaakoush, N.O., Kaarniranta, K., Kaasik, A., Kabuta, T., Kaeffer, B., Kågedal, K., Kahana, A., Kajimura, S., Kakhlon, O., Kalia, M., Kalvakolanu, D.V., Kamada, Y., Kambas, K., Kaminskyy, V.O., Kampinga, H.H., Kandouz, M., Kang, C., Kang, R., Kang, T.-C., Kanki, T., Kanneganti, T.-D., Kanno, H., Kanthasamy, A.G., Kantorow, M., Kaparakis-Liaskos, M., Kapuy, O., Karantza, V., Karim, M.R., Karmakar, P., Kaser, A., Kaushik, S., Kawula, T., Kaynar, A.M., Ke, P.-Y., Ke, Z.-J., Kehrl, J.H., Keller, K.E., Kemper, J.K., Kenworthy, A.K., Kepp, O., Kern, A., Kesari, S., Kessel, D., Ketteler, R., do Kettelhut, I.C., Khambu, B., Khan, M.M., Khandelwal, V.K., Khare, S., Kiang, J.G., Kiger, A.A., Kihara, A., Kim, A.L., Kim, C.H., Kim, D.R., Kim, D.-H., Kim, E.K., Kim, H.Y., Kim, H.-R., Kim, J.-S., Kim, J.H., Kim, J.C., Kim, J.H., Kim, K.W., Kim, M.D., Kim, M.-M., Kim, P.K., Kim, S.W., Kim, S.-Y., Kim, Y.-S., Kim, Y., Kimchi, A., Kimmelman, A.C., Kimura, T., King, J.S., Kirkegaard, K., Kirkin, V., Kirshenbaum, L.A., Kishi, S., Kitajima, Y., Kitamoto, K., Kitaoka, Y., Kitazato, K., Kley, R.A., Klimecki, W.T., Klinkenberg, M., Klucken, J., Knævelsrud, H., Knecht, E., Knuppertz, L., Ko, J.-L., Kobayashi, S., Koch, J.C., Koechlin-Ramonatxo, C., Koenig, U., Koh, Y.H., Köhler, K., Kohlwein, S.D., Koike, M., Komatsu, M., Kominami, E., Kong, D., Kong, H.J., Konstantakou, E.G., Kopp, B.T., Korcsmaros, T., Korhonen, L., Korolchuk, V.I., Koshkina, N.V., Kou, Y., Koukourakis, M.I., Koumenis, C., Kovács, A.L., Kovács, T., Kovacs, W.J., Koya, D., Kraft, C., Krainc, D., Kramer, H., Kravic-Stevovic, T., Krek, W., Kretz-Remy, C., Krick, R., Krishnamurthy, M., Kriston-Vizi, J., Kroemer, G., Kruer, M.C., Kruger, R., Ktistakis, N.T., Kuchitsu, K., Kuhn, C., Kumar, A.P., Kumar, A., Kumar, A., Kumar, D., Kumar, D., Kumar, R., Kumar, S., Kundu, M., Kung, H.-J., Kuno, A., Kuo, S.-H., Kuret, J., Kurz, T., Kwok, T., Kwon, T.K., Kwon, Y.T., Kyrmizi, I., La Spada, A.R., Lafont, F., Lahm, T., Lakkaraju, A., Lam, T., Lamark, T., Lancel, S., Landowski, T.H., Lane, D.J.R., Lane, J.D., Lanzi, C., Lapaquette, P., Lapierre, L.R., Laporte, J., Laukkarinen, J., Laurie, G.W., Lavandero, S., Lavie, L., LaVoie, M.J., Law, B.Y.K., Law, H.K.-W., Law, K.B., Layfield, R., Lazo, P.A., Le Cam, L., Le Roch, K.G., Le Stunff, H., Leardkamolkarn, V., Lecuit, M., Lee, B.-H., Lee, C.-H., Lee, E.F., Lee, G.M., Lee, H.-J., Lee, H., Lee, J.K., Lee, J., Lee, J.-H., Lee, J.H., Lee, M., Lee, M.-S., Lee, P.J., Lee, S.W., Lee, S.-J., Lee, S.-J., Lee, S.Y., Lee, S.H., Lee, S.S., Lee, S.-J., Lee, S., Lee, Y.-R., Lee, Y.J., Lee, Y.H., Leeuwenburgh, C., Lefort, S., Legouis, R., Lei, J., Lei, Q.-Y., Leib, D.A., Leibowitz, G., Lekli, I., Lemaire, S.D., Lemasters, J.J., Lemberg, M.K., Lemoine, A., Leng, S., Lenz, G., Lenzi, P., Lerman, L.O., Lettieri Barbato, D., Leu, J.I.-J., Leung, H.Y., Levine, B., Lewis, P.A., Lezoualc'h, F., Li, C.,

Li, F., Li, F.-J., Li, J., Li, K., Li, L., Li, M., Li, M., Li, Q., Li, R., Li, S., Li, W., Li, W., Li, X., Li, Y., Lian, J., Liang, C., Liang, Q., Liao, Y., Liberal, J., Liberski, P.P., Lie, P., Lieberman, A.P., Lim, H.J., Lim, K.-L., Lim, K., Lima, R.T., Lin, C.-S., Lin, C.-F., Lin, F., Lin, F., Lin, F.-C., Lin, K., Lin, K.-H., Lin, P.-H., Lin, T., Lin, W.-W., Lin, Y.-S., Lin, Y., Linden, R., Lindholm, D., Lindqvist, L.M., Lingor, P., Linkermann, A., Liotta, L.A., Lipinski, M.M., Lira, V.A., Lisanti, M.P., Liton, P.B., Liu, B., Liu, C., Liu, C.-F., Liu, F., Liu, H.-J., Liu, J., Liu, J.-J., Liu, J.-L., Liu, K., Liu, L., Liu, L., Liu, Q., Liu, R.-Y., Liu, S., Liu, S., Liu, W., Liu, X.-D., Liu, X., Liu, X.-H., Liu, X., Liu, X., Liu, X., Liu, Y., Liu, Y., Liu, Z., Liu, Z., Liuzzi, J.P., Lizard, G., Ljujic, M., Lodhi, I.J., Logue, S.E., Lokeshwar, B.L., Long, Y.C., Lonial, S., Loos, B., López-Otín, C., López-Vicario, C., Lorente, M., Lorenzi, P.L., Lõrincz, P., Los, M., Lotze, M.T., Lovat, P.E., Lu, B., Lu, B., Lu, J., Lu, Q., Lu, S.-M., Lu, S., Lu, Y., Luciano, F., Luckhart, S., Lucocq, J.M., Ludovico, P., Lugea, A., Lukacs, N.W., Lum, J.J., Lund, A.H., Luo, H., Luo, J., Luo, S., Luparello, C., Lyons, T., Ma, J., Ma, Y., Ma, Y., Ma, Z., Machado, J., Machado-Santelli, G.M., Macian, F., MacIntosh, G.C., MacKeigan, J.P., Macleod, K.F., MacMicking, J.D., MacMillan-Crow, L.A., Madeo, F., Madesh, M., Madrigal-Matute, J., Maeda, A., Maeda, T., Maegawa, G., Maellaro, E., Maes, H., Magariños, M., Maiese, K., Maiti, T.K., Maiuri, L., Maiuri, M.C., Maki, C.G., Malli, R., Malorni, W., Maloyan, A., Mami-Chouaib, F., Man, N., Mancias, J.D., Mandelkow, E.-M., Mandell, M.A., Manfredi, A.A., Manié, S.N., Manzoni, C., Mao, K., Mao, Z., Mao, Z.-W., Marambaud, P., Marconi, A.M., Marelja, Z., Marfe, G., Margeta, M., Margittai, E., Mari, M., Mariani, F.V., Marin, C., Marinelli, S., Mariño, G., Markovic, I., Marquez, R., Martelli, A.M., Martens, S., Martin, K.R., Martin, S.J., Martin, S., Martin-Acebes, M.A., Martín-Sanz, P., Martinand-Mari, C., Martinet, W., Martinez, J., Martinez-Lopez, N., Martinez-Outschoorn, U., Martínez-Velázquez, M., Martinez-Vicente, M., Martins, W.K., Mashima, H., Mastrianni, J.A., Matarese, G., Matarrese, P., Mateo, R., Matoba, S., Matsumoto, N., Matsushita, T., Matsuura, A., Matsuzawa, T., Mattson, M.P., Matus, S., Maugeri, N., Mauvezin, C., Mayer, A., Maysinger, D., Mazzolini, G.D., McBrayer, M.K., McCall, K., McCormick, C., McInerney, G.M., McIver, S.C., McKenna, S., McMahon, J.J., McNeish, I.A., Mechta-Grigoriou, F., Medema, J.P., Medina, D.L., Megyeri, K., Mehrpour, M., Mehta, J.L., Mei, Y., Meier, U.-C., Meijer, A.J., Meléndez, A., Melino, G., Melino, S., de Melo, E.J.T., Mena, M.A., Meneghini, M.D., Menendez, J.A., Menezes, R., Meng, L., Meng, L.-H., Meng, S., Menghini, R., Menko, A.S., Menna-Barreto, R.F., Menon, M.B., Meraz-Ríos, M.A., Merla, G., Merlini, L., Merlot, A.M., Meryk, A., Meschini, S., Meyer, J.N., Mi, M.-T., Miao, C.-Y., Micale, L., Michaeli, S., Michiels, C., Migliaccio, A.R., Mihailidou, A.S., Mijaljica, D., Mikoshiba, K., Milan, E., Miller-Fleming, L., Mills, G.B., Mills, I.G., Minakaki, G., Minassian, B.A., Ming, X.-F., Minibayeva, F., Minina, E.A., Mintern, J.D., Minucci, S., Miranda-Vizuete, A., Mitchell, C.H., Miyamoto, S., Miyazawa, K., Mizushima, N., Mnich, K., Mograbi, B., Mohseni, S., Moita, L.F., Molinari, M., Molinari, M., Møller, A.B., Mollereau, B., Mollinedo, F., Mongillo, M., Monick, M.M., Montagnaro, S., Montell, C., Moore, D.J., Moore, M.N., Mora-Rodriguez, R., Moreira, P.I., Morel, E., Morelli, M.B., Moreno, S., Morgan, M.J., Moris, A., Moriyasu, Y., Morrison, J.L., Morrison, L.A., Morselli, E., Moscat, J., Moseley, P.L., Mostowy, S., Motori, E., Mottet, D., Mottram, J.C., Moussa, C.E.-H., Mpakou, V.E., Mukhtar, H., Mulcahy Levy, J.M., Muller, S., Muñoz-Moreno, R., Muñoz-Pinedo, C., Münz, C., Murphy, M.E., Murray, J.T., Murthy, A., Mysorekar, I.U., Nabi, I.R., Nabissi, M., Nader, G.A., Nagahara, Y., Nagai, Y., Nagata, K., Nagelkerke, A., Nagy, P., Naidu, S.R.,

Nair, S., Nakano, H., Nakatogawa, H., Nanjundan, M., Napolitano, G., Naqvi, N.I., Nardacci, R., Narendra, D.P., Narita, M., Nascimbeni, A.C., Natarajan, R., Navegantes, L.C., Nawrocki, S.T., Nazarko, T.Y., Nazarko, V.Y., Neill, T., Neri, L.M., Netea, M.G., Netea-Maier, R.T., Neves, B.M., Ney, P.A., Nezis, I.P., Nguyen, H.T., Nguyen, H.P., Nicot, A.-S., Nilsen, H., Nilsson, P., Nishimura, M., Nishino, I., Niso-Santano, M., Niu, H., Nixon, R.A., Njar, V.C., Noda, T., Noegel, A.A., Nolte, E.M., Norberg, E., Norga, K.K., Noureini, S.K., Notomi, S., Notterpek, L., Nowikovsky, K., Nukina, N., Nürnberger, T., O'Donnell, V.B., O'Donovan, T., O'Dwyer, P.J., Oehme, I., Oeste, C.L., Ogawa, M., Ogretmen, B., Ogura, Y., Oh, Y.J., Ohmuraya, M., Ohshima, T., Ojha, R., Okamoto, K., Okazaki, T., Oliver, F.J., Ollinger, K., Olsson, S., Orban, D.P., Ordonez, P., Orhon, I., Orosz, L., O'Rourke, E.J., Orozco, H., Ortega, A.L., Ortona, E., Osellame, L.D., Oshima, J., Oshima, S., Osiewacz, H.D., Otomo, T., Otsu, K., Ou, J.-H.J., Outeiro, T.F., Ouyang, D.-Y., Ouyang, H., Overholtzer, M., Ozbun, M.A., Ozdinler, P.H., Ozpolat, B., Pacelli, C., Paganetti, P., Page, G., Pages, G., Pagnini, U., Pajak, B., Pak, S.C., Pakos-Zebrucka, K., Pakpour, N., Palková, Z., Palladino, F., Pallauf, K., Pallet, N., Palmieri, M., Paludan, S.R., Palumbo, C., Palumbo, S., Pampliega, O., Pan, H., Pan, W., Panaretakis, T., Pandey, A., Pantazopoulou, A., Papackova, Z., Papademetrio, D.L., Papassideri, I., Papini, A., Parajuli, N., Pardo, J., Parekh, V.V., Parenti, G., Park, J.-I., Park, J., Park, O.K., Parker, R., Parlato, R., Parys, J.B., Parzych, K.R., Pasquet, J.-M., Pasquier, B., Pasumarthi, K.B., Patschan, D., Patterson, C., Pattingre, S., Pattison, S., Pause, A., Pavenstädt, H., Pavone, F., Pedrozo, Z., Peña, F.J., Peñalva, M.A., Pende, M., Peng, J., Penna, F., Penninger, J.M., Pensalfini, A., Pepe, S., Pereira, G.J., Pereira, P.C., Pérez-de la Cruz, V., Pérez-Pérez, M.E., Pérez-Rodríguez, D., Pérez-Sala, D., Perier, C., Perl, A., Perlmutter, D.H., Perrotta, I., Pervaiz, S., Pesonen, M., Pessin, J.E., Peters, G.J., Petersen, M., Petrache, I., Petrof, B.J., Petrovski, G., Phang, J.M., Piacentini, M., Pierdominici, M., Pierre, P., Pierrefite-Carle, V., Pietrocola, F., Pimentel-Muiños, F.X., Pinar, M., Pineda, B., Pinkas-Kramarski, R., Pinti, M., Pinton, P., Piperdi, B., Piret, J.M., Platanias, L.C., Platta, H.W., Plowey, E.D., Pöggeler, S., Poirot, M., Polčic, P., Poletti, A., Poon, A.H., Popelka, H., Popova, B., Poprawa, I., Poulose, S.M., Poulton, J., Powers, S.K., Powers, T., Pozuelo-Rubio, M., Prak, K., Prange, R., Prescott, M., Priault, M., Prince, S., Proia, R.L., Proikas-Cezanne, T., Prokisch, H., Promponas, V.J., Przyklenk, K., Puertollano, R., Pugazhenthi, S., Puglielli, L., Pujol, A., Puyal, J., Pyeon, D., Qi, X., Qian, W.-B., Qin, Z.-H., Qiu, Y., Qu, Z., Quadrilatero, J., Quinn, F., Raben, N., Rabinowich, H., Radogna, F., Ragusa, M.J., Rahmani, M., Raina, K., Ramanadham, S., Ramesh, R., Rami, A., Randall-Demllo, S., Randow, F., Rao, H., Rao, V.A., Rasmussen, B.B., Rasse, T.M., Ratovitski, E.A., Rautou, P.-E., Ray, S.K., Razani, B., Reed, B.H., Reggiori, F., Rehm, M., Reichert, A.S., Rein, T., Reiner, D.J., Reits, E., Ren, J., Ren, X., Renna, M., Reusch, J.E., Revuelta, J.L., Reyes, L., Rezaie, A.R., Richards, R.I., Richardson, D.R., Richetta, C., Riehle, M.A., Rihn, B.H., Rikihisa, Y., Riley, B.E., Rimbach, G., Rippo, M.R., Ritis, K., Rizzi, F., Rizzo, E., Roach, P.J., Robbins, J., Roberge, M., Roca, G., Roccheri, M.C., Rocha, S., Rodrigues, C.M., Rodríguez, C.I., de Cordoba, S.R., Rodriguez-Muela, N., Roelofs, J., Rogov, V.V., Rohn, T.T., Rohrer, B., Romanelli, D., Romani, L., Romano, P.S., Roncero, M.I.G., Rosa, J.L., Rosello, A., Rosen, K.V., Rosenstiel, P., Rost-Roszkowska, M., Roth, K.A., Roué, G., Rouis, M., Rouschop, K.M., Ruan, D.T., Ruano, D., Rubinsztein, D.C., Rucker, E.B., Rudich, A., Rudolf, E., Rudolf, R., Ruegg, M.A., Ruiz-Roldan, C., Ruparelia, A.A., Rusmini, P., Russ, D.W., Russo, G.L., Russo, G., Russo, R., Rusten, T.E., Ryabovol, V.,

Tian, L., Till, A., Ting, J.P.-Y., Titorenko, V.I., Toker, L., Toldo, S., Tooze, S.A., Topisirovic, I., Torgersen, M.L., Torosantucci, L., Torriglia, A., Torrisi, M.R., Tournier, C., Towns, R., Trajkovic, V., Travassos, L.H., Triola, G., Tripathi, D.N., Trisciuoglio, D., Troncoso, R., Trougakos, I.P., Truttmann, A.C., Tsai, K.-J., Tschan, M.P., Tseng, Y.-H., Tsukuba, T., Tsung, A., Tsvetkov, A.S., Tu, S., Tuan, H.-Y., Tucci, M., Tumbarello, D.A., Turk, B., Turk, V., Turner, R.F., Tveita, A.A., Tyagi, S.C., Ubukata, M., Uchiyama, Y., Udelnow, A., Ueno, T., Umekawa, M., Umemiya-Shirafuji, R., Underwood, B.R., Ungermann, C., Ureshino, R.P., Ushioda, R., Uversky, V.N., Uzcátegui, N.L., Vaccari, T., Vaccaro, M.I., Váchová, L., Vakifahmetoglu-Norberg, H., Valdor, R., Valente, E.M., Vallette, F., Valverde, A.M., Van den Berghe, G., Van Den Bosch, L., van den Brink, G.R., van der Goot, F.G., van der Klei, I.J., van der Laan, L.J., van Doorn, W.G., van Egmond, M., van Golen, K.L., Van Kaer, L., van Lookeren Campagne, M., Vandenabeele, P., Vandenberghe, W., Vanhorebeek, I., Varela-Nieto, I., Vasconcelos, M.H., Vasko, R., Vavvas, D.G., Vega-Naredo, I., Velasco, G., Velentzas, A.D., Velentzas, P.D., Vellai, T., Vellenga, E., Vendelbo, M.H., Venkatachalam, K., Ventura, N., Ventura, S., Veras, P.S., Verdier, M., Vertessy, B.G., Viale, A., Vidal, M., Vieira, H.L.A., Vierstra, R.D., Vigneswaran, N., Vij, N., Vila, M., Villar, M., Villar, V.H., Villarroya, J., Vindis, C., Viola, G., Viscomi, M.T., Vitale, G., Vogl, D.T., Voitsekhovskaja, O.V., von Haefen, C., von Schwarzenberg, K., Voth, D.E., Vouret-Craviari, V., Vuori, K., Vyas, J.M., Waeber, C., Walker, C.L., Walker, M.J., Walter, J., Wan, L., Wan, X., Wang, B., Wang, C., Wang, C.-Y., Wang, C., Wang, C., Wang, C., Wang, D., Wang, F., Wang, F., Wang, G., Wang, H.-J., Wang, H., Wang, H.-G., Wang, H., Wang, H.-D., Wang, J., Wang, J., Wang, M., Wang, M.-Q., Wang, P.-Y., Wang, P., Wang, R.C., Wang, S., Wang, T.-F., Wang, X., Wang, X.-J., Wang, X.-W., Wang, X., Wang, X., Wang, Y., Wang, Y., Wang, Y., Wang, Y.-J., Wang, Y., Wang, Y., Wang, Y.T., Wang, Y., Wang, Z.-N., Wappner, P., Ward, C., Ward, D.M., Warnes, G., Watada, H., Watanabe, Y., Watase, K., Weaver, T.E., Weekes, C.D., Wei, J., Weide, T., Weihl, C.C., Weindl, G., Weis, S.N., Wen, L., Wen, X., Wen, Y., Westermann, B., Weyand, C.M., White, A.R., White, E., Whitton, J.L., Whitworth, A.J., Wiels, J., Wild, F., Wildenberg, M.E., Wileman, T., Wilkinson, D.S., Wilkinson, S., Willbold, D., Williams, C., Williams, K., Williamson, P.R., Winklhofer, K.F., Witkin, S.S., Wohlgemuth, S.E., Wollert, T., Wolvetang, E.J., Wong, E., Wong, G.W., Wong, R.W., Wong, V.K.W., Woodcock, E.A., Wright, K.L., Wu, C., Wu, D., Wu, G.S., Wu, J., Wu, J., Wu, M., Wu, M., Wu, S., Wu, W.K., Wu, Y., Wu, Z., Xavier, C.P., Xavier, R.J., Xia, G.-X., Xia, T., Xia, W., Xia, Y., Xiao, H., Xiao, J., Xiao, S., Xiao, W., Xie, C.-M., Xie, Z., Xie, Z., Xilouri, M., Xiong, Y., Xu, C., Xu, C., Xu, F., Xu, H., Xu, H., Xu, J., Xu, J., Xu, J., Xu, L., Xu, X., Xu, Y., Xu, Y., Xu, Z.-X., Xu, Z., Xue, Y., Yamada, T., Yamamoto, A., Yamanaka, K., Yamashina, S., Yamashiro, S., Yan, B., Yan, B., Yan, X., Yan, Z., Yanagi, Y., Yang, D.-S., Yang, J.-M., Yang, L., Yang, M., Yang, P.-M., Yang, P., Yang, Q., Yang, W., Yang, W.Y., Yang, X., Yang, Y., Yang, Y., Yang, Z., Yang, Z., Yao, M.-C., Yao, P.J., Yao, X., Yao, Z., Yao, Z., Yasui, L.S., Ye, M., Yedvobnick, B., Yeganeh, B., Yeh, E.S., Yeyati, P.L., Yi, F., Yi, L., Yin, X.-M., Yip, C.K., Yoo, Y.-M., Yoo, Y.H., Yoon, S.-Y., Yoshida, K.-I., Yoshimori, T., Young, K.H., Yu, H., Yu, J.J., Yu, J.-T., Yu, J., Yu, L., Yu, W.H., Yu, X.-F., Yu, Z., Yuan, J., Yuan, Z.-M., Yue, B.Y., Yue, J., Yue, Z., Zacks, D.N., Zacksenhaus, E., Zaffaroni, N., Zaglia, T., Zakeri, Z., Zecchini, V., Zeng, J., Zeng, M., Zeng, Q., Zervos, A.S., Zhang, D.D., Zhang, F., Zhang, G., Zhang, G.-C., Zhang, H., Zhang, H., Zhang, H., Zhang, H., Zhang, J., Zhang, J.,

Zhang, J., Zhang, J., Zhang, J.-P., Zhang, L., Zhang, L., Zhang, L., Zhang, L., Zhang, M.-Y., Zhang, X., Zhang, X.D., Zhang, Y., Zhang, Y., Zhang, Y., Zhang, Y., Zhang, Y., Zhao, M., Zhao, W.-L., Zhao, X., Zhao, Y.G., Zhao, Y., Zhao, Y., Zhao, Y.-X., Zhao, Z., Zhao, Z.J., Zheng, D., Zheng, X.-L., Zheng, X., Zhivotovsky, B., Zhong, Q., Zhou, G.-Z., Zhou, G., Zhou, H., Zhou, S.-F., Zhou, X.-J., Zhu, H., Zhu, H., Zhu, W.-G., Zhu, W., Zhu, X.-F., Zhu, Y., Zhuang, S.-M., Zhuang, X., Ziparo, E., Zois, C.E., Zoladek, T., Zong, W.-X., Zorzano, A., Zughaier, S.M., 2016. Guidelines for the use and interpretation of assays for monitoring autophagy (3rd edition). Autophagy 12, 1–222. https://doi.org/10. 1080/15548627.2015.1100356.

Krogsaeter, E.K., Biel, M., Wahl-Schott, C., Grimm, C., 2018. The protein interaction networks of mucolipins and two-pore channels. Biochim. Biophys. Acta Mol. Cell Res. 1866, 1111–1123. https://doi.org/10.1016/j.bbamcr.2018.10.020.

Lagostena, L., Festa, M., Pusch, M., Carpaneto, A., 2017. The human two-pore channel 1 is modulated by cytosolic and luminal calcium. Sci. Rep. 7, 43900. https://doi.org/10. 1038/srep43900.

LaPlante, J.M., Falardeau, J., Sun, M., Kanazirska, M., Brown, E.M., Slaugenhaupt, S.A., Vassilev, P.M., 2002. Identification and characterization of the single channel function of human mucolipin-1 implicated in mucolipidosis type IV, a disorder affecting the lysosomal pathway. FEBS Lett. 532, 183–187. https://doi.org/10.1016/S0014-5793(02) 03670-0.

Li, X., Rydzewski, N., Hider, A., Zhang, X., Yang, J., Wang, W., Gao, Q., Cheng, X., Xu, H., 2016a. A molecular mechanism to regulate lysosome motility for lysosome positioning and tubulation. Nat. Cell Biol. 18, 404–417. https://doi.org/10.1038/ncb3324.

Li, R.-J., Xu, J., Fu, C., Zhang, J., Zheng, Y.G., Jia, H., Liu, J.O., 2016b. Regulation of mTORC1 by lysosomal calcium and calmodulin. eLife 5, e19360. https://doi.org/10. 7554/eLife.19360.

Liang, C., Lee, J., Inn, K.-S., Gack, M.U., Li, Q., Roberts, E.A., Vergne, I., Deretic, V., Feng, P., Akazawa, C., Jung, J.U., 2008. Beclin1-binding UVRAG targets the class C Vps complex to coordinate autophagosome maturation and endocytic trafficking. Nat. Cell Biol. 10, 776–787. https://doi.org/10.1038/ncb1740.

Lin-Moshier, Y., Keebler, M.V., Hooper, R., Boulware, M.J., Liu, X., Churamani, D., Abood, M.E., Walseth, T.F., Brailoiu, E., Patel, S., Marchant, J.S., 2014. The two-pore channel (TPC) interactome unmasks isoform-specific roles for TPCs in endolysosomal morphology and cell pigmentation. Proc. Natl. Acad. Sci. U. S. A. 111, 13087–13092. https://doi.org/10.1073/pnas.1407004111.

Luzio, J.P., Bright, N.A., Pryor, P.R., 2007a. The role of calcium and other ions in sorting and delivery in the late endocytic pathway. Biochem. Soc. Trans. 35, 1088–1091. https://doi.org/10.1042/BST0351088.

Luzio, J.P., Pryor, P.R., Bright, N.A., 2007b. Lysosomes: fusion and function. Nat. Rev. Mol. Cell Biol. 8, 622–632. https://doi.org/10.1038/nrm2217.

Maejima, I., Takahashi, A., Omori, H., Kimura, T., Takabatake, Y., Saitoh, T., Yamamoto, A., Hamasaki, M., Noda, T., Isaka, Y., Yoshimori, T., 2013. Autophagy sequesters damaged lysosomes to control lysosomal biogenesis and kidney injury. EMBO J. 32, 2336–2347. https://doi.org/10.1038/emboj.2013.171.

Manzoni, M., Monti, E., Bresciani, R., Bozzato, A., Barlati, S., Bassi, M.T., Borsani, G., 2004. Overexpression of wild-type and mutant mucolipin proteins in mammalian cells: effects on the late endocytic compartment organization. FEBS Lett. 567, 219–224. https://doi.org/10.1016/j.febslet.2004.04.080.

Marchant, J.S., Patel, S., 2015. Two-pore channels at the intersection of endolysosomal membrane traffic. Biochem. Soc. Trans. 43, 434–441. https://doi.org/10.1042/ BST20140303.

Matsunaga, K., Saitoh, T., Tabata, K., Omori, H., Satoh, T., Kurotori, N., Maejima, I., Shirahama-Noda, K., Ichimura, T., Isobe, T., Akira, S., Noda, T., Yoshimori, T., 2009. Two Beclin 1-binding proteins, Atg14L and Rubicon, reciprocally regulate autophagy at different stages. Nat. Cell Biol. 11, 385–396. https://doi.org/10.1038/ncb1846.

Medina, D.L., Di Paola, S., Peluso, I., Armani, A., De Stefani, D., Venditti, R., Montefusco, S., Scotto-Rosato, A., Prezioso, C., Forrester, A., Settembre, C., Wang, W., Gao, Q., Xu, H., Sandri, M., Rizzuto, R., De Matteis, M.A., Ballabio, A., 2015. Lysosomal calcium signalling regulates autophagy through calcineurin and TFEB. Nat. Cell Biol. 17, 288–299. https://doi.org/10.1038/ncb3114.

Medina, D.L., Fraldi, A., Bouche, V., Annunziata, F., Mansueto, G., Spampanato, C., Puri, C., Pignata, A., Martina, J.A., Sardiello, M., Palmieri, M., Polishchuk, R., Puertollano, R., Ballabio, A., 2011. Transcriptional activation of lysosomal exocytosis promotes cellular clearance. Dev. Cell 21, 421–430. https://doi.org/10.1016/j.devcel. 2011.07.016.

Micsenyi, M.C., Dobrenis, K., Stephney, G., Pickel, J., Vanier, M.T., Slaugenhaupt, S.A., Walkley, S.U., 2009. Neuropathology of the Mcoln1(−/−) knockout mouse model of mucolipidosis type IV. J. Neuropathol. Exp. Neurol. 68, 125–135. https://doi.org/10. 1097/NEN.0b013e3181942cf0.

Monastyrska, I., Rieter, E., Klionsky, D.J., Reggiori, F., 2009. Multiple roles of the cytoskeleton in autophagy. Biol. Rev. Camb. Philos. Soc. 84, 431–448. https://doi.org/ 10.1111/j.1469-185X.2009.00082.x.

Morgan, A.J., Platt, F.M., Lloyd-Evans, E., Galione, A., 2011. Molecular mechanisms of endolysosomal Ca2+ signalling in health and disease. Biochem. J. 439, 349–374. https://doi.org/10.1042/BJ20110949.

Nakamura, S., Shigeyama, S., Minami, S., Shima, T., Akayama, S., Matsuda, T., Esposito, A., Napolitano, G., Kuma, A., Namba-Hamano, T., Nakamura, J., Yamamoto, K., Sasai, M., Tokumura, A., Miyamoto, M., Oe, Y., Fujita, T., Terawaki, S., Takahashi, A., Hamasaki, M., Yamamoto, M., Okada, Y., Komatsu, M., Nagai, T., Takabatake, Y., Xu, H., Isaka, Y., Ballabio, A., Yoshimori, T., 2020. LC3 lipidation is essential for TFEB activation during the lysosomal damage response to kidney injury. Nat. Cell Biol. 22, 1252–1263. https://doi.org/10.1038/s41556-020-00583-9.

Nazio, F., Strappazzon, F., Antonioli, M., Bielli, P., Cianfanelli, V., Bordi, M., Gretzmeier, C., Dengjel, J., Piacentini, M., Fimia, G.M., Cecconi, F., 2013. mTOR inhibits autophagy by controlling ULK1 ubiquitylation, self-association and function through AMBRA1 and TRAF6. Nat. Cell Biol. 15, 406–416. https://doi.org/10. 1038/ncb2708.

Nicklin, P., Bergman, P., Zhang, B., Triantafellow, E., Wang, H., Nyfeler, B., Yang, H., Hild, M., Kung, C., Wilson, C., Myer, V.E., MacKeigan, J.P., Porter, J.A., Wang, Y.K., Cantley, L.C., Finan, P.M., Murphy, L.O., 2009. Bidirectional transport of amino acids regulates mTOR and autophagy. Cell 136, 521–534. https://doi.org/10. 1016/j.cell.2008.11.044.

Ogunbayo, O.A., Duan, J., Xiong, J., Wang, Q., Feng, X., Ma, J., Zhu, M.X., Evans, A.M., 2018. mTORC1 controls lysosomal Ca2+ release through the two-pore channel TPC2. Sci. Signal. 11, 5775. https://doi.org/10.1126/scisignal.aao5775.

Onyenwoke, R.U., Sexton, J.Z., Yan, F., Díaz, M.C.H., Forsberg, L.J., Major, M.B., Brenman, J.E., 2015. The mucolipidosis IV Ca2+ channel TRPML1 (MCOLN1) is regulated by the TOR kinase. Biochem. J. 470, 331–342. https://doi.org/10.1042/ BJ20150219.

Orsi, A., Razi, M., Dooley, H.C., Robinson, D., Weston, A.E., Collinson, L.M., Tooze, S.A., 2012. Dynamic and transient interactions of Atg9 with autophagosomes, but not membrane integration, are required for autophagy. Mol. Biol. Cell 23, 1860–1873. https://doi.org/10.1091/mbc.E11-09-0746.

Papinski, D., Schuschnig, M., Reiter, W., Wilhelm, L., Barnes, C.A., Maiolica, A., Hansmann, I., Pfaffenwimmer, T., Kijanska, M., Stoffel, I., Lee, S.S., Brezovich, A., Lou, J.H., Turk, B.E., Aebersold, R., Ammerer, G., Peter, M., Kraft, C., 2014. Early steps in autophagy depend on direct phosphorylation of Atg9 by the Atg1 kinase. Mol. Cell 53, 471–483. https://doi.org/10.1016/j.molcel.2013.12.011.

Patel, S., Kilpatrick, B.S., 2018. Two-pore channels and disease. Biochim. Biophys. Acta Mol. Cell Res. 1865, 1678–1686. https://doi.org/10.1016/j.bbamcr.2018.05.004.

Pereira, G.J.S., Antonioli, M., Hirata, H., Ureshino, R.P., Nascimento, A.R., Bincoletto, C., Vescovo, T., Piacentini, M., Fimia, G.M., Smaili, S.S., 2017. Glutamate induces autophagy via the two-pore channels in neural cells. Oncotarget 8, 12730–12740. https://doi.org/10.18632/oncotarget.14404.

Pereira, G.J.S., Hirata, H., Fimia, G.M., do Carmo, L.G., Bincoletto, C., Han, S.W., Stilhano, R.S., Ureshino, R.P., Bloor-Young, D., Churchill, G., Piacentini, M., Patel, S., Smaili, S.S., 2011. Nicotinic acid adenine dinucleotide phosphate (NAADP) regulates autophagy in cultured astrocytes. J. Biol. Chem. 286, 27875–27881. https://doi.org/10.1074/jbc.C110.216580.

Plesch, E., Chen, C.-C., Butz, E., Scotto Rosato, A., Krogsaeter, E.K., Yinan, H., Bartel, K., Keller, M., Robaa, D., Teupser, D., Holdt, L.M., Vollmar, A.M., Sippl, W., Puertollano, R., Medina, D., Biel, M., Wahl-Schott, C., Bracher, F., Grimm, C., 2018. Selective agonist of TRPML2 reveals direct role in chemokine release from innate immune cells. eLife 7, e9720. https://doi.org/10.7554/eLife.39720.

Proikas-Cezanne, T., Takacs, Z., Dönnes, P., Kohlbacher, O., 2015. WIPI proteins: essential PtdIns3P effectors at the nascent autophagosome. J. Cell Sci. 128, 207–217. https://doi.org/10.1242/jcs.146258.

Pryor, P.R., Mullock, B.M., Bright, N.A., Gray, S.R., Luzio, J.P., 2000. The role of intra-organellar Ca(2+) in late endosome-lysosome heterotypic fusion and in the reformation of lysosomes from hybrid organelles. J. Cell Biol. 149, 1053–1062. https://doi.org/10.1083/jcb.149.5.1053.

Pryor, P.R., Reimann, F., Gribble, F.M., Luzio, J.P., 2006. Mucolipin-1 is a lysosomal membrane protein required for intracellular lactosylceramide traffic. Traffic 7, 1388–1398. https://doi.org/10.1111/j.1600-0854.2006.00475.x.

Rodríguez, A., Webster, P., Ortego, J., Andrews, N.W., 1997. Lysosomes behave as Ca2+−regulated exocytic vesicles in fibroblasts and epithelial cells. J. Cell Biol. 137, 93–104. https://doi.org/10.1083/jcb.137.1.93.

Rosato, A.S., Tang, R., Grimm, C., 2020. Two-pore and TRPML cation channels: regulators of phagocytosis, autophagy and lysosomal exocytosis. Pharmacol. Ther. 220, 107713. https://doi.org/10.1016/j.pharmthera.2020.107713.

Ruas, M., Rietdorf, K., Arredouani, A., Davis, L.C., Lloyd-Evans, E., Koegel, H., Funnell, T.M., Morgan, A.J., Ward, J.A., Watanabe, K., Cheng, X., Churchill, G.C., Zhu, M.X., Platt, F.M., Wessel, G.M., Parrington, J., Galione, A., 2010. Purified TPC isoforms form NAADP receptors with distinct roles for Ca(2+) signaling and endolysosomal trafficking. Curr. Biol. 20, 703–709. https://doi.org/10.1016/j.cub.2010.02.049.

Sakurai, Y., Kolokoltsov, A.A., Chen, C.-C., Tidwell, M.W., Bauta, W.E., Klugbauer, N., Grimm, C., Wahl-Schott, C., Biel, M., Davey, R.A., 2015. Two-pore channels control Ebola virus host cell entry and are drug targets for disease treatment. Science 347, 995–998. https://doi.org/10.1126/science.1258758.

Schaheen, L., Dang, H., Fares, H., 2006. Basis of lethality in C. elegans lacking CUP-5, the Mucolipidosis type IV orthologue. Dev. Biol. 293, 382–391. https://doi.org/10.1016/j.ydbio.2006.02.008.

Scotto Rosato, A., Montefusco, S., Soldati, C., Di Paola, S., Capuozzo, A., Monfregola, J., Polishchuk, E., Amabile, A., Grimm, C., Lombardo, A., De Matteis, M.A., Ballabio, A.,

Medina, D.L., 2019. TRPML1 links lysosomal calcium to autophagosome biogenesis through the activation of the CaMKKβ/VPS34 pathway. Nat. Commun. 10, 5630. https://doi.org/10.1038/s41467-019-13572-w.

Settembre, C., Di Malta, C., Polito, V.A., Garcia Arencibia, M., Vetrini, F., Erdin, S., Erdin, S.U., Huynh, T., Medina, D., Colella, P., Sardiello, M., Rubinsztein, D.C., Ballabio, A., 2011. TFEB links autophagy to lysosomal biogenesis. Science 332, 1429–1433. https://doi.org/10.1126/science.1204592.

Settembre, C., Zoncu, R., Medina, D.L., Vetrini, F., Erdin, S., Erdin, S.U., Huynh, T., Ferron, M., Karsenty, G., Vellard, M.C., Facchinetti, V., Sabatini, D.M., Ballabio, A., 2012. A lysosome-to-nucleus signalling mechanism senses and regulates the lysosome via mTOR and TFEB. EMBO J. 31, 1095–1108. https://doi.org/10.1038/emboj.2012.32.

Stanley, R.E., Ragusa, M.J., Hurley, J.H., 2014. The beginning of the end: how scaffolds nucleate autophagosome biogenesis. Trends Cell Biol. 24, 73–81. https://doi.org/10.1016/j.tcb.2013.07.008.

Stokłosa, P., Probst, D., Reymond, J.-L., Peinelt, C., 2020. The name tells the story: two-pore channels. Cell Calcium 89, 102215. https://doi.org/10.1016/j.ceca.2020.102215.

Sun, Q., Fan, W., Chen, K., Ding, X., Chen, S., Zhong, Q., 2008. Identification of Barkor as a mammalian autophagy-specific factor for Beclin 1 and class III phosphatidylinositol 3-kinase. Proc. Natl. Acad. Sci. U. S. A. 105, 19211–19216. https://doi.org/10.1073/pnas.0810452105.

Sun, T., Wang, X., Lu, Q., Ren, H., Zhang, H., 2011. CUP-5, the C. elegans ortholog of the mammalian lysosomal channel protein MLN1/TRPML1, is required for proteolytic degradation in autolysosomes. Autophagy 7, 1308–1315. https://doi.org/10.4161/auto.7.11.17759.

Sun, X., Yang, Y., Zhong, X.Z., Cao, Q., Zhu, X.-H., Zhu, X., Dong, X.-P., 2018. A negative feedback regulation of MTORC1 activity by the lysosomal Ca2+ channel MCOLN1 (mucolipin 1) using a CALM (calmodulin)-dependent mechanism. Autophagy 14, 38–52. https://doi.org/10.1080/15548627.2017.1389822.

Sun, W., Yue, J., 2018. TPC2 mediates autophagy progression and extracellular vesicle secretion in cancer cells. Exp. Cell Res. 370, 478–489. https://doi.org/10.1016/j.yexcr.2018.07.013.

Venkatachalam, K., Long, A.A., Elsaesser, R., Nikolaeva, D., Broadie, K., Montell, C., 2008. Motor deficit in a Drosophila model of mucolipidosis type IV due to defective clearance of apoptotic cells. Cell 135, 838–851. https://doi.org/10.1016/j.cell.2008.09.041.

Venugopal, B., Mesires, N.T., Kennedy, J.C., Curcio-Morelli, C., Laplante, J.M., Dice, J.F., Slaugenhaupt, S.A., 2009. Chaperone-mediated autophagy is defective in mucolipidosis type IV. J. Cell. Physiol. 219, 344–353. https://doi.org/10.1002/jcp.21676.

Vergarajauregui, S., Connelly, P.S., Daniels, M.P., Puertollano, R., 2008. Autophagic dysfunction in mucolipidosis type IV patients. Hum. Mol. Genet. 17, 2723–2737. https://doi.org/10.1093/hmg/ddn174.

Vergarajauregui, S., Martina, J.A., Puertollano, R., 2009. Identification of the penta-EF-hand protein ALG-2 as a Ca2+-dependent interactor of mucolipin-1. J. Biol. Chem. 284, 36357–36366. https://doi.org/10.1074/jbc.M109.047241.

Vergarajauregui, S., Puertollano, R., 2006. Two di-leucine motifs regulate trafficking of mucolipin-1 to lysosomes. Traffic 7, 337–353. https://doi.org/10.1111/j.1600-0854.2006.00387.x.

Wang, W., Gao, Q., Yang, M., Zhang, X., Yu, L., Lawas, M., Li, X., Bryant-Genevier, M., Southall, N.T., Marugan, J., Ferrer, M., Xu, H., 2015. Up-regulation of lysosomal TRPML1 channels is essential for lysosomal adaptation to nutrient starvation. Proc. Natl. Acad. Sci. U. S. A. 112, E1373–E1381. https://doi.org/10.1073/pnas.1419669112.

Wang, W., Zhang, X., Gao, Q., Lawas, M., Yu, L., Cheng, X., Gu, M., Sahoo, N., Li, X., Li, P., Ireland, S., Meredith, A., Xu, H., 2017. A voltage-dependent K+ channel in the lysosome is required for refilling lysosomal Ca2+ stores. J. Cell Biol. 216, 1715–1730. https://doi.org/10.1083/jcb.201612123.

Wong, C.-O., Li, R., Montell, C., Venkatachalam, K., 2012. Drosophila TRPML is required for TORC1 activation. Curr. Biol. 22, 1616–1621. https://doi.org/10.1016/j.cub.2012.06.055.

Xu, H., Ren, D., 2015. Lysosomal physiology. Annu. Rev. Physiol. 77, 57–80. https://doi.org/10.1146/annurev-physiol-021014-071649.

Yang, J., Zhao, Z., Gu, M., Feng, X., Xu, H., 2019. Release and uptake mechanisms of vesicular Ca2+ stores. Protein Cell 10, 8–19. https://doi.org/10.1007/s13238-018-0523-x.

Yu, L., McPhee, C.K., Zheng, L., Mardones, G.A., Rong, Y., Peng, J., Mi, N., Zhao, Y., Liu, Z., Wan, F., Hailey, D.W., Oorschot, V., Klumperman, J., Baehrecke, E.H., Lenardo, M.J., 2010. Termination of autophagy and reformation of lysosomes regulated by mTOR. Nature 465, 942–946. https://doi.org/10.1038/nature09076.

Zhang, X., Chen, W., Gao, Q., Yang, J., Yan, X., Zhao, H., Su, L., Yang, M., Gao, C., Yao, Y., Inoki, K., Li, D., Shao, R., Wang, S., Sahoo, N., Kudo, F., Eguchi, T., Ruan, B., Xu, H., 2019. Rapamycin directly activates lysosomal mucolipin TRP channels independent of mTOR. PLoS Biol. 17, e3000252. https://doi.org/10.1371/journal.pbio.3000252.

Zhang, X., Cheng, X., Yu, L., Yang, J., Calvo, R., Patnaik, S., Hu, X., Gao, Q., Yang, M., Lawas, M., Delling, M., Marugan, J., Ferrer, M., Xu, H., 2016a. MCOLN1 is a ROS sensor in lysosomes that regulates autophagy. Nat. Commun. 7, 12109. https://doi.org/10.1038/ncomms12109.

Zhang, X., Li, X., Xu, H., 2012. Phosphoinositide isoforms determine compartment-specific ion channel activity. Proc. Natl. Acad. Sci. U. S. A. 109, 11384–11389. https://doi.org/10.1073/pnas.1202194109.

Zhang, J., Zhou, W., Lin, J., Wei, P., Zhang, Y., Jin, P., Chen, M., Man, N., Wen, L., 2016b. Autophagic lysosomal reformation depends on mTOR reactivation in H2O2-induced autophagy. Int. J. Biochem. Cell Biol. 70, 76–81. https://doi.org/10.1016/j.biocel.2015.11.009.

Zhong, Y., Wang, Q.J., Li, X., Yan, Y., Backer, J.M., Chait, B.T., Heintz, N., Yue, Z., 2009. Distinct regulation of autophagic activity by Atg14L and Rubicon associated with Beclin 1-phosphatidylinositol-3-kinase complex. Nat. Cell Biol. 11, 468–476. https://doi.org/10.1038/ncb1854.

CHAPTER FIVE

Mitochondrial Ca^{2+} and cell cycle regulation

Haixin Zhao[a] and Xin Pan[b,*]

[a]State Key Laboratory of Experimental Haematology, Institute of Hematology, Fifth Medical Center of Chinese PLA General Hospital, Beijing, China
[b]State Key Laboratory of Proteomics, Institute of Basic Medical Sciences, National Center of Biomedical Analysis, Beijing, China
*Corresponding author: e-mail address: xpan@ncba.ac.cn

Contents

Abstract

It has been demonstrated for more than 40 years that intracellular calcium (Ca^{2+}) controls a variety of cellular functions, including mitochondrial metabolism and cell proliferation. Cytosolic Ca^{2+} fluctuation during key stages of the cell cycle can lead to mitochondrial Ca^{2+} uptake and subsequent activation of mitochondrial oxidative

International Review of Cell and Molecular Biology, Volume 362
ISSN 1937-6448
https://doi.org/10.1016/bs.ircmb.2021.02.015

phosphorylation and a range of signaling. However, the relationship between mitochondrial Ca^{2+} and cell cycle progression has long been neglected because the molecule responsible for Ca^{2+} uptake has been unknown. Recently, the identification of the mitochondrial Ca^{2+} uniporter (MCU) has led to key advances. With improved Ca^{2+} imaging and detection, effects of MCU-mediated mitochondrial Ca^{2+} have been observed at different stages of the cell cycle. Elevated Ca^{2+} signaling boosts ATP and ROS production, remodels cytosolic Ca^{2+} pathways and reprograms cell fate-determining networks. These findings suggest that manipulating mitochondrial Ca^{2+} signaling may serve as a potential strategy in the control of many crucial biological events, such as tumor development and cell division in hematopoietic stem cells (HSCs). In this review, we summarize the current understanding of the role of mitochondrial Ca^{2+} signaling during different stages of the cell cycle and highlight the potential physiological and pathological significance of mitochondrial Ca^{2+} signaling.

1. Introduction

The cell cycle is a tightly controlled cellular process (Bloom and Cross, 2007; Hochegger et al., 2008; Murray, 2004). During the cell cycle, a cell replicates its DNA in S phase and segregates the newly duplicated genomes into two daughter cells in M phase. Prior to S phase, cells prepare enzymes required for DNA synthesis during G1, while cells prepare to enter mitosis during G2. Although it has been well established that cell cycle checkpoints precisely regulate cell cycle progression, there is growing evidence that other signals such as Ca^{2+} signaling and metabolic pathways are related to cell cycle progression (Humeau et al., 2018; Lee and Finkel, 2013; Machaca, 2011; Salazar-Roa and Malumbres, 2017).

Mitochondria, previous considered "the powerhouse of the cell," have emerged as a membrane platform for many signaling processes (Bornstein et al., 2020; Dasgupta, 2019; Spinelli and Haigis, 2018). Growing evidence demonstrates that this dynamic organelle participates in cell cycle regulation (Carlton et al., 2020; Mishra and Chan, 2014; Spurlock et al., 2020). Early studies showed that mitochondria change in number, shape, size and cellular location throughout the cell cycle (Kanfer et al., 2015; Mitra et al., 2009; Sarraf et al., 2019; Taguchi et al., 2007). These findings were further supported by the observation of increased oxidative phosphorylation (OXPHOS) during S and G2/M phases (Montemurro et al., 2017; Skog et al., 1982; Webster and Hof, 1969). New evidences revealed that cell cycle regulators, such as cyclin-dependent kinase 1 (CDK1), can phosphorylate a cluster of mitochondrial proteins, including the mitochondrial protein import channel component Tom6 (Harbauer et al., 2014) and 5 complex I subunits in the respiratory chain (Wang et al., 2014), thus boosting mitochondrial OXPHOS when cells enter mitosis. These studies provide evidence that mitochondrial signaling is regulated and executed with high spatial and temporal control to ensure proper cell cycle progression.

Calcium (Ca^{2+}) signals play a central role in cell cycle regulation (Lu and Means, 1993; Machaca, 2011; Pinto et al., 2015), and it is well established that changes in Ca^{2+} levels affect cell cycle progression. Over the past 30 years, rapid and temporary rises in intracellular Ca^{2+}, termed Ca^{2+} transients, have been observed during many stages of the cell cycle in a variety of cell types (Baker and Warner, 1972; Kao et al., 1990; Poenie et al., 1985). The connection between Ca^{2+} signaling and the cell cycle has focused on specific cytoplasmic Ca^{2+} binding proteins such as calmodulin and CaMKII (Chafouleas et al., 1982; Kahl and Means, 2003; Skelding et al., 2011). However, given the versatile nature of Ca^{2+} signals, the downstream effectors are not merely limited to the Ca^{2+}-dependent modes of action.

The elevation of cytosolic Ca^{2+} levels typically leads to rapid mitochondrial Ca^{2+} uptake. However, researchers have long neglected the role of mitochondrial Ca^{2+} in cell cycle progression because the identity of the molecule responsible for Ca^{2+} was unknown. The discovery of the mitochondrial Ca^{2+} uniporter (MCU) and its main regulators means that the role of mitochondrial Ca^{2+} homeostasis in cell cycle can now be directly interrogated.

In this review, we will first summarize previously-established knowledge about the role of intracellular Ca^{2+} signaling in regulating the cell cycle. We will then describe recent advances in our understanding of the MCU and its regulators and describe the evidence connecting cell cycle progression with Ca^{2+} influx into the mitochondrial matrix, as well as what is known about the mechanisms underpinning mitochondrial Ca^{2+} control of cell cycle progression. Finally, we will characterize the physiological and pathological importance of the uniporter, particularly in the context of hematopoietic reconstruction and cancer development.

2. Cytosolic calcium signaling during the cell cycle

Calcium (Ca^{2+}) is one of the most important biological cations, and is implicated in almost every processes from cell survival to cell death (Berridge et al., 1998). Ca^{2+} acts as a versatile intracellular messenger that controls numerous processes such as fertilization, proliferation and differentiation

(Pinto et al., 2015; Snoeck, 2020). In the 1980s and 1990s, numerous studies indicated that cytosolic Ca^{2+} levels fluctuate across stages of the cell cycle and play an important regulatory role in cell proliferation (Poenie et al., 1985, 1986; Ratan et al., 1988). Following the development of Ca^{2+} fluorescent indicators such as Fura-2, many research groups reported on the phenomenon of rapid, temporary rises in Ca^{2+} levels, termed Ca^{2+} transients, during specific cell cycle stages. For example, during the early cell cycles after fertilization of sea urchin embryos, transient increases in Ca^{2+} levels occur during the nuclear envelop breakdown (NEB), the metaphase/anaphase transition and the cleavage of daughter cells (Clothier and Timourian, 1972; Poenie et al., 1985). Similar Ca^{2+} transients have been observed during meiosis and mitosis—especially the stage of metaphase/anaphase transition—in a range of species, including oocyte and embryo from *Xenopus laevis* (Baker and Warner, 1972; Grandin and Charbonneau, 1991), Swiss 3T3 fibroblasts from mice (Kao et al., 1990; Tombes and Borisy, 1989), and kidney epithelial Ptk1 or Ptk2 cells from rat kangaroo (Poenie et al., 1986, Ratan et al., 1988). In addition, cells in other stages also exhibit Ca^{2+} transients. For example, in G1/S-synchronized Cos-7 cells, multiple spontaneous Ca^{2+} oscillations that last for up to 40 min have been observed; a similar phenomenon was observed in some HeLa cells undergoing G1/S transition (Russa et al., 2009).

Although Ca^{2+} transients have been observed in a number of studies, results are variable, with transients not always detected in cells and varying in amplitude. This may be because while some Ca^{2+} signals are easy to observe because they are intense and have long durations, brief or low intensity Ca^{2+} transients may be missed in some of the cells during imaging, leading to the conclusion that some cell cycles take place without Ca^{2+} transients (Tombes and Borisy, 1989; Whitaker, 2006). As a result of this inconsistent reporting, the necessity of Ca^{2+} fluctuation in cell cycle regulation has been challenged for years. One possible explanation is based on the characteristics of the spatial organization of Ca^{2+} release channels, such as inositol 1,4,5-trisphosphate receptor (IP3R) in endoplasmic reticulum (ER), which can create a Ca^{2+} signaling microdomain (Parry et al., 2005; Torok et al., 1998). Due to the initial methodological limitations, such as the low-resolution of microscopy and low sensitivity of Ca^{2+} indicators, small and rapid Ca^{2+} signals inside the microdomain could be so localized as to be undetectable (Whitaker, 2006). This possibility is in keeping with the Parry et al. finding that ER cluster around the nucleus and become intimately associated with the mitotic spindle when cells enter mitosis (Bobinnec et al., 2003). Thus, the distribution of ER may generate a Ca^{2+} signal microdomain in which Ca^{2+} release is under very tight spatial control (Parry et al., 2005, 2006; Whitaker, 2006).

Although Ca^{2+} signals cannot always be detected, there is a consensus that Ca^{2+} signaling is involved in cell cycle regulation. Microinjection of Ca^{2+} or Ca^{2+} chelators leads to accelerated cell division or arrested cell cycle respectively, indicating the regulatory relationship between these two events. For example, the injection of EGTA or BAPTA preceding NEB chelated the potential Ca^{2+} signals and stopped mitosis in the sea urchin embryo. Relieving the effect of chelators by reinjection of high Ca^{2+}-containing rescue buffer, the Ca^{2+} transients re-emerged and cellular mitotic progression continued (Steinhardt and Alderton, 1988). Similarly, in mammalian cells, such as PtK1 and Swiss 3T3 cells, microinjection of Ca^{2+}-containing solutions stimulated the transition from metaphase to anaphase and promoted precocious chromatid separation (Izant, 1983). In contrast, injection of EGTA and BAPTA was effective in both blocking anaphase and reducing elevated Ca^{2+} levels (Izant, 1983; Tombes and Borisy, 1989). These results, together with the observation of Ca^{2+} signaling at specific cell cycle stages, have led to the conclusion that Ca^{2+} signals are essential for cell cycle progression.

Research on the downstream targets of Ca^{2+} signals has mostly focused on specific cytoplasmic Ca^{2+} binding proteins, such as calmodulin (CaM) and calcium/calmodulin-stimulated protein kinase (CaMK) (Skelding et al., 2011). It has been observed that the expression of CaM is related to cell cycle progression, with higher levels during G1/S transition (Chafouleas et al., 1982) and G2/M phase (Lu et al., 1993). Knockdown of CaM can lead to cell cycle arrest, including G1 and mitosis (Rasmussen and Means, 1989), indicating the requirement of CaM and the associated kinases CaMKI and CaMKII in these two phases. Consistent with the effect of CaM, it has been demonstrated that CaMKI and CaMKII mediate G1/S and G2/M phases, respectively. Mechanically, during the late G1 phase, CaMKI activates Cdk4/cyclin D1, resulting in phosphorylation of retinoblastoma protein (Rb) and its dissociation from E2F transcription factors. The activated E2F then transcripts genes responsible for promoting entry into S phase (Kahl and Means, 2004). Others have also reported that CaMKII can stimulate cyclin D1 expression through nuclear factor kappaB (NF-κB) (Torricelli et al., 2008). During S phase, CaMKII activates MEK/ERK, which in turn leads to phosphorylation and degradation of p27 and alleviates the inhibition of Cdk2, thus promoting S-G2/M progression (Li et al., 2009). A series of findings demonstrated that CaMKII is essential for G2 and M phase progression. For example, CaMKII directly targets cdc25c, which activates Cdk1/cyclin B and promotes G2/M transition (Patel et al., 1999). When cells enter mitosis, CaMKII is reported to be localized on the centrosomes and spindle pole (Ohta et al., 1990). The distribution of the kinase may benefit the microtubule dynamics through the regulation of MCAK, a microtubule depolymerase, which is essential for spindle maintenance during mitosis (Holmfeldt et al., 2005). In the metaphase of meiosis II, a similar stage compared with mitosis, CaMKII can phosphorylate Emi2 and trigger a polo-like kinase 1 (Plx1)-dependent degradation signal of Emi2 (Lorca et al., 1993; Rauh et al., 2005). Emi2 is an inhibitor of the anaphase-promoting complex/cyclosome (APC/C), whose activation leads to anaphase onset and subsequent mitotic or meiotic exit. In sum, CaM- and Ca^{2+}-mediated activation of CaMKI and CaMKII play an important role during cell cycle progression.

However, given that Ca^{2+} signals may be involved in almost all intracellular signaling, it is difficult to establish a complete understanding of the mechanism underlying the relationship between Ca^{2+} and cell cycle regulation. Ca^{2+} may influence the cell cycle through many other mechanisms, such as mitochondrial signaling and metabolic pathways. The related directions are still worth further exploring in the future.

3. Mitochondrial Ca^{2+} uniporter complex (MCUC) and its regulatory mechanisms

Previous evidence has shown that purified mitochondria can take up and buffer large amounts of Ca^{2+} (Deluca and Engstrom, 1961; Gunter and Pfeiffer, 1990; Vasington and Murphy, 1962), providing evidence of a close relationship between Ca^{2+} and mitochondria. However, the identity of the molecule which transfers Ca^{2+} from cytosol into the matrix has remained a mystery for more than 50 years.

Mitochondria are enclosed by a double membranous structure, which is equipped with numerous channels responsible for ion permeation. It has been well established that the voltage dependent anion channel (VDAC) controls Ca^{2+} crossing the outer mitochondrial membrane (OMM) from the cytosol (Shoshan-Barmatz and Mizrachi, 2012). However, little was known about the Ca^{2+} transporter located on the inner mitochondrial membrane (IMM), preventing a complete understanding of the physiological and pathological role of mitochondrial Ca^{2+} uptake. This situation changed in 2011, when two groups identified CCDC109A as the gene encoding the mitochondrial Ca^{2+} uniporter (MCU) (Baughman et al., 2011; De Stefani et al., 2011). Within the next several years a series of regulators that control the uniporter's activity were identified along with the intracellular functions of Ca^{2+} influx via the MCU. This recent research and subsequent studies have opened up a new "molecular era" of mitochondrial Ca^{2+} research (Feno et al., 2020; Giorgi et al., 2018; Kamer and Mootha, 2015).

3.1 Extramitochondrial Ca²⁺ elevation as the driving signal

It has been well established that Ca^{2+} uptake through mitochondria is cytosolic Ca^{2+} level-dependent, or rather, extramitochondrial Ca^{2+} level-dependent (Moreau et al., 2006). The MCU is a channel with a very low Ca^{2+} affinity of about 5–10 µmol/L (Carafoli, 2012). However, the basal cytosolic Ca^{2+} level is only about 100 nM. Even when being stimulated, the average concentration of Ca^{2+} in the cytosol is less than 3 µmol/L. These conditions suggest it would be difficult to activate the MCU under physiological conditions. In order to overcome the low Ca^{2+} affinity of MCU, a mitochondrial Ca^{2+} microdomain exists between the Ca^{2+} release channel and subsequent uptaking machinery constructed with VDAC and MCU (Raturi and Simmen, 2013). For example, ER is a major site of Ca^{2+} storage, and the concentration of Ca^{2+} in ER may reach 0.5 mM, which is 1000 times higher than that in the cytosol. It has been reported that the ER membrane is often very close, about 10–60 nm distance, to the outer membrane of adjacent mitochondria, allowing for the formation of so-called mitochondria-associated membrane (MAM) contacts (Gaigg et al., 1995; Rizzuto et al., 1998). The construction of MAM contacts improves the stability of the interaction between ER and mitochondria, forming Ca^{2+} hotspots locally once Ca^{2+} is released by IP3R and ultimately facilitating efficient Ca^{2+} transfer through the IP3R-VDAC-MCU axis. Similarly, there also exist plasma membrane-associated mitochondria (PAM) contacts, where mitochondria form associations with the voltage-gated Ca^{2+} channels in the plasma membrane (Suski et al., 2014). Disruption of the interaction between ER and mitochondria may lead to dysregulated Ca^{2+} transfer and decreased mitochondrial energy production, thus inducing diseases (Paillusson et al., 2017). As a result, Ca^{2+} fluctuation outside the mitochondria acts as a perquisite for the opening of the MCUC.

3.2 Mitochondrial Ca²⁺ uniporter complex and its Ca²⁺ sensing mechanism

According to the latest studies, the pore-forming proteins MCU and its paralog MCUb interact with the regulatory subunits MICU1, MICU2, and EMRE to construct the IMM Ca²⁺ uniporter complex (Belosludtsev et al., 2019) (Fig. 1). MCU is a 40 kDa protein, containing two transmembrane domains responsible for IMM localization, and an intervening loop region responsible for Ca²⁺ transport (Baughman et al., 2011; De Stefani et al., 2011). Four or five MCU subunits, dependent on species, aggregate together and form the core channel of the complex (Fan et al., 2018; Nguyen et al., 2018b; Oxenoid et al., 2016; Raffaello et al., 2013; Yoo et al., 2018). In addition, MCUb, a dominant–negative form of MCU,

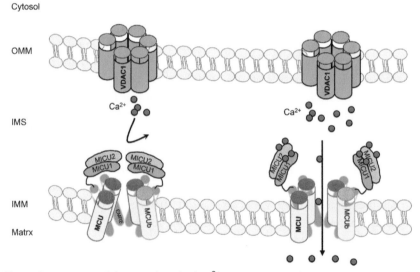

Fig. 1 Components of the mitochondrial Ca²⁺ uniporter complex (MCUC). Mitochondria are organelles equipped with two membranes: the outer mitochondrial membrane (OMM) and the inner mitochondrial membrane (IMM). The membranes thus separates compartments or regions, including cytosol, intermembrane space (IMS) and the matrix. Voltage dependent anion channel 1 (VDAC1) controls the Ca²⁺ across the outer mitochondrial membrane (OMM) from cytosol, while MCUC transfers Ca²⁺ from IMS to the matrix. The uniporter complex consists of the pore-forming proteins, mitochondrial Ca²⁺ uniporter (MCU), MCU dominant-negative regulatory subunit b (MCUb), and essential MCU regulator (EMRE), together with two IMS-resident proteins mitochondrial Ca²⁺ uptake protein 1 (MICU1) and mitochondrial Ca²⁺ uptake protein 2 (MICU2). Under resting conditions, MICU1/2 dimers dock on MCU and act as gatekeepers of the uniporter. When Ca²⁺ level in IMS rises, MICU1/2 move away from the pore. The conformational rearrangements of MICU1/2 thus facilitate the Ca²⁺ influx into the matrix through MCU.

participates in the hetero-oligomer construction of the pore, which determines the Ca^{2+} uptake capacity, depending on the relative ratio of MCU and MCUb (Raffaello et al., 2013).

Neither MCU nor MCUb is able to open and close the channel according to the surrounding Ca^{2+} concentration, because of a lack of Ca^{2+} binding domains. However, there is evidence for a sigmoidal dependence of the Ca^{2+} uptake rate on Ca^{2+} concentration (Csordas et al., 2013), indicating that there is a mechanism for sensing extramitochondrial Ca^{2+} concentration.

MICU1 was the first regulatory molecule of MCU to be identified (Perocchi et al., 2010). Structure analysis revealed that MICU1 contains EF-hand domains, which are responsible for its Ca^{2+} binding ability (Perocchi et al., 2010). Recent evidence indicates that MICU1, together with its homolog MICU2, forms a heterodimer through disulfide bonds and performs both positive and negative roles in the regulation of MCUM. In brief, under resting conditions (0.1 µM Ca^{2+}), MICU1/2 act as uniporter gatekeepers (Csordas et al., 2013; Kamer et al., 2017), keeping tightly docked on the MCU to seal the pore. Depletion of MICU1 often results in increased basal matrix Ca^{2+} levels (Mallilankaraman et al., 2012). Upon Ca^{2+} elevation, Ca^{2+} binds to the MICU1-MICU2 heterodimer cooperatively, inducing conformational changes and crippling MCU-MICU1 interactions (Fan et al., 2020; Wu et al., 2020). The MICU1/2 heterodimer then moves away from the pore, leading to the Ca^{2+}-dependent activation of the uniporter (Fig. 1). There is evidence that the interaction of the MICU1/MICU2 heterodimer with the MCU is mediated by another regulatory subunit, EMRE. EMRE is a 10 kDa IMM protein which is essential for MCU activity. Depletion of EMRE reduces mitochondrial Ca^{2+} uptake significantly (Sancak et al., 2013). In addition to its linker functions, recent studies have suggested an expanded role of EMRE in Ca^{2+} sensing and uniporter gatekeeping. In a surprising finding, Vais et al. reported that MCU activity is also regulated by the matrix Ca^{2+} concentration, which is sensed by EMRE (Vais et al., 2016). Other groups also reported that EMRE is responsible for the assembly, opening and maintenance of the MCU core complex (Van Keuren et al., 2020; Wang et al., 2019). Although the underlying mechanisms are becoming clearer, new challenges have also emerged. For example, the molecular relationship and the interactions among these regulatory subunits appear to be more complex than initially believed (Fan et al., 2020; Liu et al., 2020; Wu et al., 2020; Zhuo et al., 2020). Further research is needed to understand the characteristics of Ca^{2+} sensing signaling on both sides of the mitochondrial.

3.3 Post-translational modification of MCU

In addition to the regulatory model mentioned above, there are other layers of regulation, such as post-translational modification of MCU.

The first identified post-translational modification of MCU was phosphorylation. 1 year after the identification of MCU, Joiner et al. found that a CaMKII inhibitory protein CaMKIIN, in particular mitochondrially-localized CaMKIIN, significantly inhibits the MCU's Ca^{2+} uptake capacity (Joiner et al., 2012), indicating that CaMKII has a positive mitochondrial Ca^{2+} regulatory function. This study also demonstrated that CaMKII could directly phosphorylate the NTD domain of the MCU at two potential sites, Ser57 and Ser92. Moreover, CaMKII activation did not increase the mitochondrial Ca^{2+} uptake capacity of the $MCU^{S57A/S92A}$ mutant. Although the regulatory role of CaMKII was questioned in a subsequent electrophysiological study (Fieni et al., 2014), further evidence confirmed the initial conclusion with the demonstration that the phosphorylation of MCU at Ser92 induces changes in structure and charge distribution in the loop structure of NTD, which in turn leads to the dimerization of MCU-EMRE tetramers and modulation of mitochondrial Ca^{2+} uptake (Lee et al., 2015, 2020). Later, using a phosphor-specific antibody to recognize Ser92-phosphorylated MCU, the phosphorylation of MCU mediated by CaMKII was observed when vascular smooth muscle cells (VSMC) were stimulated by platelet-derived growth factor (PDGF) (Nguyen et al., 2018a), an agonist inducing VSMC migration and neointimal hyperplasia. Taken together, these results underscore the critical role of the phosphorylation of the MCU NTD domain in Ca^{2+} transport activity.

Similarly, the phosphorylation of MCU at a site nearby, Ser57, is associated with increased Ca^{2+} uptake capacity. When cells enter mitosis, the energy deficits caused by the numerous ATP-consuming events activates the cellular energy sensor, AMPK. It has been shown that AMPK can translocate into the mitochondria (Cai et al., 2020; Zhao et al., 2019) and phosphorylate MCU at Ser57, facilitating Ca^{2+} transfer (Zhao et al., 2019). However, it seems that the kinases that phosphorylate MCU at Ser57 are not limited to AMPK, as deletion of AMPK does not affect the MCU phosphorylation level under resting conditions, indicating that basal MCU phosphorylation at Ser57 may be mediated by other kinases, such as CaMKII. By comparing the amino acid sequences around the two potential phosphorylated sites, Ser57 and Ser92, both of which reside in a conserved RXXS motif. The result suggests that these two sites may be modified by multiple kinases under different conditions. It is worth noting that the modification of MCU does not affect the Ca^{2+} release signal in the cytosol. It seems that the modification occurs when cells are experiencing energy deficits or stimulations from environmental fluctuations, suggesting that the modification of MCU may act as a priming event which facilitates or alleviates the subsequent mitochondrial Ca^{2+} response to cope with the forthcoming stimulation. A similar effect was also exhibited in the Pyk2-mediated phosphorylation of MCU (Hirschler-Laszkiewicz et al., 2018; Jin et al., 2014). Further studies are needed to confirm this hypothesis.

In addition to the activity regulation through phosphorylation, during hypoxia and inflammatory-mediated oxidative stress the MCU can be targeted by reactive oxygen species (ROS), which promotes S-glutathionylation of the Cys97 in MCU protein subunits. This modification does not affect the interaction of MCU with other uniporter subunits, but increases the stability of the complex, thereby promoting Ca^{2+} accumulation in mitochondria and increasing susceptibility to cell death (Dong et al., 2017).

4. Mitochondrial calcium signaling during the cell cycle

The Ca^{2+} which is taken up into mitochondria regulates several enzymes as a cofactor in the matrix, enhancing mitochondrial activity including ATP generation, ROS production, metabolic reprogramming and mPTP opening. Additionally, mitochondrial Ca^{2+} uptake buffers cytosolic Ca^{2+} and contributes to the termination of signaling which affects events in the cytosol (Fig. 2). Detailed mechanisms are discussed below.

4.1 Facilitating energy production and mitochondrial metabolism

It is widely believed that the entry of mitochondrial Ca^{2+} boosts mitochondrial ATP production in response to the increased energy demand. Indeed, biochemistry studies from the 1970s demonstrated that Ca^{2+} in the mitochondrial matrix activates three key enzymes of the tricarboxylic acid (TCA) cycle, α-ketoglutarate dehydrogenase (αKGDH), isocitrate dehydrogenase (IDH), and pyruvate dehydrogenase (PDH) (Denton et al., 1972, 1978; McCormack and Denton, 1979). In addition, the activation of several complexes of the electron transport chain (ETC) and mitochondrial F$_1$F$_0$ ATP synthase (Complex V) also seem to be Ca^{2+}-dependent (Glancy et al., 2013; Murphy et al., 1990; Territo et al., 2000). With improved methodologies, the direct regulation of Ca^{2+} on mitochondrial ATP synthesis was visualized at the single cell level (Imamura et al., 2009; Nakano et al., 2011)

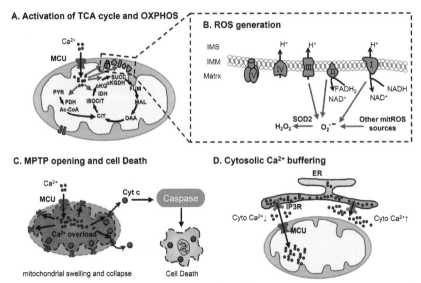

Fig. 2 Intracellular functions of MCU-mediated Ca^{2+} signaling. (A) Mitochondrial Ca^{2+} activates three key dehydrogenase in matrix, including α-ketoglutarate dehydrogenase (αKGDH), isocitrate dehydrogenase (IDH), pyruvate dehydrogenase (PDH), several complexes of electron transport chain (ETC) and mitochondrial F$_1$F$_0$ ATP synthase (Complex V) on the IMM, thus promoting TCA cycle, oxidative phosphorylation (OXPHOS) and subsequent ATP production. (B) Mitochondrial Ca^{2+}-induced activation of ETC may lead to the generation of ROS. Superoxide (O$_2^-$) leaked from complex I, II and III, converts to hydrogen peroxide (H$_2$O$_2$) through superoxide dismutase SOD2. These molecules can be released into IMS and cytosol, not only triggering redox signaling in cells, but also leading to damage stresses. (C) Ca^{2+} overload in matrix mediates opening of the mitochondrial permeability transition pore (mPTP). Once opened, the mPTP results in the membrane potential collapse, mitochondrial swelling, release of cytochrome *c* and subsequent cell death. (D) Mitochondria could uptake amounts of Ca^{2+} from the cytosol, thus modulating the duration of cytosolic Ca^{2+} signals and preventing excess Ca^{2+} accumulation. When the Ca^{2+} uptake activity is disrupted, the elevated cytosolic Ca^{2+} level results in cellular functions' change.

(Fig. 2A). Moreover, Ca^{2+} uptake through mitochondria may reprogram cellular metabolism and affect cell functions and fates through multiple pathways, given that Ca^{2+}-induced OXPHOS activation promotes the generation of intermediaries which are associated with epigenetic modification, including acetyl-CoA, α-ketoglutarate (α-KG) and NADH (Carey et al., 2015; Imai et al., 2000; Moussaieff et al., 2015).

4.2 Reactive oxygen species (ROS) production

Mitochondria are the major source of intracellular ROS. When electrons are transferred through a series of electron transporters, ROS is generated from the electron leakage (Brand, 2016; Mailloux, 2015). ROS has been thought of as a harmful by-product of the ETC for many years. The presence of ROS has been linked to factors causing oxidative damage of nucleic acids, proteins and lipids. However, initial assumptions have been replaced in light of recent evidence of ROS' function as a second messenger and signaling transducer (Holmstrom and Finkel, 2014). For example, ROS can directly affect many enzymes' activity through a post-translational modification of its cysteine residues (Meng et al., 2002). The multiple functions of ROS have led to its association with an ever-widening range of physiological and pathological processes, from stem cell survival to aging and eventually cell death (Holmstrom and Finkel, 2014; Sies and Jones, 2020) (Fig. 2B). Given that elevated Ca^{2+} inside mitochondria activates the matrix dehydrogenase and respiratory chain, it is not surprising that there is a Ca^{2+} dependent ROS generation mechanism (Bertero and Maack, 2018; Hempel and Trebak, 2017). Recent studies have supported this hypothesis with the finding that MCU-silenced breast cancer cells exhibited reduced mitochondrial ROS level, which led to a decreased expression of HIF1α and thus hampered tumor growth and invasion (Tosatto et al., 2016).

4.3 Mitochondrial permeability transition pore (mPTP) opening

Despite the beneficial metabolic effects of Ca^{2+} uptake, there are some studies demonstrating that mitochondrial Ca^{2+} overload is associated with cell death leading to diseases such as heart failure and neurodegeneration (Joiner et al., 2012; Luo and Anderson, 2013; Sheng and Cai, 2012). The underlying mechanism involves a mitochondrial Ca^{2+} dependent mPTP opening, which results in membrane potential collapse, mitochondrial swelling, and release of cytochrome c and ROS (Blattner et al., 2001) (Fig. 2C). Given that the pore component of the mPTP complex has not been fully characterized and that use of the existing inhibitor, cyclosporine A (CsA), produces off-target effects and low therapeutic efficacy (Briston et al., 2019; Cung et al., 2015; Monassier et al., 2016), depletion or blockage of MCU might be considered as a promising strategy to block injury caused by mPTP opening (Kon et al., 2017; Luongo et al., 2015).

4.4 Cytosolic Ca^{2+} buffering and signaling

Acting as one of the major Ca^{2+} buffering organelles, mitochondria rapidly take up Ca^{2+} from the cytosol, modulate the duration of cytosolic Ca^{2+}

signals and prevent excess Ca^{2+} accumulation (Fig. 2D). These functions have been demonstrated in studies of various cell types, especially non-muscle cells (De Stefani et al., 2016). When the capacity of mitochondrial Ca^{2+} uptake was increased by MCU overexpression, a significant reduction in the cytosolic Ca^{2+} signal evoked by histamine was observed (De Stefani et al., 2011). Likewise, when Ca^{2+} entry was blocked by MCU silencing or disrupting ER-mitochondria tethering, although the effects are not always significant there is evidence that both the resting cytosolic Ca^{2+} level and the peak amplitude of the stimulated cytosolic Ca^{2+} rises are elevated, probably due to the reduced mitochondrial buffering capacity (De Stefani et al., 2011; Drago et al., 2012; Hirabayashi et al., 2017; Koval et al., 2019).

4.5 *In vivo* evidence of mitochondrial Ca^{2+} mediated cell cycle regulation

The relationship between Ca^{2+} signaling and cell cycle progression has been investigated for more than 40 years. However, considering the ubiquitous and versatile nature of Ca^{2+}, the underlying mechanism requires further enquiry. Due to historically limited knowledge of the Ca^{2+}-uptaking machinery, there was previously no way to explore the physiological and pathological roles of mitochondrial Ca^{2+} signaling. As such, the potential contribution of mitochondrial Ca^{2+} to the cell cycle was also under-examined. However, with recent improvements in the understanding of MCU and its regulatory mechanisms, the issue is beginning to receive considerable attention. The first relevant cases were reported in mice with knockout of MCU and its regulatory subunits. MCU and EMRE knockout mice in C57BL/6 background exhibit embryonic lethality (Liu et al., 2020; Murphy et al., 2014; Pan et al., 2013). The $MCU^{-/-}$ embryos exhibited a mild size reduction and die around E11.5–E13.5, which was later found to be partially due to the mitotic defect in fetal liver (Zhao et al., 2019). Liver sections from $MCU^{-/-}$ embryos showed an increased mitotic index, coincident with increased dead cells (Zhao et al., 2019). $MICU1^{-/-}$ mice exhibit a slightly different phenotype, probably because of dual function of MICU1 in regulating Ca^{2+} uptake. Due to the impaired gatekeeping effect, MICU1-deleted cells exhibited resting mitochondrial Ca^{2+} overload, which may result in a high, but not complete, perinatal death of $MICU1^{-/-}$ mice (Liu et al., 2016). The surviving $MICU1^{-/-}$ mice showed slight retardation of development with a smaller body weight. Reducing the overloaded Ca^{2+} level in mitochondria by deleting one allele of EMRE almost completely rescues the perinatal mortality of $MICU1^{-/-}$ mice (Liu et al., 2016). On the other hand, MICU1 also positively regulates mitochondrial Ca^{2+} uptake, and MICU1-deleted liver cells exhibit a lower rate of mitochondrial Ca^{2+} uptake. After partial hepatectomy (PHx), liver regeneration was delayed in $MICU^{-/-}$ liver. In addition, hepatocyte proliferation, measured by 5-bromodeoxyuridine (BrdU) incorporation, was abolished in remnant $MICU1^{-/-}$ livers, in which the expression of cyclin D1 was decreased (Antony et al., 2016). These results suggest that MICU1-mediated Ca^{2+} uptake may promote the transition of primed hepatocytes to the proliferative state and subsequent proliferation through an unknown mechanism. Interestingly, in a study where Pan et al. constructed viable $MCU^{-/-}$ mice in an outbred CD1 background (Pan et al., 2013), they found there was still significant embryonic lethality in mice, as they are born at a lower ratio than the expected 25% (Murphy et al., 2014; Zhao et al., 2019). Similar viable $EMRE^{-/-}$ mice were reported recently (Liu et al., 2020), where surviving mice also exhibited relatively lower body and organ weight compared to wild-type (WT) (Koval et al., 2019; Liu et al., 2020; Pan et al., 2013). More importantly, proliferating cell nuclear antigen (PCNA)-labeled proliferating hepatocytes were lower in liver sections from the postnatal alive $MCU^{-/-}$ mice (Koval et al., 2019). The knockout mice also exhibited a significant delay of wound healing after cutaneous punch biopsy (Koval et al., 2019). Taken together, these findings suggest that mitochondrial Ca^{2+} uptake via the MCU may play a critical role in cell proliferation, organ development and body growth.

4.6 Mitochondrial Ca^{2+} signaling imaging during the cell cycle

The level of mitochondrial Ca^{2+} usually oscillates rapidly and in synchrony with the Ca^{2+} oscillations in the cytosol. The existence of MAM contacts further increases sensitivity to the mitochondrial uptaking signaling. As a result, cytosolic Ca^{2+} fluctuations during certain stages of the cell cycle may lead to synchronous mitochondrial Ca^{2+} changes. However, due to the lack of proper mitochondrial Ca^{2+} indicators, this phenomenon had not been observed until recently. The mitochondrial Ca^{2+} dynamics during the cell cycle and the underlying mechanisms will be described below (Table 1).

Table 1 Mitochondrial Ca^{2+} dynamics during the cell cycle.

Phenotype	Mitochondrial Ca^{2+} indicator	Cell type	Species	Phases of the cell cycle	Roles of mitochondrial Ca^{2+} in the cell cycle	References
Elevated mitochondrial Ca^{2+} level	Mito-Pericam	INS 832/13 cells	Rat	Synchronized S and G2/M	The elevated mitochondrial Ca^{2+} level activates the TCA cycle and mitochondrial respiration	Montemurro et al. (2017)
Mitochondrial Ca^{2+} transients	4mt-GCaMP6	HeLa, PtK1, NIH3T3, HepG2, LM3	Human, rat kangaroo, mouse	Mitosis	The mitochondrial Ca^{2+} transients boost mitochondrial respiration, restore energy homeostasis, regulate microtubule depolymerization and promote proper mitotic progression	Zhao et al. (2019)
Mitochondrial Ca^{2+} transients	4mt-GCaMP6	HeLa	Human	Interphase	Not defined	Zhao et al. (2019)
RuR treatment led to early cell division arrest	N/A	Embryos	Xenopus	Not defined	Fertilization-induced mitochondrial Ca^{2+} dynamics increase ROS production and cell cycle progression	Han et al. (2018)
MCU silencing or RU360 treatment delayed cell cycle progression	N/A	VSMCs	Mouse	G1/S transition	MCU-mediated Ca^{2+} influx regulates activity of CaMKII, mitochondrial morphology and energy production during G1/S transition	Koval et al. (2019)
Elevated mitochondrial Ca^{2+} level	Rhod-2	HSCs	Mouse	Cell cycle initiation	Not defined	Umemoto et al. (2018)

In 2017, using mitochondrial localized Pericam, a genetic encoded ratiometric Ca^{2+} indicator, Montemurro et al. found that synchronized S and G2/M cells exhibited relatively higher mitochondrial Ca^{2+} levels compared with G1/S cells, coinciding with the increased OCR (Montemurro et al., 2017). Due to the low Ca^{2+} affinity of the indicator and the lack of live-cell time-lapse imaging, details of the connection between mitochondrial Ca^{2+} and cell cycle progression remain to be clarified.

In 2019, using the improved mitochondrial Ca^{2+} indicator 4mt-GCaMP6, another group found that a rapid Ca^{2+} burst occurs in some cells that had entered mitosis (Zhao et al., 2019). The transients in the mitochondria occurred almost simultaneously with the cytosolic Ca^{2+} transients. As the rapid and small mitochondrial Ca^{2+} transients were not easily captured, a Na^{+}/Ca^{2+} exchanger inhibitor CGP37157 was added to increase the amplitude and duration of Ca^{2+} signaling without affecting mitosis. Under these conditions, more mitochondrial Ca^{2+} transients were captured using time-lapse imaging. However, only about 60% of cells exhibited mitochondrial Ca^{2+} transients during mitosis. Although it is likely that some brief and small transients were missed, the authors reported that the mitotic Ca^{2+} signaling acts as a response to the cellular energetic status. Based on their observations, they suggest that mitochondrial Ca^{2+} transients tend to occur in cells with low resting ATP levels. Because mitosis represents a highly energy-consuming cellular process, the acute energy shortage may appear when these cells enter mitosis, thus leading to mitochondrial Ca^{2+} transients through AMPK activation. In contrast, there is no requirement to turn on the Ca^{2+} signals in cells with high energy storage.

It has been demonstrated that the mitochondrial Ca^{2+} transients that occur during mitosis are mediated by the MCU in a study where MCU depletion dramatically decreased the frequency and amplitude of mitochondrial Ca^{2+} transients and delayed mitotic progression (Zhao et al., 2019). Importantly, mitochondrial Ca^{2+} transients were also observed in some interphase cells, which were likely in other key stages of the cell cycle, such as the G1/S transition. Additional studies found that MCU-mediated Ca^{2+} signaling is involved in a range of cell cycle stages, including fertilization-induced egg activation (Han et al., 2018), G1/S transition (Koval et al., 2019) and mitotic progression (Zhao et al., 2019) (Fig. 3). This regulation may contribute to many physiological and pathological processes, such as embryonic development and postnatal growth.

A. Mitochondrial Ca²⁺ signaling in fertilization induced early embryonic cell division

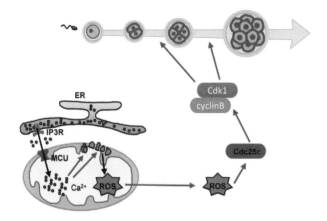

B. Mitochondrial Ca²⁺ signaling during G1/S transition and mitosis

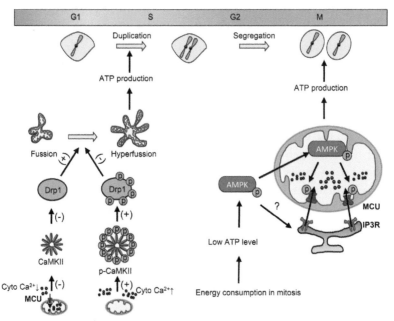

Fig. 3 Mitochondrial Ca²⁺ signaling in cell cycle regulation. (A) Fertilization in *Xenopus* embryos triggers Ca²⁺ wave, which results in mitochondrial Ca²⁺ uptake and elevated ROS production. The released ROS from mitochondria regulates Cdc25C activity and ensures successful cell cycle progression through early embryonic development. (B) Mitochondrial Ca²⁺ level fluctuates along the cell cycle, together with the

5. The mitochondrial calcium uptake mechanism

5.1 Fertilization-induced egg activation

It has been widely found that cytosolic Ca^{2+} is involved in fertilization-induced egg activation, cell division and subsequent embryonic development. Early studies reported that Ca^{2+} transients appeared following the fertilization of eggs in model animals, such as sea urchin and *Xenopus laevis* (Grandin and Charbonneau, 1991; Steinhardt and Alderton, 1988). Additionally, it has been reported that mitochondrial respiration increased simultaneously with the occurrence of Ca^{2+} transients (Nakazawa et al., 1970). A more recent study also reported increased ROS levels following fertilization (Foerder et al., 1978). These phenomena helped elucidate the complex connections between rises in Ca^{2+}, mitochondrial ATP synthesis and ROS production.

Using a transgenic *Xenopus* line expressing Hyper, an H_2O_2 indicator, a rapid rise of ROS production was observed following fertilization *in vivo* (Han et al., 2018). It was further demonstrated that increased ROS production was dependent on Ca^{2+} signaling. Abolishing Ca^{2+} signaling through the Ca^{2+} chelator EGTA or overexpression of IP3 phosphatase completely arrested ROS production. ROS is mainly generated from mitochondria, because mitochondrial ETC inhibitors also attenuate ROS production, indicating a potential mitochondrial Ca^{2+} influx-mediated ROS mechanism after fertilization. It has been found that inhibiting mitochondrial Ca^{2+} uptake via ruthenium red (RuR) injection or MCUb overexpression led to significantly reduced ROS production and early cell division arrest

---

morphology changes of the dynamic organelle. During G1/S transition, PDGF-induced elevated Ca^{2+} stimulates Ca^{2+} transfer from cytosol to mitochondria via MCU. MCU-mediated cytosolic Ca^{2+} clearance prevents excessive activation of CaMKII and maintains the hyperfused mitochondrial morphology, which correlates with the high ATP output. Once the Ca^{2+} transfer pathway is disrupted, sustained activation of CaMKII promotes Drp1 phosphorylation, which leads to mitochondrial fragment and cell cycle arrest. On the other hand, spontaneous mitochondrial Ca^{2+} transients occur during mitosis. The Ca^{2+} signaling acts as an adaptive response to overcome the acute energy shortage during this stage. When cells enter mitosis, cytosolic ATP level decreases probably due to many energy-consuming processes. As a result, AMPK, the intracellular energy sensor, promotes Ca^{2+} release from ER through a yet-to-be characterized mechanism, translocates into the mitochondria and phosphorylates MCU. The modification of MCU facilitates Ca^{2+} uptake, boosts mitochondrial ATP production and contributes to proper mitotic progression.

(Han et al., 2018) (Fig. 3A). Interestingly, inhibition of mitochondrial ATP synthesis via oligomycin did not affect ROS generation and embryonic development at least up to the blastula stage, ruling out the involvement of Ca^{2+} induced ATP generation in early stage development (Han et al., 2018).

A potential point of confusion is that although $MCU^{-/-}$ mice experience embryonic lethality, they can survive until at least E11.5 in a C57BL/6 background (Zhao et al., 2019). The developmental arrest does not occur up to the blastula stage, as the embryos are normal except for their small size. One explanation is that while MCU-mediated metabolic changes play a key role in early developmental processes in lower animals such as *Xenopus laevis*, higher animals may evolve mechanisms to overcome MCU deficiency-induced defects in early development. Further studies are needed to address this issue.

5.2 Mitosis

Similar to the fertilization-induced cell division, Ca^{2+} transients have been reported occurring in association with nuclear envelope breakdown (NEB), metaphase to anaphase transition and cytokinesis (Kao et al., 1990; Poenie et al., 1986; Ratan et al., 1988). As described above, mitochondrial Ca^{2+} transients during mitotic progression were recently observed in multiple cell types, including normal cells, such as PtK1, NIH3t3 cells and cancer cells, such as HeLa, HepG2, LM3 cells (Zhao et al., 2019). Analyzing hundreds of cells for evidence of mitochondrial Ca^{2+} transients during mitosis has revealed that the appearance of transients does not correlate temporally with NEB or anaphase onset. Rather, most of the transients occurred in metaphase, in accordance with morphological chromosomal changes, suggesting a potential role for mitochondrial Ca^{2+} in the metaphase to anaphase transition. Indeed, siRNA mediated MCU silencing not only inhibits spontaneous mitochondrial Ca^{2+} transients but also prolongs the duration between NEB and the separation of the sister chromosome. This phenomenon has been validated in a variety of cell types. For example, similar results have been found in MCU knockout embryonic liver cells. In addition, more mitotic delay-induced cell death was detected when MCU levels were depleted. The mitotic delay accompanied by the consequent cell death, known as mitotic catastrophe, may partially account for the embryonic lethality in E11.5–13.5. A recent study also demonstrated that inhibition of Ca^{2+} transfer from ER to mitochondria may induce a mitotic catastrophe due to the compromised bioenergetics (Cardenas et al., 2016). Taken together these results suggest a regulatory role of mitochondrial Ca^{2+} in mitosis *in vitro* and *in vivo*.

Progression through G2 and mitotic entry has been thought of as an energy-demanding process and includes spindle formation, checkpoint silencing, microtubule depolymerization, chromosome segregation and poleward movement (Bershadsky and Gelfand, 1981; Hartman and Vale, 1999; Miniowitz-Shemtov et al., 2010; Nakada et al., 2010; Salazar-Roa and Malumbres, 2017). It has been found that the oxygen consumption rate fluctuates in a Cdk1-dependent manner in response to the increased energy demand at these stages. At G2/M transition, cyclin B1/Cdk1 translocates to the mitochondria, phosphorylates 5 complex I (CI) subunits, and boosts mitochondrial ATP production (Wang et al., 2014). Similarly, when yeast cells enter mitosis, Cdk1 can also stimulate mitochondrial respiration activity through the phosphorylation of a protein import channel subunit, Tom6 (Harbauer et al., 2014). However, beyond the cell cycle dependent regulation of mitochondrial biogenesis, mitochondrial ATP production is also mediated by the spontaneous Ca^{2+} transients occurring in mitosis (Zhao et al., 2019). MCU depletion significantly inhibits the elevated mitochondrial activity during mitotic progression. Inhibition of the mitochondrial ATP synthase through oligomycin treatment was found to prolong the duration of mitosis in a manner similar to the reported effect of MCU silencing. Furthermore, the delay caused by oligomycin is not further prolonged by MCU depletion, indicating that mitochondrial respiration is impaired in MCU knockdown mitotic cells. As a result, the energy released from MCU-depleted mitochondria cannot replenish the ATP pool to meet the needs of the increased ATP consumption during mitosis. ATP fluctuation during mitosis has been observed using a cytosolic ATP fluorescent indicator, ATeam1.03. Results indicate that intracellular ATP levels drop in early mitosis and recover before anaphase onset. In line with the finding of prolonged mitosis, ATP recovery is also delayed in MCU-depleted mitotic cells. Interestingly, adding ATP extracellularly can increase the intracellular ATP level through micropinocytosis and rescue the MCU knockdown-induced mitotic delay. Taken together, these results demonstrate that MCU-mediated mitochondrial Ca^{2+} transients are required for ATP production and proper mitotic progression.

ATP, the basic energy source, is responsible for many cellular and molecular events during mitosis. However, MCU-mediated acute ATP production seems to be required for certain mitotic processes, including

microtubule depolymerization and tension establishment between kineto-
chores (Zhao et al., 2019). Adding ATP extracellularly can rescue both
the impaired microtubule depolymerization and weakened tension in
MCU knockdown cells. Importantly, spindle formation and checkpoint
silencing, which are widely believed as energy-demanding processes
(Hartman and Vale, 1999; Miniowitz-Shemtov et al., 2010), are not
MCU-ATP axis-dependent. MCU depletion does not affect the disassoci-
ation of MAD2 from CDC20, a critical step for checkpoint silencing. The
spindles in MCU-depleted cells also seem intact. The role of MCU-
mediated rapid ATP production may partially account for the localized
high concentration of ATP released from mitochondria. Recent studies
showed that Miro-Cenp-F interaction connected mitochondria to the tips
of dynamic microtubules and caused tip-tracking mitochondrial movements
along the microtubule (Kanfer et al., 2015). This distribution of mitochondria may account for localized ATP production and generation of the pulling
forces involved in microtubule dynamics. Further studies are needed to
detail the mechanisms by which mitochondrial ATP production or metabolic signaling specifically affects mitosis.

The spontaneous Ca^{2+} transients in mitochondria originate from those in
the cytosol. As described before, although the cytosolic Ca^{2+} transients have
been observed for more than 40 years, the upstream events that trigger these
Ca^{2+} transients have remained unclear. It has been suggested that the cytosolic Ca^{2+} transients were induced by the second messenger IP3 (Ciapa et al.,
1994; Twigg et al., 1988). Different from the membrane signaling transduction pathways, the endogenous phosphoinositide messenger system fluctuates with a rhythm corresponding to the cell cycle progression in the sea
urchin, triggering Ca^{2+} release during the cell cycle from intracellular stores,
such as ER (Ciapa et al., 1994). Further evidence also demonstrated that
the IP3 cycles occur independent of the cyclin oscillators, as blocking cyclin
synthesis through inhibitors does not affect the periodic activation of
IP3 (Ciapa et al., 1994; Han et al., 1992). However, for some time there
are no further clues about the triggers that generate Ca^{2+} release.
Interestingly, the use of fluorescent indicators to monitor cytosolic ATP
levels and mitochondrial Ca^{2+} dynamics indicates that the Ca^{2+} transients
occur in an energy-dependent manner, controlled by the AMP/ATP levels
and the energy sensor, AMPK (Zhao et al., 2019). Due to the many energy-
consuming events that occur, ATP levels drop quickly after cells enter
mitosis. When the ATP level drops to a critical level and cells experience
an acute energy crisis, mitochondrial Ca^{2+} transients occur and restore

energetic homeostasis. In addition, based on their resting cytosolic ATP levels, cells can be divided into groups. Cells with low ATP levels are more likely to produce mitochondrial Ca^{2+} transients than other groups. Similarly, mimicking a low energy state by AMP pretreatment, the amplitude and occurrence of both cytosolic and mitochondrial Ca^{2+} transients increase. In contrast, incubation of ATP, mimicking a high energy state, significantly inhibits Ca^{2+} transients. These results indicate that Ca^{2+} transients function as an adaptive response to cellular energy status, especially during mitosis. The energy state, given as the AMP/ATP ratio, is usually detected by the crucial energy sensor, AMPK (Hardie et al., 2012). Knocking out AMPK inhibits both cytosol and mitochondrial Ca^{2+} signaling, suggesting that mitotic energy defect-induced AMPK activation may serve as the potential upstream trigger for IP3 fluctuation and subsequent Ca^{2+} transients (Fig. 3B). However, further research is needed to fully understand the underlying mechanisms.

5.3 G1/S transition

In addition to the regulatory role of MCU-mediated Ca^{2+} influx during mitosis and fertilization, MCU is also required for fulfilling the metabolic demands during G1/S transition, a key event accompanied by the activation of cytosolic Ca^{2+}/CaMK and the hyperfusion of mitochondria (Koval et al., 2019). Inhibition of MCU by siRNA silencing or RU360 treatment strongly reduces the proliferation rate of VSMCs stimulated by PDGF. These results are further supported by studies with cultured skin fibroblasts, isolated from WT and MCU$^{-/-}$ mice. For example, it has been reported that MCU deficiency delayed cell cycle progression at the G1/S transition. MCU$^{-/-}$ cells exhibited an increased G0/G1 fraction and decreased S phase cell fraction compared with WT cells after release from cell cycle arrest. Similarly, the expression of cyclin D and cyclin E, two proteins required for the transition from G1 to S phase, decreased in MCU$^{-/-}$ cells (Koval et al., 2019).

As described before, accurate regulation of intracellular Ca^{2+} concentration plays a crucial role in ensuring normal cellular functions. During G1/S transition, Ca^{2+} transients have been observed in association with increased concentration of cytosolic Ca^{2+}(Montemurro et al., 2017; Russa et al., 2009). On one hand, the elevated Ca^{2+} stimulates Ca^{2+} transfer from cytosol to mitochondria via MCU. At the same time, to match the increased energetic demand, mitochondria undergo hyperfused, tubular morphological changes (Mitra et al., 2009), which increase the mitochondrial capacity to take up Ca^{2+}. On the other hand, the activation of MCU facilitates cytosolic Ca^{2+} clearance, prevents excessive activation of CaMKII and maintains the hyperfused morphology of mitochondria (Koval et al., 2019). When MCU is inhibited, delayed cytosolic Ca^{2+} clearance increases the activation of CaMKII, which promotes PDGF-dependent Drp1 phosphorylation and mitochondrial fission (Fig. 3B). Although Ca^{2+} uptake fuels mitochondrial energy production, mitochondrial morphology changes between fission and fusion play a decisive role in ATP synthesis activity, especially in PDGF-induced cell proliferation, because inhibition of CaMKII can restore Drp1 phosphorylation, mitochondrial fragment and OCR efficiently even in the absence of MCU. However, because CaMKII would have been activated during G1/S transition, questions remain about how cells activate or deactivate it during the cell cycle. The specific connections between CaMKII, mitochondrial Ca^{2+} and morphological changes also need to be further investigated.

5.4 Initiation of hematopoietic stem cells division

Stem cells, due to their potential biomedical applications, have recently received worldwide attention. Unlike most cell types, stem cells usually reside in a specialized microenvironment, called a "stem cell niche" to maintain their pluripotent status. For example, hematopoietic stem cells (HSCs), exist within the bone marrow (BM), which provides a hypoxia condition for the maintenance of HSC quiescence (Crane et al., 2017). When facing stresses such as interferon treatment or 5-fluorouracil (5-FU) induced BM suppression, HSCs are forced to exit quiescence and undergo the cell division cycle (Baldridge et al., 2010; Harrison and Lerner, 1991). In this situation, elevated intracellular Ca^{2+} concentration is observed before HSCs enter the cell cycle (Umemoto et al., 2018). Meanwhile, mitochondrial Ca^{2+} levels also increase, accompanied by an enhancement of mitochondrial membrane potential, increased mitochondrial superoxide level and increased intracellular ATP content, indicating a potential mitochondrial Ca^{2+} uptake process during the initiation of HSC division. Inhibiting the increase of intracellular Ca^{2+} via Nifedipine treatment, an antagonist of L-type voltage-gated Ca^{2+} channels (LTCCs), significantly reduces mitochondrial functions and prolongs HSC division intervals.

Compared to quiescent HSCs, the metabolic program of the cycling HSCs is quite different, especially represented by the activation of mitochondrial OXPHOS (Nakamura-Ishizu et al., 2020; Zheng et al., 2020). It is therefore plausible that mitochondrial Ca^{2+} uptake may serve as a metabolic switch that activates the cell cycle and determines the cells' fate. Culturing HSCs in a high Ca^{2+} condition *in vitro* increases mitochondrial respiration and compromises their stem cell functions, while promoting gene expression of a series of differentiation markers (Luchsinger et al., 2019). Conversely, culturing HSCs in a low Ca^{2+} condition increases their maintenance of stem cell features (Luchsinger et al., 2019), indicating that manipulating mitochondrial Ca^{2+} signaling may alter the cell fate of HSCs. Similarly, quiescent HSCs have been shown to display a relatively low intracellular Ca^{2+} level in both cytosol and mitochondria *in vivo*, which is partially achieved via low plasma membrane Ca^{2+} influx activity (Umemoto et al., 2018). The Ca^{2+} influx in HSCs is inhibited, probably due to the adenosine provided by the surrounding myeloid progenitor cells. Suppressing plasma membrane Ca^{2+} influx, which affects mitochondrial Ca^{2+} level, leads to up-regulation in the expression of genes associated with quiescence, self-renewal and HSC markers (Umemoto et al., 2018), resulting in the maintenance of HSC stem cell features. However, as there is no direct evidence of targeting mitochondrial Ca^{2+} signaling in hematopoietic system, it is still to be determined whether MCU or mitochondrial Ca^{2+} signaling participates in initiating cell proliferation and HSC cell fate decisions.

In addition, the mechanisms by which mitochondrial Ca^{2+} regulates cell cycle- and stem cell-related gene expression also needs further exploration. One potential mechanism involves epigenetic regulation. Aside from increased ATP synthesis and ROS production, Ca^{2+} activates three mitochondrial dehydrogenases (Denton et al., 1972, 1978; McCormack and Denton, 1979) and regulates the epigenome through promoting the formation of acetyl-CoA and α-ketoglutarate (α-KG) (Lombardi et al., 2019). Acetyl-CoA is an essential substrate for histone acetylation. The level of α-KG regulates the activity of histone demethylase, including histone lysine demethylases (KDMs) and ten-eleven translocated (TET) enzymes. In addition, Ca^{2+} also accelerates NADH shuttling from cytosol to mitochondria, regenerating cytosolic NAD^+ which in turn affects a class of histone/protein deacetylases, Sirtuins (Marcu et al., 2014). Recently, Lombardi et al. demonstrated that loss of mitochondrial Ca^{2+} uptake promotes myofibroblast differentiation and fibrosis. They found that MICU1-mediated MCU gating leads to a metabolic switch. The resulting activation of α-KG dependent histone demethylases and epigenetic reprogramming then determines the cell fate transition (Lombardi et al., 2019). A similar regulatory model may occur in multiple cellular processes, such as cell proliferation, stem cell differentiation and cell fate decision.

6. Mitochondrial Ca^{2+} dependent cell cycle regulation in tumorigenesis

Cancer cells are characterized by infinite proliferation and therapeutic resistance due to their remarkable adaptability and robustness. One of the noteworthy hallmarks of cancer is deregulating cellular energetics (Hanahan and Weinberg, 2011). Although cancer cells metabolize mainly though aerobic glycolysis and exhibit the "Warburg effect,"(Liberti and Locasale, 2016), the additional contribution of mitochondria to the growth and survival of tumor cells cannot be neglected (Vasan et al., 2020). Mitochondria not only provide metabolic intermediates and energy, but also integrate signaling from cytosol or nuclei to determine cell fate and function. Increasing evidence shows that mitochondria play important roles in promoting cancer. The mechanism may involve Ca^{2+} and transporters localized on the mitochondrial membrane. Targeting this Ca^{2+} driven metabolism may represent a new strategy for cancer therapy.

Mitochondrial Ca^{2+} signaling contributes to malignant transformation through multiple mechanisms, including cell death response, migration, invasion and proliferation. In this section, the discussion is primarily focused on the cell cycle regulatory role of mitochondrial Ca^{2+} signaling in cancer development. For more comprehensive description, we refer the interested reader to other reviews on this topic (Marchi et al., 2019; Sterea and El Hiani, 2020; Vultur et al., 2018).

In order to sustain intracellular metabolic demand and ensure growth and survival, mitochondrial Ca^{2+} signaling pathways have been reported as being remodeled in certain cancer cells. For example, in triple negative breast cancer, MCU expression was positively correlated with tumor size and degree of lymph node infiltration (Tosatto et al., 2016). MCU silencing significantly inhibits tumor growth and metastasis formation. According to The Human Protein Atlas (www.proteinatlas.org), most cancer tissues show moderate to strong MCU immunostaining, including lung and ovarian cancer. However, in other cancer cells, mitochondrial Ca^{2+} may have the opposite effect on tumor development. In human colon cancers, the expression of MCU is deregulated due to the overexpression of an MCU-targeting microRNA, miR-25, which favors cancer cell survival and tumorigenesis

(Marchi et al., 2013). Thus, the mitochondrial Ca^{2+} signaling profile may vary depending on the cell type and stages of cancer development.

Likewise, mitochondrial Ca^{2+}-dependent cell cycle control may play different roles at specific stages of tumorigenesis. On one hand, mitochondrial Ca^{2+} activation acts as a pro-oncogenic signal. Ca^{2+} transfer from the ER to mitochondria initiates the cell cycle and provides energetic output for survival. Inhibition of InsP3R activity strongly suppresses tumor growth *in vivo* and leads to mitotic catastrophe induced cell death (Cardenas et al., 2016). On the other hand, defective cell cycle progression, especially abnormal mitosis (Zhao et al., 2019), caused by MCU inhibition can result in chromosomal instability, which is conducive to tumor formation. As such, it is difficult to conclude whether mitochondrial Ca^{2+} signaling has a pro- or anti-cancer effect. Of course, considering the versatile nature of Ca^{2+} signaling, the relation between mitochondrial Ca^{2+} and tumorigenesis is more complex than described here. However, a better understanding of the underlying mechanisms may help us in developing targeted interventions to halt tumor progression and affect other related diseases.

7. Conclusions

With the discovery of MCU and the progress in Ca^{2+} imaging techniques, recent findings have revealed a crucial role of MCU-mediated mitochondrial Ca^{2+} signaling during cell cycle progression. Via the activation of mitochondrial energetic output and ROS production, MCU-mediated Ca^{2+} influx occurs in response to the metabolic demand during mitosis and fertilization-induced cell division. In PDGF-induced cell proliferation, MCU is required to balance Ca^{2+} concentration in cytosol and mitochondria, thus maintaining mitochondrial morphology of hyperfusion during G1/S transition. Despite recent progress, the research in this area is still in its infancy. Central questions remain: How is the spontaneous Ca^{2+} fluctuation upstream achieved? Is there any priming intracellular or extracellular signals that trigger the Ca^{2+} fluctuation? The answer to these questions may help us fully understand the self-adaptation mechanisms of cells when facing extracellular environmental fluctuations and intracellular metabolic needs. The discovery of AMPK-mediated MCU phosphorylation and subsequent mitochondrial Ca^{2+} uptake may provide a new perspective for understanding the relationship between Ca^{2+} signaling, metabolism and energy-consuming events.

In addition, we have yet to elucidate the biological relevance or specific mechanisms by which mitochondrial Ca^{2+} signaling regulates the cell cycle. The regulation is likely to be involved in many physiological and pathological processes, such as the initiation of HSCs division and tumor progression, both of which are characterized by robust cell division ability. Recent studies have shown that low Ca^{2+} media is associated with decreased mitochondrial activity and enhanced HSC maintenance *in vitro*, and that silencing MCU or blocking Ca^{2+} transfer from ER to mitochondria also produce a considerable antitumor effect. Therefore, manipulating mitochondrial Ca^{2+} signaling is emerging as a potential strategy in tumor treatment, hematopoiesis regeneration and interventions for many other diseases. A better understanding of the underlying mechanisms will provide guidance for future clinical applications.

Acknowledgments

The authors would like to thank all the members from National Center of Biomedical Analysis (NCBA) and L&L lab for kindly discussion and critical reading of the manuscript.

Conflict of interest

The authors declare no conflict of interest.

References

Antony, A.N., Paillard, M., Moffat, C., Juskeviciute, E., Correnti, J., Bolon, B., Rubin, E., Csordas, G., Seifert, E.L., Hoek, J.B., Hajnoczky, G., 2016. MICU1 regulation of mitochondrial Ca(2+) uptake dictates survival and tissue regeneration. Nat. Commun. 7, 10955.

Baker, P.F., Warner, A.E., 1972. Intracellular calcium and cell cleavage in early embryos of Xenopus laevis. J. Cell Biol. 53, 579–581.

Baldridge, M.T., King, K.Y., Boles, N.C., Weksberg, D.C., Goodell, M.A., 2010. Quiescent haematopoietic stem cells are activated by IFN-gamma in response to chronic infection. Nature 465, 793–797.

Baughman, J.M., Perocchi, F., Girgis, H.S., Plovanich, M., Belcher-Timme, C.A., Sancak, Y., Bao, X.R., Strittmatter, L., Goldberger, O., Bogorad, R.L., Koteliansky, V., Mootha, V.K., 2011. Integrative genomics identifies MCU as an essential component of the mitochondrial calcium uniporter. Nature 476, 341–345.

Belosludtsev, K.N., Dubinin, M.V., Belosludtseva, N.V., Mironova, G.D., 2019. Mitochondrial Ca2+ transport: mechanisms, molecular structures, and role in cells. Biochemistry (Mosc.) 84, 593–607.

Berridge, M.J., Bootman, M.D., LIPP, P., 1998. Calcium—a life and death signal. Nature 395, 645–648.

Bershadsky, A.D., Gelfand, V.I., 1981. ATP-dependent regulation of cytoplasmic microtubule disassembly. Proc. Natl. Acad. Sci. U. S. A. 78, 3610–3613.

Bertero, E., Maack, C., 2018. Calcium signaling and reactive oxygen species in mitochondria. Circ. Res. 122, 1460–1478.

Blattner, J.R., He, L., Lemasters, J.J., 2001. Screening assays for the mitochondrial permeability transition using a fluorescence multiwell plate reader. Anal. Biochem. 295, 220–226.

Bloom, J., Cross, F.R., 2007. Multiple levels of cyclin specificity in cell-cycle control. Nat. Rev. Mol. Cell Biol. 8, 149–160.

Bobinnec, Y., Marcaillou, C., Morin, X., Debec, A., 2003. Dynamics of the endoplasmic reticulum during early development of Drosophila melanogaster. Cell Motil. Cytoskeleton 54, 217–225.

Bornstein, R., Gonzalez, B., Johnson, S.C., 2020. Mitochondrial pathways in human health and aging. Mitochondrion 54, 72–84.

Brand, M.D., 2016. Mitochondrial generation of superoxide and hydrogen peroxide as the source of mitochondrial redox signaling. Free Radic. Biol. Med. 100, 14–31.

Briston, T., Selwood, D.L., Szabadkai, G., Duchen, M.R., 2019. Mitochondrial permeability transition: a molecular lesion with multiple drug targets. Trends Pharmacol. Sci. 40, 50–70.

Cai, Z., Li, C.F., Han, F., Liu, C., Zhang, A., Hsu, C.C., Peng, D., Zhang, X., Jin, G., Rezaeian, A.H., Wang, G., Zhang, W., Pan, B.S., Wang, C.Y., Wang, Y.H., Wu, S.Y., Yang, S.C., Hsu, F.C., D'agostino Jr., R.B., Furdui, C.M., Kucera, G.L., Parks, J.S., Chilton, F.H., Huang, C.Y., Tsai, F.J., Pasche, B., Watabe, K., Lin, H.K., 2020. Phosphorylation of PDHA by AMPK drives TCA cycle to promote cancer metastasis. Mol. Cell 80, 263–278 e7.

Carafoli, E., 2012. The interplay of mitochondria with calcium: an historical appraisal. Cell Calcium 52, 1–8.

Cardenas, C., Muller, M., McNeal, A., Lovy, A., Jana, F., Bustos, G., Urra, F., Smith, N., Molgo, J., Diehl, J.A., Ridky, T.W., Foskett, J.K., 2016. Selective vulnerability of cancer cells by inhibition of Ca(2+) transfer from endoplasmic reticulum to mitochondria. Cell Rep. 14, 2313–2324.

Carey, B.W., Finley, L.W., Cross, J.R., Allis, C.D., Thompson, C.B., 2015. Intracellular alpha-ketoglutarate maintains the pluripotency of embryonic stem cells. Nature 518, 413–416.

Carlton, J.G., Jones, H., Eggert, U.S., 2020. Membrane and organelle dynamics during cell division. Nat. Rev. Mol. Cell Biol. 21, 151–166.

Chafouleas, J.G., Bolton, W.E., Hidaka, H., Boyd 3rd, A.E., Means, A.R., 1982. Calmodulin and the cell cycle: involvement in regulation of cell-cycle progression. Cell 28, 41–50.

Ciapa, B., Pesando, D., Wilding, M., Whitaker, M., 1994. Cell-cycle calcium transients driven by cyclic changes in inositol trisphosphate levels. Nature 368, 875–878.

Clothier, G., Timourian, H., 1972. Calcium uptake and release by dividing sea urchin eggs. Exp. Cell Res. 75, 105–110.

Crane, G.M., Jeffery, E., Morrison, S.J., 2017. Adult haematopoietic stem cell niches. Nat. Rev. Immunol. 17, 573–590.

Csordas, G., Golenar, T., Seifert, E.L., Kamer, K.J., Sancak, Y., Perocchi, F., Moffat, C., Weaver, D., De La Fuente Perez, S., Bogorad, R., Koteliansky, V., Adijanto, J., Mootha, V.K., Hajnoczky, G., 2013. MICU1 controls both the threshold and cooperative activation of the mitochondrial Ca(2)(+) uniporter. Cell Metab. 17, 976–987.

Cung, T.T., Morel, O., Cayla, G., Rioufol, G., Garcia-Dorado, D., Angoulvant, D., Bonnefoy-Cudraz, E., Guerin, P., Elbaz, M., Delarche, N., Coste, P., Vanzetto, G., Metge, M., Aupetit, J.F., Jouve, B., Motreff, P., Tron, C., Labeque, J.N., Steg, P.G., Cottin, Y., Range, G., Clerc, J., Claeys, M.J., Coussement, P., Prunier, F., Moulin, F., Roth, O., Belle, L., Dubois, P., Barragan, P., Gilard, M., Piot, C., Colin, P., De Poli, F., Morice, M.C., Ider, O., Dubois-Rande, J.L., Unterseeh, T.,

Le Breton, H., Beard, T., Blanchard, D., Grollier, G., Malquarti, V., Staat, P., Sudre, A., Elmer, E., Hansson, M.J., Bergerot, C., Boussaha, I., Jossan, C., Derumeaux, G., Mewton, N., Ovize, M., 2015. Cyclosporine before PCI in patients with acute myocardial infarction. N. Engl. J. Med. 373, 1021–1031.

Dasgupta, S., 2019. Mitochondrion: i am more than a fuel server. Ann Transl Med 7, 594.

De Stefani, D., Raffaello, A., Teardo, E., Szabo, I., Rizzuto, R., 2011. A forty-kilodalton protein of the inner membrane is the mitochondrial calcium uniporter. Nature 476, 336–340.

De Stefani, D., Rizzuto, R., Pozzan, T., 2016. Enjoy the trip: calcium in mitochondria back and forth. Annu. Rev. Biochem. 85, 161–192.

Deluca, H.F., Engstrom, G.W., 1961. Calcium uptake by rat kidney mitochondria. Proc. Natl. Acad. Sci. U. S. A. 47, 1744–1750.

Denton, R.M., Randle, P.J., Martin, B.R., 1972. Stimulation by calcium ions of pyruvate dehydrogenase phosphate phosphatase. Biochem. J. 128, 161–163.

Denton, R.M., Richards, D.A., Chin, J.G., 1978. Calcium ions and the regulation of NAD+-linked isocitrate dehydrogenase from the mitochondria of rat heart and other tissues. Biochem. J. 176, 899–906.

Dong, Z., Shanmughapriya, S., Tomar, D., Siddiqui, N., Lynch, S., Nemani, N., Breves, S.L., Zhang, X., Tripathi, A., Palaniappan, P., Riitano, M.F., Worth, A.M., Seelam, A., Carvalho, E., Subbiah, R., Jana, F., Soboloff, J., Peng, Y., Cheung, J.Y., Joseph, S.K., Caplan, J., Rajan, S., Stathopulos, P.B., Madesh, M., 2017. Mitochondrial Ca2+ uniporter is a mitochondrial luminal redox sensor that augments MCU channel activity. Mol. Cell 65, 1014–1028 e7.

Drago, I., De Stefani, D., Rizzuto, R., Pozzan, T., 2012. Mitochondrial Ca2+ uptake contributes to buffering cytoplasmic Ca2+ peaks in cardiomyocytes. Proc. Natl. Acad. Sci. U. S. A. 109, 12986–12991.

Fan, C., Fan, M., Orlando, B.J., Fastman, N.M., Zhang, J., Xu, Y., Chambers, M.G., Xu, X., Perry, K., Liao, M., Feng, L., 2018. X-ray and cryo-EM structures of the mitochondrial calcium uniporter. Nature 559, 575–579.

Fan, M., Zhang, J., Tsai, C.W., Orlando, B.J., Rodriguez, M., Xu, Y., Liao, M., Tsai, M.F., Feng, L., 2020. Structure and mechanism of the mitochondrial Ca(2+) uniporter holocomplex. Nature 582, 129–133.

Feno, S., Rizzuto, R., Raffaello, A., Vecellio Reane, D., 2020. The molecular complexity of the mitochondrial calcium uniporter. Cell Calcium 93, 102322.

Fieni, F., Johnson, D.E., Hudmon, A., Kirichok, Y., 2014. Mitochondrial Ca2+ uniporter and CaMKII in heart. Nature 513, E1–E2.

Foerder, C.A., Klebanoff, S.J., Shapiro, B.M., 1978. Hydrogen peroxide production, chemiluminescence, and the respiratory burst of fertilization: interrelated events in early sea urchin development. Proc. Natl. Acad. Sci. U. S. A. 75, 3183–3187.

Gaigg, B., Simbeni, R., Hrastnik, C., Paltauf, F., Daum, G., 1995. Characterization of a microsomal subfraction associated with mitochondria of the yeast, Saccharomyces cerevisiae. Involvement in synthesis and import of phospholipids into mitochondria. Biochim. Biophys. Acta 1234, 214–220.

Giorgi, C., Marchi, S., Pinton, P., 2018. The machineries, regulation and cellular functions of mitochondrial calcium. Nat. Rev. Mol. Cell Biol. 19, 713–730.

Glancy, B., Willis, W.T., Chess, D.J., Balaban, R.S., 2013. Effect of calcium on the oxidative phosphorylation cascade in skeletal muscle mitochondria. Biochemistry 52, 2793–2809.

Grandin, N., Charbonneau, M., 1991. Intracellular free calcium oscillates during cell division of Xenopus embryos. J. Cell Biol. 112, 711–718.

Gunter, T.E., Pfeiffer, D.R., 1990. Mechanisms by which mitochondria transport calcium. Am. J. Physiol. 258, C755–C786.

Han, J.K., Fukami, K., Nuccitelli, R., 1992. Reducing inositol lipid hydrolysis, Ins(1,4,5)P3 receptor availability, or Ca2+ gradients lengthens the duration of the cell cycle in Xenopus laevis blastomeres. J. Cell Biol. 116, 147–156.

Han, Y., Ishibashi, S., Iglesias-Gonzalez, J., Chen, Y., Love, N.R., Amaya, E., 2018. Ca(2+)-induced mitochondrial ROS regulate the early embryonic cell cycle. Cell Rep. 22, 218–231.

Hanahan, D., Weinberg, R.A., 2011. Hallmarks of cancer: the next generation. Cell 144, 646–674.

Harbauer, A.B., Opalinska, M., Gerbeth, C., Herman, J.S., Rao, S., Schonfisch, B., Guiard, B., Schmidt, O., Pfanner, N., Meisinger, C., 2014. Mitochondria. Cell cycle-dependent regulation of mitochondrial preprotein translocase. Science 346, 1109–1113.

Hardie, D.G., Ross, F.A., Hawley, S.A., 2012. AMPK: a nutrient and energy sensor that maintains energy homeostasis. Nat. Rev. Mol. Cell Biol. 13, 251–262.

Harrison, D.E., Lerner, C.P., 1991. Most primitive hematopoietic stem cells are stimulated to cycle rapidly after treatment with 5-fluorouracil. Blood 78, 1237–1240.

Hartman, J.J., Vale, R.D., 1999. Microtubule disassembly by ATP-dependent oligomerization of the AAA enzyme katanin. Science 286, 782–785.

Hempel, N., Trebak, M., 2017. Crosstalk between calcium and reactive oxygen species signaling in cancer. Cell Calcium 63, 70–96.

Hirabayashi, Y., Kwon, S.K., Paek, H., Pernice, W.M., Paul, M.A., Lee, J., Erfani, P., Raczkowski, A., Petrey, D.S., Pon, L.A., Polleux, F., 2017. ER-mitochondria tethering by PDZD8 regulates Ca(2+) dynamics in mammalian neurons. Science 358, 623–630.

Hirschler-Laszkiewicz, I., Chen, S.J., Bao, L., Wang, J., Zhang, X.Q., Shanmughapriya, S., Keefer, K., Madesh, M., Cheung, J.Y., Miller, B.A., 2018. The human ion channel TRPM2 modulates neuroblastoma cell survival and mitochondrial function through Pyk2, CREB, and MCU activation. Am. J. Physiol. Cell Physiol. 315, C571–C586.

Hochegger, H., Takeda, S., Hunt, T., 2008. Cyclin-dependent kinases and cell-cycle transitions: does one fit all? Nat. Rev. Mol. Cell Biol. 9, 910–916.

Holmfeldt, P., Zhang, X., Stenmark, S., Walczak, C.E., Gullberg, M., 2005. CaMKIIgamma-mediated inactivation of the Kin I kinesin MCAK is essential for bipolar spindle formation. EMBO J. 24, 1256–1266.

Holmstrom, K.M., Finkel, T., 2014. Cellular mechanisms and physiological consequences of redox-dependent signalling. Nat. Rev. Mol. Cell Biol. 15, 411–421.

Humeau, J., Bravo-San Pedro, J.M., Vitale, I., Nunez, L., Villalobos, C., Kroemer, G., Senovilla, L., 2018. Calcium signaling and cell cycle: progression or death. Cell Calcium 70, 3–15.

Imai, S., Armstrong, C.M., Kaeberlein, M., Guarente, L., 2000. Transcriptional silencing and longevity protein Sir2 is an NAD-dependent histone deacetylase. Nature 403, 795–800.

Imamura, H., Nhat, K.P., Togawa, H., Saito, K., Iino, R., Kato-Yamada, Y., Nagai, T., Noji, H., 2009. Visualization of ATP levels inside single living cells with fluorescence resonance energy transfer-based genetically encoded indicators. Proc. Natl. Acad. Sci. U. S. A. 106, 15651–15656.

Izant, J.G., 1983. The role of calcium ions during mitosis. Calcium participates in the anaphase trigger. Chromosoma 88, 1–10.

Jin, O.U., Jhun, B.S., Xu, S., Hurst, S., Raffaello, A., Liu, X., Yi, B., Zhang, H., Gross, P., Mishra, J., Ainbinder, A., Kettlewell, S., Smith, G.L., Dirksen, R.T., Wang, W., Rizzuto, R., Sheu, S.S., 2014. Adrenergic signaling regulates mitochondrial Ca2+ uptake through Pyk2-dependent tyrosine phosphorylation of the mitochondrial Ca2+ uniporter. Antioxid. Redox Signal. 21, 863–879.

Joiner, M.L., Koval, O.M., Li, J., He, B.J., Allamargot, C., Gao, Z., Luczak, E.D., Hall, D.D., Fink, B.D., Chen, B., Yang, J., Moore, S.A., Scholz, T.D., Strack, S., Mohler, P.J., Sivitz, W.I., Song, L.S., Anderson, M.E., 2012. Camkii determines mitochondrial stress responses in heart. Nature 491, 269–273.

Kahl, C.R., Means, A.R., 2003. Regulation of cell cycle progression by calcium/calmodulin-dependent pathways. Endocr. Rev. 24, 719–736.

Kahl, C.R., Means, A.R., 2004. Regulation of cyclin D1/Cdk4 complexes by calcium/calmodulin-dependent protein kinase I. J. Biol. Chem. 279, 15411–15419.

Kamer, K.J., Mootha, V.K., 2015. The molecular era of the mitochondrial calcium uniporter. Nat. Rev. Mol. Cell Biol. 16, 545–553.

Kamer, K.J., Grabarek, Z., Mootha, V.K., 2017. High-affinity cooperative Ca(2+) binding by MICU1-MICU2 serves as an on-off switch for the uniporter. EMBO Rep. 18, 1397–1411.

Kanfer, G., Courtheoux, T., Peterka, M., Meier, S., Soste, M., Melnik, A., Reis, K., Aspenstrom, P., Peter, M., Picotti, P., Kornmann, B., 2015. Mitotic redistribution of the mitochondrial network by Miro and Cenp-F. Nat. Commun. 6, 8015.

Kao, J.P., Alderton, J.M., Tsien, R.Y., Steinhardt, R.A., 1990. Active involvement of Ca2+ in mitotic progression of Swiss 3T3 fibroblasts. J. Cell Biol. 111, 183–196.

Kon, N., Satoh, A., Miyoshi, N., 2017. A small-molecule DS44170716 inhibits Ca(2+)-induced mitochondrial permeability transition. Sci. Rep. 7, 3864.

Koval, O.M., Nguyen, E.K., Santhana, V., Fidler, T.P., Sebag, S.C., Rasmussen, T.P., Mittauer, D.J., Strack, S., Goswami, P.C., Abel, E.D., Grumbach, I.M., 2019. Loss of MCU prevents mitochondrial fusion in G1-S phase and blocks cell cycle progression and proliferation. Sci. Signal. 12 (579), eaav1439.

Lee, I.H., Finkel, T., 2013. Metabolic regulation of the cell cycle. Curr. Opin. Cell Biol. 25, 724–729.

Lee, Y., Min, C.K., Kim, T.G., Song, H.K., Lim, Y., Kim, D., Shin, K., Kang, M., Kang, J.Y., Youn, H.S., Lee, J.G., An, J.Y., Park, K.R., Lim, J.J., Kim, J.H., Kim, J.H., Park, Z.Y., Kim, Y.S., Wang, J., Kim, D.H., Eom, S.H., 2015. Structure and function of the N-terminal domain of the human mitochondrial calcium uniporter. EMBO Rep. 16, 1318–1333.

Lee, Y., Park, J., Lee, G., Yoon, S., Min, C.K., Kim, T.G., Yamamoto, T., Kim, D.H., Lee, K.W., Eom, S.H., 2020. S92 phosphorylation induces structural changes in the N-terminus domain of human mitochondrial calcium uniporter. Sci. Rep. 10, 9131.

Li, N., Wang, C., Wu, Y., Liu, X., Cao, X., 2009. Ca(2+)/calmodulin-dependent protein kinase II promotes cell cycle progression by directly activating MEK1 and subsequently modulating p27 phosphorylation. J. Biol. Chem. 284, 3021–3027.

Liberti, M.V., Locasale, J.W., 2016. The warburg effect: how does it benefit cancer cells? Trends Biochem. Sci. 41, 211–218.

Liu, J.C., Liu, J., Holmstrom, K.M., Menazza, S., Parks, R.J., Fergusson, M.M., Yu, Z.X., Springer, D.A., Halsey, C., Liu, C., Murphy, E., Finkel, T., 2016. MICU1 serves as a molecular gatekeeper to prevent in vivo mitochondrial calcium overload. Cell Rep. 16, 1561–1573.

Liu, J.C., Syder, N.C., Ghorashi, N.S., Willingham, T.B., Parks, R.J., Sun, J., Fergusson, M.M., Liu, J., Holmstrom, K.M., Menazza, S., Springer, D.A., Liu, C., Glancy, B., Finkel, T., Murphy, E., 2020. EMRE is essential for mitochondrial calcium uniporter activity in a mouse model. JCI Insight 5 (4), e134063.

Lombardi, A.A., Gibb, A.A., Arif, E., Kolmetzky, D.W., Tomar, D., Luongo, T.S., Jadiya, P., Murray, E.K., Lorkiewicz, P.K., Hajnoczky, G., Murphy, E., Arany, Z.P., Kelly, D.P., Margulies, K.B., Hill, B.G., Elrod, J.W., 2019. Mitochondrial calcium exchange links metabolism with the epigenome to control cellular differentiation. Nat. Commun. 10, 4509.

Lorca, T., Cruzalegui, F.H., Fesquet, D., Cavadore, J.C., Mery, J., Means, A., Doree, M., 1993. Calmodulin-dependent protein kinase II mediates inactivation of MPF and CSF upon fertilization of Xenopus eggs. Nature 366, 270–273.

Lu, K.P., Means, A.R., 1993. Regulation of the cell cycle by calcium and calmodulin. Endocr. Rev. 14, 40–58.

Lu, K.P., Osmani, S.A., Osmani, A.H., Means, A.R., 1993. Essential roles for calcium and calmodulin in G2/M progression in Aspergillus nidulans. J. Cell Biol. 121, 621–630.

Luchsinger, L.L., Strikoudis, A., Danzl, N.M., Bush, E.C., Finlayson, M.O., Satwani, P., Sykes, M., Yazawa, M., Snoeck, H.W., 2019. Harnessing hematopoietic stem cell low intracellular calcium improves their maintenance in vitro. Cell Stem Cell 25, 225–240 e7.

Luo, M., Anderson, M.E., 2013. Mechanisms of altered Ca(2)(+) handling in heart failure. Circ. Res. 113, 690–708.

Luongo, T.S., Lambert, J.P., Yuan, A., Zhang, X., Gross, P., Song, J., Shanmughapriya, S., Gao, E., Jain, M., Houser, S.R., Koch, W.J., Cheung, J.Y., Madesh, M., Elrod, J.W., 2015. The mitochondrial calcium uniporter matches energetic supply with cardiac workload during stress and modulates permeability transition. Cell Rep. 12, 23–34.

Machaca, K., 2011. Ca(2+) signaling, genes and the cell cycle. Cell Calcium 49, 323–330.

Mailloux, R.J., 2015. Teaching the fundamentals of electron transfer reactions in mitochondria and the production and detection of reactive oxygen species. Redox Biol. 4, 381–398.

Mallilankaraman, K., Doonan, P., Cardenas, C., Chandramoorthy, H.C., Muller, M., Miller, R., Hoffman, N.E., Gandhirajan, R.K., Molgo, J., Birnbaum, M.J., Rothberg, B.S., Mak, D.O., Foskett, J.K., Madesh, M., 2012. MICU1 is an essential gatekeeper for MCU-mediated mitochondrial Ca(2+) uptake that regulates cell survival. Cell 151, 630–644.

Marchi, S., Lupini, L., Patergnani, S., Rimessi, A., Missiroli, S., Bonora, M., Bononi, A., Corra, F., Giorgi, C., De Marchi, E., Poletti, F., Gafa, R., Lanza, G., Negrini, M., Rizzuto, R., Pinton, P., 2013. Downregulation of the mitochondrial calcium uniporter by cancer-related miR-25. Curr. Biol. 23, 58–63.

Marchi, S., Vitto, V.A.M., Danese, A., Wieckowski, M.R., Giorgi, C., Pinton, P., 2019. Mitochondrial calcium uniporter complex modulation in cancerogenesis. Cell Cycle 18, 1068–1083.

Marcu, R., Wiczer, B.M., Neeley, C.K., Hawkins, B.J., 2014. Mitochondrial matrix Ca(2)(+) accumulation regulates cytosolic NAD(+)/NADH metabolism, protein acetylation, and sirtuin expression. Mol. Cell. Biol. 34, 2890–2902.

McCormack, J.G., Denton, R.M., 1979. The effects of calcium ions and adenine nucleotides on the activity of pig heart 2-oxoglutarate dehydrogenase complex. Biochem. J. 180, 533–544.

Meng, T.C., Fukada, T., Tonks, N.K., 2002. Reversible oxidation and inactivation of protein tyrosine phosphatases in vivo. Mol. Cell 9, 387–399.

Miniowitz-Shemtov, S., Teichner, A., Sitry-Shevah, D., Hershko, A., 2010. ATP is required for the release of the anaphase-promoting complex/cyclosome from inhibition by the mitotic checkpoint. Proc. Natl. Acad. Sci. U. S. A. 107, 5351–5356.

Mishra, P., Chan, D.C., 2014. Mitochondrial dynamics and inheritance during cell division, development and disease. Nat. Rev. Mol. Cell Biol. 15, 634–646.

Mitra, K., Wunder, C., Roysam, B., Lin, G., Lippincott-Schwartz, J., 2009. A hyperfused mitochondrial state achieved at G1-S regulates cyclin E buildup and entry into S phase. Proc. Natl. Acad. Sci. U. S. A. 106, 11960–11965.

Monassier, L., Ayme-Dietrich, E., Aubertin-Kirch, G., Pathak, A., 2016. Targeting myocardial reperfusion injuries with cyclosporine in the CIRCUS Trial—pharmacological reasons for failure. Fundam. Clin. Pharmacol. 30, 191–193.

Montemurro, C., Vadrevu, S., Gurlo, T., Butler, A.E., Vongbunyong, K.E., Petcherski, A., Shirihai, O.S., Satin, L.S., Braas, D., Butler, P.C., Tudzarova, S., 2017. Cell cycle-related metabolism and mitochondrial dynamics in a replication-competent pancreatic beta-cell line. Cell Cycle 16, 2086–2099.

Moreau, B., Nelson, C., Parekh, A.B., 2006. Biphasic regulation of mitochondrial Ca2+ uptake by cytosolic Ca2+ concentration. Curr. Biol. 16, 1672–1677.

Moussaieff, A., Rouleau, M., Kitsberg, D., Cohen, M., Levy, G., Barasch, D., Nemirovski, A., Shen-Orr, S., Laevsky, I., Amit, M., Bomze, D., Elena-Herrmann, B., Scherf, T., Nissim-Rafinia, M., Kempa, S., Itskovitz-Eldor, J., Meshorer, E., Aberdam, D., Nahmias, Y., 2015. Glycolysis-mediated changes in acetyl-CoA and histone acetylation control the early differentiation of embryonic stem cells. Cell Metab. 21, 392–402.

Murphy, A.N., Kelleher, J.K., Fiskum, G., 1990. Submicromolar Ca2+ regulates phosphorylating respiration by normal rat liver and AS-30D hepatoma mitochondria by different mechanisms. J. Biol. Chem. 265, 10527–10534.

Murphy, E., Pan, X., Nguyen, T., Liu, J., Holmstrom, K.M., Finkel, T., 2014. Unresolved questions from the analysis of mice lacking MCU expression. Biochem. Biophys. Res. Commun. 449, 384–385.

Murray, A.W., 2004. Recycling the cell cycle: cyclins revisited. Cell 116, 221–234.

Nakada, D., Saunders, T.L., Morrison, S.J., 2010. Lkb1 regulates cell cycle and energy metabolism in haematopoietic stem cells. Nature 468, 653–658.

Nakamura-Ishizu, A., Ito, K., Suda, T., 2020. Hematopoietic stem cell metabolism during development and aging. Dev. Cell 54, 239–255.

Nakano, M., Imamura, H., Nagai, T., Noji, H., 2011. Ca(2)(+) regulation of mitochondrial ATP synthesis visualized at the single cell level. ACS Chem. Biol. 6, 709–715.

Nakazawa, T., Asami, K., Shoger, R., Fujiwara, A., Yasumasu, I., 1970. Ca2+ uptake H+ ejection and respiration in sea urchin eggs on fertilization. Exp. Cell Res. 63, 143–146.

Nguyen, E.K., Koval, O.M., Noble, P., Broadhurst, K., Allamargot, C., Wu, M., Strack, S., Thiel, W.H., Grumbach, I.M., 2018a. CaMKII (Ca(2+)/calmodulin-dependent kinase II) in mitochondria of smooth muscle cells controls mitochondrial mobility, migration, and neointima formation. Arterioscler. Thromb. Vasc. Biol. 38, 1333–1345.

Nguyen, N.X., Armache, J.P., Lee, C., Yang, Y., Zeng, W., Mootha, V.K., Cheng, Y., Bai, X.C., Jiang, Y., 2018b. Cryo-EM structure of a fungal mitochondrial calcium uniporter. Nature 559, 570–574.

Ohta, Y., Ohba, T., Miyamoto, E., 1990. Ca2+/calmodulin-dependent protein kinase II: localization in the interphase nucleus and the mitotic apparatus of mammalian cells. Proc. Natl. Acad. Sci. U. S. A. 87, 5341–5345.

Oxenoid, K., Dong, Y., Cao, C., Cui, T., Sancak, Y., Markhard, A.L., Grabarek, Z., Kong, L., Liu, Z., Ouyang, B., Cong, Y., Mootha, V.K., Chou, J.J., 2016. Architecture of the mitochondrial calcium uniporter. Nature 533, 269–273.

Paillusson, S., Gomez-Suaga, P., Stoica, R., Little, D., Gissen, P., Devine, M.J., Noble, W., Hanger, D.P., Miller, C.C.J., 2017. alpha-Synuclein binds to the ER-mitochondria tethering protein VAPB to disrupt Ca(2+) homeostasis and mitochondrial ATP production. Acta Neuropathol. 134, 129–149.

Pan, X., Liu, J., Nguyen, T., Liu, C., Sun, J., Teng, Y., Fergusson, M.M., Rovira, I., Allen, M., Springer, D.A., Aponte, A.M., Gucek, M., Balaban, R.S., Murphy, E., Finkel, T., 2013. The physiological role of mitochondrial calcium revealed by mice lacking the mitochondrial calcium uniporter. Nat. Cell Biol. 15, 1464–1472.

Parry, H., McDougall, A., Whitaker, M., 2005. Microdomains bounded by endoplasmic reticulum segregate cell cycle calcium transients in syncytial Drosophila embryos. J. Cell Biol. 171, 47–59.

Parry, H., McDougall, A., Whitaker, M., 2006. Endoplasmic reticulum generates calcium signalling microdomains around the nucleus and spindle in syncytial Drosophila embryos. Biochem. Soc. Trans. 34, 385–388.

Patel, R., Holt, M., Philipova, R., Moss, S., Schulman, H., Hidaka, H., Whitaker, M., 1999. Calcium/calmodulin-dependent phosphorylation and activation of human Cdc25-C at the G2/M phase transition in HeLa cells. J. Biol. Chem. 274, 7958–7968.

Perocchi, F., Gohil, V.M., Girgis, H.S., Bao, X.R., McCombs, J.E., Palmer, A.E., Mootha, V.K., 2010. MICU1 encodes a mitochondrial EF hand protein required for Ca(2+) uptake. Nature 467, 291–296.

Pinto, M.C., Kihara, A.H., Goulart, V.A., Tonelli, F.M., Gomes, K.N., Ulrich, H., Resende, R.R., 2015. Calcium signaling and cell proliferation. Cell. Signal. 27, 2139–2149.

Poenie, M., Alderton, J., Tsien, R.Y., Steinhardt, R.A., 1985. Changes of free calcium levels with stages of the cell division cycle. Nature 315, 147–149.

Poenie, M., Alderton, J., Steinhardt, R., Tsien, R., 1986. Calcium rises abruptly and briefly throughout the cell at the onset of anaphase. Science 233, 886–889.

Raffaello, A., De Stefani, D., Sabbadin, D., Teardo, E., Merli, G., Picard, A., Checchetto, V., Moro, S., Szabo, I., Rizzuto, R., 2013. The mitochondrial calcium uniporter is a multimer that can include a dominant-negative pore-forming subunit. EMBO J. 32, 2362–2376.

Rasmussen, C.D., Means, A.R., 1989. Calmodulin is required for cell-cycle progression during G1 and mitosis. EMBO J. 8, 73–82.

Ratan, R.R., Maxfield, F.R., Shelanski, M.L., 1988. Long-lasting and rapid calcium changes during mitosis. J. Cell Biol. 107, 993–999.

Raturi, A., Simmen, T., 2013. Where the endoplasmic reticulum and the mitochondrion tie the knot: the mitochondria-associated membrane (MAM). Biochim. Biophys. Acta 1833, 213–224.

Rauh, N.R., Schmidt, A., Bormann, J., Nigg, E.A., Mayer, T.U., 2005. Calcium triggers exit from meiosis II by targeting the APC/C inhibitor XErp1 for degradation. Nature 437, 1048–1052.

Rizzuto, R., Pinton, P., Carrington, W., Fay, F.S., Fogarty, K.E., Lifshitz, L.M., Tuft, R.A., Pozzan, T., 1998. Close contacts with the endoplasmic reticulum as determinants of mitochondrial Ca2+ responses. Science 280, 1763–1766.

Russa, A.D., Maesawa, C., Satoh, Y., 2009. Spontaneous [Ca2+]i oscillations in G1/S phase-synchronized cells. J. Electron Microsc. (Tokyo) 58, 321–329.

Salazar-Roa, M., Malumbres, M., 2017. Fueling the cell division cycle. Trends Cell Biol. 27, 69–81.

Sancak, Y., Markhard, A.L., Kitami, T., Kovacs-Bogdan, E., Kamer, K.J., Udeshi, N.D., Carr, S.A., Chaudhuri, D., Clapham, D.E., Li, A.A., Calvo, S.E., Goldberger, O., Mootha, V.K., 2013. EMRE is an essential component of the mitochondrial calcium uniporter complex. Science 342, 1379–1382.

Sarraf, S.A., Sideris, D.P., Giagtzoglou, N., Ni, L., Kankel, M.W., Sen, A., Bochicchio, L.E., Huang, C.H., Nussenzweig, S.C., Worley, S.H., Morton, P.D., Artavanis-Tsakonas, S., Youle, R.J., Pickrell, A.M., 2019. PINK1/parkin influences cell cycle by sequestering TBK1 at damaged mitochondria, inhibiting mitosis. Cell Rep. 29, 225–235 e5.

Sheng, Z.H., Cai, Q., 2012. Mitochondrial transport in neurons: impact on synaptic homeostasis and neurodegeneration. Nat. Rev. Neurosci. 13, 77–93.

Shoshan-Barmatz, V., Mizrachi, D., 2012. VDAC1: from structure to cancer therapy. Front. Oncol. 2, 164.

Sies, H., Jones, D.P., 2020. Reactive oxygen species (ROS) as pleiotropic physiological signalling agents. Nat. Rev. Mol. Cell Biol. 21, 363–383.

Skelding, K.A., Rostas, J.A., Verrills, N.M., 2011. Controlling the cell cycle: the role of calcium/calmodulin-stimulated protein kinases I and II. Cell Cycle 10, 631–639.

Skog, S., Tribukait, B., Sundius, G., 1982. Energy metabolism and ATP turnover time during the cell cycle of Ehrlich ascites tumour cells. Exp. Cell Res. 141, 23–29.

Snoeck, H.W., 2020. Calcium regulation of stem cells. EMBO Rep. 21, e50028.

Spinelli, J.B., Haigis, M.C., 2018. The multifaceted contributions of mitochondria to cellular metabolism. Nat. Cell Biol. 20, 745–754.

Spurlock, B., Tullet, J., Hartman, J.L.T., Mitra, K., 2020. Interplay of mitochondrial fission-fusion with cell cycle regulation: possible impacts on stem cell and organismal aging. Exp. Gerontol. 135, 110919.

Steinhardt, R.A., Alderton, J., 1988. Intracellular free calcium rise triggers nuclear envelope breakdown in the sea urchin embryo. Nature 332, 364–366.

Sterea, A.M., El Hiani, Y., 2020. The role of mitochondrial calcium signaling in the pathophysiology of cancer cells. Adv. Exp. Med. Biol. 1131, 747–770.

Suski, J.M., Lebiedzinska, M., Wojtala, A., Duszynski, J., Giorgi, C., Pinton, P., Wieckowski, M.R., 2014. Isolation of plasma membrane-associated membranes from rat liver. Nat. Protoc. 9, 312–322.

Taguchi, N., Ishihara, N., Jofuku, A., Oka, T., Mihara, K., 2007. Mitotic phosphorylation of dynamin-related GTPase Drp1 participates in mitochondrial fission. J. Biol. Chem. 282, 11521–11529.

Territo, P.R., Mootha, V.K., French, S.A., Balaban, R.S., 2000. Ca(2+) activation of heart mitochondrial oxidative phosphorylation: role of the F(0)/F(1)-ATPase. Am. J. Physiol. Cell Physiol. 278, C423–C435.

Tombes, R.M., Borisy, G.G., 1989. Intracellular free calcium and mitosis in mammalian cells: anaphase onset is calcium modulated, but is not triggered by a brief transient. J. Cell Biol. 109, 627–636.

Torok, K., Wilding, M., Groigno, L., Patel, R., Whitaker, M., 1998. Imaging the spatial dynamics of calmodulin activation during mitosis. Curr. Biol. 8, 692–699.

Torricelli, C., Fortino, V., Capurro, E., Valacchi, G., Pacini, A., Muscettola, M., Soucek, K., Maioli, E., 2008. Rottlerin inhibits the nuclear factor kappaB/cyclin-D1 cascade in MCF-7 breast cancer cells. Life Sci. 82, 638–643.

Tosatto, A., Sommaggio, R., Kummerow, C., Bentham, R.B., Blacker, T.S., Berecz, T., Duchen, M.R., Rosato, A., Bogeski, I., Szabadkai, G., Rizzuto, R., Mammucari, C., 2016. The mitochondrial calcium uniporter regulates breast cancer progression via HIF-1alpha. EMBO Mol. Med. 8, 569–585.

Twigg, J., Patel, R., Whitaker, M., 1988. Translational control of InsP3-induced chromatin condensation during the early cell cycles of sea urchin embryos. Nature 332, 366–369.

Umemoto, T., Hashimoto, M., Matsumura, T., Nakamura-Ishizu, A., Suda, T., 2018. Ca(2+)-mitochondria axis drives cell division in hematopoietic stem cells. J. Exp. Med. 215, 2097–2113.

Vais, H., Mallilankaraman, K., Mak, D.D., Hoff, H., Payne, R., Tanis, J.E., Foskett, J.K., 2016. EMRE is a matrix Ca(2+) sensor that governs gatekeeping of the mitochondrial Ca(2+) uniporter. Cell Rep. 14, 403–410.

Van Keuren, A.M., Tsai, C.W., Balderas, E., Rodriguez, M.X., Chaudhuri, D., Tsai, M.F., 2020. Mechanisms of EMRE-dependent MCU opening in the mitochondrial calcium uniporter complex. Cell Rep. 33, 108486.

Vasan, K., Werner, M., Chandel, N.S., 2020. Mitochondrial metabolism as a target for cancer therapy. Cell Metab. 32, 341–352.

Vasington, F.D., Murphy, J., 1962. Ca ion uptake by rat kidney mitochondria and its dependence on respiration and phosphorylation. J. Biol. Chem. 237, 2670–2677.

Vultur, A., Gibhardt, C.S., Stanisz, H., Bogeski, I., 2018. The role of the mitochondrial calcium uniporter (MCU) complex in cancer. Pflugers Arch. 470, 1149–1163.

Wang, Z., Fan, M., Candas, D., Zhang, T.Q., Qin, L., Eldridge, A., Wachsmann-Hogiu, S., Ahmed, K.M., Chromy, B.A., Nantajit, D., Duru, N., He, F., Chen, M., Finkel, T., Weinstein, L.S., Li, J.J., 2014. Cyclin B1/Cdk1 coordinates mitochondrial respiration for cell-cycle G2/M progression. Dev. Cell 29, 217–232.

Wang, Y., Nguyen, N.X., She, J., Zeng, W., Yang, Y., Bai, X.C., Jiang, Y., 2019. Structural mechanism of EMRE-dependent gating of the human mitochondrial calcium uniporter. Cell 177, 1252–1261 e13.

Webster, P.L., Hof, J.V., 1969. Dependence of energy and aerobic metabolism of initiation of DNA synthesis and mitosis by G1 and G2 cells. Exp. Cell Res. 55, 88–94.

Whitaker, M., 2006. Calcium microdomains and cell cycle control. Cell Calcium 40, 585–592.

Wu, W., Shen, Q., Zhang, R., Qiu, Z., Wang, Y., Zheng, J., Jia, Z., 2020. The structure of the MICU1-MICU2 complex unveils the regulation of the mitochondrial calcium uniporter. EMBO J. 39 (19), e104285.

Yoo, J., Wu, M., Yin, Y., Herzik Jr., M.A., Lander, G.C., Lee, S.Y., 2018. Cryo-EM structure of a mitochondrial calcium uniporter. Science 361, 506–511.

Zhao, H., Li, T., Wang, K., Zhao, F., Chen, J., Xu, G., Zhao, J., Li, T., Chen, L., Li, L., Xia, Q., Zhou, T., Li, H.Y., Li, A.L., Finkel, T., Zhang, X.M., Pan, X., 2019. AMPK-mediated activation of MCU stimulates mitochondrial Ca(2+) entry to promote mitotic progression. Nat. Cell Biol. 21, 476–486.

Zheng, C.X., Sui, B.D., Qiu, X.Y., Hu, C.H., Jin, Y., 2020. Mitochondrial regulation of stem cells in bone homeostasis. Trends Mol. Med. 26, 89–104.

Zhuo, W., Zhou, H., Guo, R., Yi, J., Zhang, L., Yu, L., Sui, Y., Zeng, W., Wang, P., Yang, M., 2020. Structure of intact human MCU supercomplex with the auxiliary MICU subunits. Protein Cell 12 (3), 220–229.

CHAPTER SIX

The mitochondrial calcium homeostasis orchestra plays its symphony: Skeletal muscle is the guest of honor

Gaia Gherardi[†], Agnese De Mario[†], and Cristina Mammucari*

Department of Biomedical Sciences, University of Padua, Padua, Italy
*Corresponding author: e-mail address: cristina.mammucari@unipd.it

Contents

[†] Equal contribution.

International Review of Cell and Molecular Biology, Volume 362
ISSN 1937-6448
https://doi.org/10.1016/bs.ircmb.2021.03.005

Abstract

Skeletal muscle mitochondria are placed in close proximity of the sarcoplasmic reticulum (SR), the main intracellular Ca^{2+} store. During muscle activity, excitation of sarcolemma and of T-tubule triggers the release of Ca^{2+} from the SR initiating myofiber contraction. The rise in cytosolic Ca^{2+} determines the opening of the mitochondrial calcium uniporter (MCU), the highly selective channel of the inner mitochondrial membrane (IMM), causing a robust increase in mitochondrial Ca^{2+} uptake. The Ca^{2+}-dependent activation of TCA cycle enzymes increases the synthesis of ATP required for SERCA activity. Thus, Ca^{2+} is transported back into the SR and cytosolic $[Ca^{2+}]$ returns to resting levels eventually leading to muscle relaxation. In recent years, thanks to the molecular identification of MCU complex components, the role of mitochondrial Ca^{2+} uptake in the pathophysiology of skeletal muscle has been uncovered. In this chapter, we will introduce the reader to a general overview of mitochondrial Ca^{2+} accumulation. We will tackle the key molecular players and the cellular and pathophysiological consequences of mitochondrial Ca^{2+} dyshomeostasis. In the second part of the chapter, we will discuss novel findings on the physiological role of mitochondrial Ca^{2+} uptake in skeletal muscle. Finally, we will examine the involvement of mitochondrial Ca^{2+} signaling in muscle diseases.

1. Introduction

The second messenger Ca^{2+} decodes extracellular stimuli to modulate different tissue-specific functions including gene expression, endocrine secretion, muscle contraction, synaptic transmission and others (Rizzuto et al., 2012). The effectiveness of Ca^{2+} signaling relies on the maintenance of a $[Ca^{2+}]$ gradient between the extracellular space and the intracellular milieu, that is tightly controlled by the presence of pumps and transporters placed both at the plasma membrane and at organelles (Rizzuto et al., 2012). The low intracellular $[Ca^{2+}]$ allows the fine-tuned regulation of Ca^{2+}-sensitive proteins, i.e., enzymes, channels, and transcription factors (Berridge et al., 2000). In this context, mitochondria play an important role in regulating and decoding Ca^{2+} inputs. Because of their intrinsically dynamic nature, mitochondria localize at specific positions within the cell where, by buffering cytosolic $[Ca^{2+}]$ ($[Ca^{2+}]c$), regulate Ca^{2+}-dependent processes. In physiological conditions, mitochondrial Ca^{2+} regulates the activity of TCA cycle enzymes, thus exerting a positive role on oxidative metabolism. However, in response to pathological stimuli, excessive mitochondrial Ca^{2+} entry triggers the opening of the mitochondrial permeability transition pore (mPTP) and the release of pro-apoptotic factors, eventually leading to cell death.

2. Physiological roles of mitochondrial calcium uptake

Mitochondrial Ca^{2+} uptake regulates the activity of pyruvate dehydrogenase (PDH), isocitrate dehydrogenase (IDH), and a-ketoglutarate dehydrogenase (a-KGDH) (McCormack and Denton, 1984), thus increasing the rate of NADH production. While IDH and a-KGDH are directly activated by Ca^{2+} (Denton et al., 1978), PDH is indirectly activated by the Ca^{2+}-dependent pyruvate dehydrogenase phosphatase 1 (PDP1) (Denton et al., 1972). The rise in NADH levels consequent to Ca^{2+}-mediated TCA cycle stimulation increases respiratory chain activity and eventually ATP production. Accordingly, mitochondrial ATP levels increase in response to cell stimulation with endoplasmic reticulum (ER) Ca^{2+}-releasing agonists. On the contrary, when Ca^{2+} transients are dampened (by ER Ca^{2+} depletion or by Ca^{2+} chelators) mitochondrial ATP rise is proportionally reduced (Jouaville et al., 1999).

In addition to the regulation of TCA cycle enzymes activity, Ca^{2+} can directly modulate the electron transport chain (ETC) and the F_1F_0-ATP synthase (Glancy et al., 2013; Territo et al., 2000). Furthermore, Ca^{2+} boosts mitochondrial metabolism through local activation of cyclic AMP (cAMP) production, which in turn increases ATP generation through a mitochondrial PKA-dependent pathway (Di Benedetto et al., 2013).

In addition to mitochondrial Ca^{2+} uptake, other factors participate to the regulation of bioenergetics. The availability of TCA cycle substrates is crucial to sustain ATP production and it impacts on mitochondrial Ca^{2+} accumulation itself (Bricker et al., 2012; Herzig et al., 2012). Indeed, the inhibition of pyruvate entry into mitochondria reduces matrix Ca^{2+} accumulation as a consequence of increased MICU1 protein levels (Nemani et al., 2020). Bioenergetics is additionally controlled by extra-mitochondrial Ca^{2+}, which impinges on different substrates transporters including the malate-aspartate shuttle and the mitochondrial glutamate-aspartate carrier aralar1 (Palmieri et al., 2001; Satrústegui et al., 2007; Szibor et al., 2020). The physiological role of mitochondrial Ca^{2+} in the control of bioenergetics has been demonstrated both in several settings. In hepatocytes, high frequency cytosolic $[Ca^{2+}]c$ oscillations are decoded by mitochondria. In turn, mitochondrial $[Ca^{2+}]$ ($[Ca^{2+}]mt$) oscillations cause an increase in NADH levels due to activation of matrix dehydrogenases (Hajnóczky et al., 1995). Reduced ER-mitochondria Ca^{2+} transfer triggers PDH phosphorylation, decreased TCA cycle activity and lower ATP production.

The increase in AMP/ATP ratio is sensed by AMPK eventually leading to autophagy induction (Cárdenas et al., 2010b). Direct coupling of mitochondrial Ca^{2+} transients to oxidative metabolism has also been confirmed by studies conducted in patients' fibroblasts carrying the 13514A > G mutation of the ND5 subunit of NADH dehydrogenase. Compared to fibroblasts of healthy subjects, mitochondrial Ca^{2+} uptake is reduced in patients' cells, leading to increased AMPK activity and pro-survival autophagy (Granatiero et al., 2016). Several other studies, conducted in different cellular models, including cardiomyocytes, rat cortical astrocytes, and pancreatic acinar cells, support this evidence (Balaban, 2009; Boitier et al., 1999; Voronina et al., 2010). In vivo studies corroborating the role of mitochondria Ca^{2+} uptake in bioenergetics regulation are discussed in following sections of this chapter.

As mentioned above, mitochondrial Ca^{2+} uptake not only controls intra-organelle functions, but also shapes cytosolic Ca^{2+} signaling. By buffering $[Ca^{2+}]c$ increases, mitochondria spatially remodel intracellular $[Ca^{2+}]$ eventually impinging on cytosolic Ca^{2+}-dependent activities. Energized mitochondria, located in close proximity of Ca^{2+} channels of the ER or of the plasma membrane (PM), promptly remove Ca^{2+} ions. Since Ca^{2+}, in turn, modulates the activity of the channels, mitochondrial Ca^{2+} accumulation plays a crucial role in controlling Ca^{2+} dynamics. This mechanism is particularly relevant in IP3R-mediated ER Ca^{2+} release. The initial cytosolic Ca^{2+} rise, due to the opening of IP3Rs, favors ER Ca^{2+} release. However, in the absence of mitochondria Ca^{2+} buffering, the local elevation of $[Ca^{2+}]c$ triggers the inhibition of IP3Rs. On the other hand, efficient mitochondrial Ca^{2+} uptake hampers the negative feedback exerted by the sustained rise of cytosolic Ca^{2+} on IP3Rs (Bezprozvanny et al., 1991). This general mechanism has peculiar tissue-specific variations. In hepatocytes, mitochondrial Ca^{2+} uptake suppresses the local positive stimulation of Ca^{2+} on IP3R providing a subcellular heterogeneous IP3 sensitivity, therefore setting the threshold for IP3-dependent $[Ca^{2+}]c$ signaling (Hajnóczky et al., 1999). In astrocytes IP3R2, which is the most expressed isoform, is only positively regulated by $[Ca^{2+}]$, thus the buffering capacity of mitochondria negatively controls the Ca^{2+} wave propagation (Boitier et al., 1999). Finally, in cardiomyocytes, in which the Ca^{2+} release is operated by ryanodine receptor (RyR), mitochondrial Ca^{2+} accumulation diminishes the frequency and the duration of cytosolic Ca^{2+} transients eventually control the excitation-contraction (EC) coupling, the mechanism whereby an action potential causes the contraction of a myocyte (Pacher et al., 2002).

[Ca^{2+}]c shaping is also influenced by the subcellular localization of mitochondria. In excitable cells, mitochondria localized in the proximity of voltage operated Ca^{2+} channels (VOCs) at the PM decrease the magnitude of local Ca^{2+} microdomains generated around the open channels (De Mario et al., 2017). As a consequence, the magnitude of exocytosis is diminished (Montero et al., 2000). In pancreatic acinar cells, mitochondria act as a "firewall" that prevent the propagation of the cytosolic Ca^{2+} waves generated in the apical area of the cell, therefore splitting the cell in two functional compartments able to generate distinct cytosolic Ca^{2+} signals (Park et al., 2001).

[Ca^{2+}]c shaping depends not only on mitochondrial Ca^{2+} uptake, but also on the activity of the efflux mechanisms. Indeed, mitochondrial Ca^{2+} efflux favors the fast refilling of the ER and contributes to the increase in local [Ca^{2+}] close to the IP3R, eventually impinging on the onset of the subsequent ER Ca^{2+} release event (Ishii et al., 2006).

Altogether these data define mitochondria as efficient Ca^{2+} buffers able to shape cytosolic Ca^{2+} transients by modulating Ca^{2+} channels activity and by limiting Ca^{2+} waves propagation.

3. Calcium transport through the outer mitochondrial membrane

In response to agonist-induced activation of PM receptors, opening of 1,4,5-trisphosphate receptors (IP3Rs) or ryanodine receptors (RyRs) triggers Ca^{2+} release from the ER or from the sarcoplasmic reticulum (SR). Subsequently, at ER-mitochondria contact sites, Ca^{2+} is taken up by mitochondria (Csordás et al., 1999). To reach the mitochondrial matrix, cytosolic Ca^{2+} must cross both the outer mitochondrial membrane (OMM) and the inner mitochondrial membrane (IMM) (Fig. 1). In the OMM, voltage-dependent anion channels (VDACs) are highly permeable to Ca^{2+} allowing its transport into the intermembrane space (IMS). Only the abundance of VDAC channels modulates OMM Ca^{2+} transfer capacity (Rizzuto et al., 2012). VDACs assume different conformations, and the transition between open (diameter of the channel pore: 2.5 nm) and closed (pore size: 0.9 nm) states occurs in a voltage-dependent manner: low transmembrane potentials determine a high-conductance, anion selective state, whereas increased voltages (20–40 mV) promote lower conducting conformations which are impermeable to nucleotides but permeable to Ca^{2+} (Colombini, 2012). It has been postulated that the voltage potential across the IMM might influence the OMM pores. As a consequence, while VDACs located at the

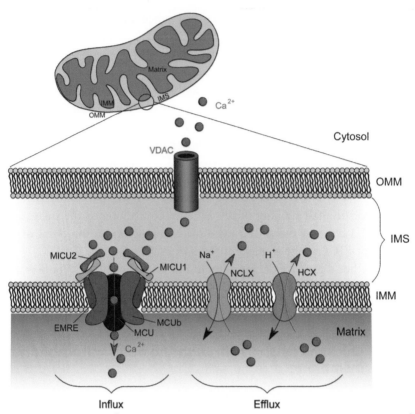

Fig. 1 Mitochondrial Ca^{2+} transport proteins. To reach the mitochondrial matrix, Ca^{2+} ions cross both the outer (OMM) and the inner (IMM) mitochondrial membranes. Ca^{2+} passes the OMM thanks to the presence of a large conductance channel, the voltage-dependent anion channel (VDAC). Ca^{2+} uptake into the mitochondrial matrix is mediated by the mitochondrial Ca^{2+} uniporter (MCU), a highly Ca^{2+} selective channel located at the IMM. MCU complex is composed of channel-forming subunits, i.e., MCU, MCUb, and EMRE, and regulatory interactors, i.e., MICU1 and MICU2. MICU3, which is highly expressed in the brain, and MICU1.1, which is specifically expressed in the skeletal muscle, are not represented here. Ca^{2+} is extruded by the mitochondrial sodium/calcium exchanger (NCLX) or by the proton/calcium exchanger (HCX), which are both located at the IMM.

ER-mitochondria contact sites might assume a closed state, VDACs positioned between the contacts should stay in the high-conducting open conformation (Adams et al., 1989).

Three VDAC isoforms have been identified (VDAC1, VDAC2, and VDAC3), VDAC1 being considered the main Ca^{2+} transport channel (Messina et al., 2012). VDAC1, by interacting with the IP3R located at the ER, senses low amplitude pro-apoptotic Ca^{2+} signals (De Stefani et al., 2012). VDAC1 silencing reduces mitochondrial Ca^{2+} uptake and decreases ceramide-induced cell death (De Stefani et al., 2012; Rapizzi et al., 2002). Surprisingly, despite similar effects on mitochondrial Ca^{2+} accumulation compared to VDAC1, VDAC2 silencing enhances cell death, possibly because it modulates Bak oligomerization on the OMM (De Stefani et al., 2012).

4. Calcium transport through the inner mitochondrial membrane

The IMM is less permeable to ions and small molecules than the OMM. Respiratory chain activity generates an electrochemical gradient across the IMM ($-180\,$mV inside) which drives the synthesis of ATP and allows Ca^{2+} uptake by mitochondria (Rizzuto et al., 2012; Stock et al., 1999). Accordingly, treatment with an uncoupler such as p-(trifluoromethoxy)-phenyl-hydrazone (FCCP), that collapses the $\Delta\Psi$m, abolishes mitochondrial Ca^{2+} accumulation.

Based on the low affinity for Ca^{2+} of the mitochondrial uniporter, it was originally assumed that mitochondria would not impinge on Ca^{2+} dynamics under physiological conditions, but rather act as a low affinity buffer in case of pathological Ca^{2+} overload (Pozzan and Rizzuto, 2000). A novel role of mitochondria in Ca^{2+} homeostasis emerged when the [Ca^{2+}]m was directly measured in living cells by means of a mitochondria-targeted Ca^{2+}-sensitive probe based on the photoprotein aequorin (Rizzuto et al., 1992). In contrast with previous predictions, a rapid rise in [Ca^{2+}]m was detected in response to physiological increases in [Ca^{2+}]c triggered by agonist-induced ER Ca^{2+} release. This apparent discrepancy was explained by the demonstration that microdomains of high [Ca^{2+}] are generated in sites of close proximity between mitochondria and the ER and induce the opening of the uniporter (Rizzuto et al., 1998).

5. The mitochondrial calcium uniporter complex

The mitochondrial calcium uniporter complex is composed of pore-forming subunits (i.e., MCU, MCUb and essential MCU regulator-EMRE), and of regulatory elements (mitochondrial Ca^{2+} uptake MICU family proteins).

The mitochondrial calcium uniporter (MCU) gene, originally known as CCDC109A, encodes a 40 kDa protein that contains two transmembrane domains linked by a highly conserved loop that faces the IMS, while both the C- and the N-termini face the mitochondrial matrix (Baradaran et al., 2018; Baughman et al., 2011; De Stefani et al., 2011). The two transmembrane domains are separated by a short loop, the "DIME" motif, characterized by negatively charged amino acids that are essential to confer Ca^{2+} selectivity to the channel (De Stefani et al., 2011), as confirmed by molecular structures (Lee et al., 2015). High-resolution cryo-EM structural analyses indicate that the channel is organized into tetramers (Baradaran et al., 2018; Fan et al., 2018; Nguyen et al., 2018; Yoo et al., 2018), as originally proposed (Raffaello et al., 2013). MCU forms tetramers with a non-obvious symmetry: the transmembrane domain displays a fourfold symmetry, while the N-terminal domain on the matrix side shows a twofold symmetry axis. Recently, the structure of the human MCU-EMRE-MICU1-MICU2 supercomplex has been solved (Fan et al., 2020; Zhuo et al., 2020). Within the supercomplex, MICU1 does not directly contact MCU, but its presence on the MCU complex is mediated by the interaction with EMRE, suggesting a critical function of EMRE in coupling MICUs with the MCU channel (Zhuo et al., 2020). In addition, MICU1–EMRE interaction helps to keep MICU1 in place to seal the MCU pore at low $[Ca^{2+}]$, and to prevent MICU1 dissociation from the complex when $[Ca^{2+}]$ increases (Fan et al., 2020).

5.1 MCU

Genetic manipulation of MCU in different animal models highlights the role of mitochondrial Ca^{2+} uptake in energy production. In *Trypanosoma brucei*, MCU silencing impairs mitochondrial Ca^{2+} uptake and increases AMP/ATP ratio, resulting in a severe growth inhibition and reduced infectivity (Huang et al., 2013). In *Danio rerio*, MCU silencing causes major developmental defects, impinging on RhoA signaling and F-actin dynamics (Prudent et al., 2013). In *Drosophila melanogaster*, overexpression of a dominant-negative isoform of MCU in mushroom body neurons during pupation causes memory impairment, decreased number of synaptic vesicles and increased axon length (Drago and Davis, 2016). However, *Caenorhabditis elegans* MCU null mutants are viable and fertile (Xu and Chisholm, 2014). Nonetheless, impaired mitochondrial Ca^{2+} uptake negatively impact on ROS-dependent wound healing.

In *Mus musculus*, the study of MCU deletion is biased by the fact that the resulting phenotype strictly depends on the mouse strain. The first MCU$^{-/-}$ mouse generated in a mixed background (C57/BL6 and CD1) showed a relatively mild phenotype, characterized by increased plasma lactate levels upon starvation and impaired exercise performance accompanied by a reduction of skeletal muscle pyruvate dehydrogenase activity (Pan et al., 2013). However, in a pure C57BL/6 background MCU deletion is embryonically lethal (Murphy et al., 2014).

Here, we will summarize the role of MCU in the heart and in the brain, while mitochondrial Ca^{2+} uptake in skeletal muscle will be discussed below.

Despite the importance of mitochondrial Ca^{2+} buffering in cardiac physiology, the heart phenotype of the constitutive MCU knockout mouse was surprisingly mild (Holmström and Finkel, 2014; Pan et al., 2013). As expected, mitochondria isolated from MCU$^{-/-}$ cardiomyocytes do not take up Ca^{2+} and have lower resting Ca^{2+} levels compared to controls. However, basal ATP levels are unaffected by MCU deletion, demonstrating a preserved basal mitochondrial energetics (Holmström et al., 2015). In addition, mice lacking MCU show normal basal cardiac function in terms of ejection fraction, fractional shortening, stroke volume, and chamber size, both in adulthood and aging (Holmström et al., 2015). Similar results were obtained in a transgenic mice expressing a dominant-negative MCU isoform, MCU^{D260Q,E263Q} (DN-MCU), in the same genetic background of the constitutive MCU$^{-/-}$ model. DN-MCU mice have normal heart rate; however, they display an impairment in the "fight or flight" response (Wu et al., 2015). Overall, these data suggest that mitochondrial Ca^{2+} accumulation is dispensable both for basal cardiac function and during increased workload. However, as opposed to the total MCU$^{-/-}$ and DN-MCU mouse models (Holmström et al., 2015; Pan et al., 2013; Wu et al., 2015), in a cardiac-specific tamoxifen-inducible model, MCU ablation strongly protects from ischemic-reperfusion injury (IR injury) of the heart. In addition, despite the normal baseline phenotype, the cardiac-specific inducible MCU$^{-/-}$ mice show lack of responsiveness to β-adrenergic stimulation and decreased running capacity, suggesting that MCU is required to stimulate Ca^{2+}-dependent metabolism upon increased workload (Kwong et al., 2015; Luongo et al., 2015).

Compared to wild type counterparts, brains of inducible neuron-specific MCU knockout mice and MCU-silenced primary cortical neurons show a significant reduction of the hypoxic/ischemic damage and decreased cell death without impairment of neuronal mitochondria metabolism

(Nichols et al., 2018). However, brain mitochondria isolated from total $MCU^{-/-}$ mice are not protected from hypoxic/ischemic injury (Nichols et al., 2017). Thus, the brain responds to chronic MCU ablation by establishing adaptive metabolic mechanisms in order to cope with impaired mitochondrial Ca^{2+} signaling.

5.2 MCUb

A second component of the MCU channel-forming subunits is the 33 kDa protein MCUb, also known as CCDC109B. MCUb topology is similar to the one of MCU, with two transmembrane domains and both N- and C-termini protruding into the matrix. However, despite the huge sequence similarity with MCU in the transmembrane domains, MCUb displays altered ion permeation (Raffaello et al., 2013). This is due to two amino acid substitutions (the murine Arg251 and Glu256 residues are mutated into Trp and Val, respectively), located in close proximity of the DIME motif and conserved among species, that drastically reduce channel conductivity (Raffaello et al., 2013). In living cells, MCUb overexpression reduces the amplitude of $[Ca^{2+}]$m transients evoked by agonist stimulation, and MCUb silencing elicits the opposite effect, indicating that this protein acts as a dominant-negative subunit of the channel. The expression and stoichiometry of MCU and MCUb vary significantly among mammalian tissues. For example, the heart exhibits a high MCU:MCUb ratio, while skeletal muscle displays a lower ratio. This accounts for tissue-specific variations in channel activity and in mitochondrial Ca^{2+} currents (Fieni et al., 2012). To investigate the role of MCUb in the heart, both MCUb knockout and transgenic mice were developed. MCUb deletion exacerbates the pathological consequences of IR injury (Huo et al., 2020) while MCUb exerts a protective effect when overexpressed from at least 1 month before the induction of left coronary artery ligation (Huo et al., 2020; Lambert et al., 2019). Strikingly, when MCUb was overexpressed for 1 week before ligation, increased mortality after IR injury was observed possibly due to acute bioenergetics alterations (Lambert et al., 2019).

5.3 EMRE

The third channel-forming subunit is EMRE, a single-pass transmembrane protein of 10 kDa. EMRE N-terminus is located in the mitochondrial matrix, where it interacts with MCU, while its C-terminus interacts in the IMS with MICU1 (Sancak et al., 2013). The acidic C-terminal domain has been proposed as a matrix Ca^{2+} sensor that governs the MCU activity. Thus, EMRE acts as a regulatory complex able to sense [Ca^{2+}] at both sides of IMM (Vais et al., 2016).

In yeast cells, EMRE is essential to assemble a functional channel with human MCU (Kovács-Bogdán et al., 2014). However, in planar lipid bilayer MCU and the regulatory subunits MICU1 and MICU2 display channel activity without the presence of EMRE (Patron et al., 2014).

To determine the relevance of EMRE in vivo, an EMRE$^{-/-}$ mouse model was generated (Liu et al., 2020). EMRE is necessary for MCU activity, as EMRE$^{-/-}$ isolated mitochondria do not accumulate Ca^{2+}. Similarly to the total MCU knockout mouse model (Pan et al., 2013), EMRE deletion results in almost complete embryonic lethality in inbred C57Bl6/N mice. However, EMRE$^{-/-}$ animals of mixed genetic background are viable, although smaller and born with less frequency. No differences were detected in terms of basal metabolic functions, running capacity and cardiac functionality in the hybrid Bl6/N-CD1 EMRE$^{-/-}$ animals compared to controls. Thus, similarly to the MCU knockout mouse model (Murphy et al., 2014; Pan et al., 2013), adaptations occur to sustain mitochondrial activity in the absence of a functional MCU channel.

5.4 MICU proteins

A critical aspect of mitochondria Ca^{2+} uptake is the sigmoidal response to cytosolic Ca^{2+} levels. At low [Ca^{2+}]c mitochondrial Ca^{2+} uptake is negligible, despite the huge driving force ensured by the mitochondrial membrane potential (MMP). When [Ca^{2+}]c rises above a certain threshold upon agonist stimulation, the speed and amplitude of mitochondrial Ca^{2+} uptake exponentially increase. This represent a fundamental property to protect the organelle from a vain energy expenditure due to cations vicious cycle across the IMM and to Ca^{2+} overload. MCU protein lacks Ca^{2+} binding domains. Thus, regulatory proteins located in the IMS, which interact with the pore-forming subunits, control the fine-tuned regulation of channel opening. The proteins responsible for the sigmoidal response to extra-mitochondrial Ca^{2+} belong to the MICU family. MICU1, MICU1.1, MICU2 and MICU3 are mitochondrial EF-hand Ca^{2+}-binding domain containing proteins, each of them characterized by different tissue expression profiles and peculiar functions.

MICU1, previously known as CBARA, is a 54 kDa protein containing two EF-hand Ca^{2+} binding domains, and it was the first member of MICU's family to be identified (Csordás et al., 2013; Perocchi et al., 2010). MICU1 has a dual-mode control on MCU. When extra-mitochondrial $[Ca^{2+}]$ is low, MICU1 keeps the channel close to avoid constant Ca^{2+} entry. On the other hand, when extra-mitochondrial $[Ca^{2+}]$ exceeds a certain threshold, MICU1 acts as a cooperative activator of MCU warranting high Ca^{2+} conductivity (Csordás et al., 2013; Mallilankaraman et al., 2012; Patron et al., 2014). The recent resolution of the human holocomplex structure allows to elucidate the molecular mechanism of MICU1 Ca^{2+} sensing (Fan et al., 2020). MICU1 is able to dock directly the IMS surface of MCU forming a cap on the pore entrance. When the $[Ca^{2+}]$ rises, MICU1 loses most of the interaction with MCU leaving the pore uncovered (Fan et al., 2020).

MICU2, formerly known as EFHA1, is a vertebrate paralog of MICU1. Similarly to MICU1, MICU2 is located in the IMS and has a similar tissue expression pattern (Plovanich et al., 2013). In many cell types, such as HeLa cells and patient-derived fibroblasts carrying a MICU1 deletion, the stability of MICU2 depends on the presence of MICU1, since MICU1 knockdown destabilizes MICU2 protein without affecting its mRNA levels (Debattisti et al., 2019; Patron et al., 2014). However, in MICU1 knockout MEFs and in MICU1 knockout skeletal muscle, MICU2 protein levels are unchanged (Antony et al., 2016; Debattisti et al., 2019). While MICU1 is able to form homodimers, MICU2 can only form heterodimers with MICU1. These heterodimers are stabilized by a disulfide bond through two conserved cysteine residues to regulate MCU channel activity (Patron et al., 2014). In addition, electrophysiological studies performed in planar lipid bilayer showed that MICU2 inhibits MCU activity at low $[Ca^{2+}]$, acting as the genuine gatekeeper of the channel (Patron et al., 2014). However, other hypotheses have been proposed regarding the regulation of the MCU activity exerted by MICU1-MICU2. Mootha and coworkers proposed that both MICU1 and MICU2 are gatekeepers of the channel, since only the simultaneous lack of MICU1 and MICU2 triggers an overt dysregulation of the channel. On the contrary, in the absence of MICU2 alone, mitochondrial Ca^{2+} uptake is still inhibited by MICU1 (Kamer and Mootha, 2014). On the other hand, other studies indicate that MICU1 alone acts both as a gatekeeper and as a cooperative activator of the channel, whereas MICU2 regulates the channel activation threshold (Payne et al., 2017). Finally, the stoichiometry of MCU components is tissue-specific and correlates with the distinct mitochondrial Ca^{2+} uptake properties of each tissue (Paillard et al., 2017). In mouse liver and skeletal muscle MICU1 is highly expressed compare to its expression levels in the heart, while MICU2 is abundant in the heart. This specific feature confers to the liver and muscle high cooperative activation of the channel and increases the threshold of activation, which is functional to stimulate oxidative metabolism in these tissues. On the other hand, the high amount of MICU2 in cardiac mitochondria is instrumental to reduce mitochondrial Ca^{2+} uptake upon cytosolic Ca^{2+} elevation, allowing the decoding of the repetitive cytosolic Ca^{2+} spikes of the beating heart into a graded increase of matrix Ca^{2+} (Paillard et al., 2017). Although discrepancies remain, as mention above the recent structural analyses have helped clarify and characterize the MICU1-MICU2 gatekeeping activity in response to Ca^{2+} transients. At low [Ca^{2+}] MICU1 acts as a cap on the channel, while upon Ca^{2+}-binding to MICU1-MICU2 heterodimer, a conformational change weakens MICU1 and MCU interaction eventually opening the channel (Fan et al., 2020; Wu et al., 2020).

A few years ago a MICU1 splice variant, i.e., MICU1.1, that is highly expressed in skeletal muscle was characterized (Vecellio Reane et al., 2016). Compared to MICU1, MICU1.1 contains an extra micro-exon between exon 5 and exon 6 that encodes four amino acids (EFWQ). MICU1.1 forms dimers with MICU2, and plays a gatekeeping function similar to that of MICU1. However, MICU1.1 binds Ca^{2+} one order of magnitude more efficiently than MICU1 and activates MCU at lower [Ca^{2+}]i than MICU1 (Vecellio Reane et al., 2016). Thus, in skeletal muscle, the MICU1.1-MICU2 dimer enhances the ability of mitochondria to take up Ca^{2+}, and increases ATP production.

Concerning MICU3, its sequence similarity to MICU1 suggested that the two proteins may have shared a common biological function (Plovanich et al., 2013). Nonetheless, MICU3 displays a reduced gatekeeper activity compared to MICU1, and it mediates a more rapid response to [Ca^{2+}] i changes, allowing a shorter delay between the [Ca^{2+}]i rise and the increase in mitochondrial Ca^{2+} uptake (Patron et al., 2019). MICU3 is involved in the metabolic flexibility of nerve terminals. Presynaptic mitochondria are able to sustain axonal ATP synthesis taking up Ca^{2+} in response of small cytosolic Ca^{2+} peaks thanks to the fine tuning of Ca^{2+} sensitivity exerted by MICU3 (Ashrafi et al., 2020).

6. Mitochondrial calcium extrusion

Two major systems have been postulated to extrude Ca^{2+} from the mitochondrial matrix in response to physiological mitochondrial [Ca^{2+}]

transients: the mitochondrial Na^+/Ca^{2+} exchanger (mNCX) and the mito-chondrial H^+/Ca^{2+} exchanger (mHCX) (Fig. 1). A third mechanism, the mPTP opening, is activated under specific pathophysiological conditions, characterized by prolonged mitochondrial Ca^{2+} overload (Bernardi et al., 1992). A discussion of the mPTP can be found in Section 8. Here, we will discuss the role of mNCX and the discovery of NCLX as mNCX encoding gene. mNCX is the predominant antiporter in excitable tissues (i.e., heart and brain), whereas mHCX is particularly active in not excitable tissues (liver and kidney). The stoichiometry of mNCX is defined as electrogenic, with three (or four) Na^+ exchanged for one Ca^{2+} (Jung et al., 1995), whereas the exchange ratio of mHCX is electroneutral (two H^+ for one Ca^{2+}) (Gunter et al., 1991). Recently, mNCX function was ascribed to NCLX (Palty et al., 2010). NCLX mediates not only Na^+/Ca^{2+} exchange but also Li^+-dependent Ca^{2+} transport.

Electron microscopy and cell fractionation experiments clearly showed that NCLX is targeted to the IMM. Na^+-dependent Ca^{2+} release was strongly reduced in NCLX knockdown cells, whereas it was enhanced upon NCLX overexpression (Palty et al., 2010). That NCLX encodes the mNCX is corroborated the discovery that NCLX-driven Ca^{2+} extrusion is inhibited by 7-chloro-5-(2-chlorophenyl)-1,5-dihydro-4,1-benzothiazepin-2(3H)-one (CGP-37157), the most selective inhibitor of mitochondrial Na^+/Ca^{2+} exchange (Palty et al., 2010). Moreover, NCLX exerts fundamental function in controlling $[Ca^{2+}]m$ in excitable cells, as demonstrated by studies conducted in a conditional cardiomyocyte-specific NCLX knockout mouse (Luongo et al., 2017). NCLX ablation causes sudden death due to acute myocardial dysfunction and fulminant heart failure. NCLX knockout hearts display sarcomere disorganization, necrotic cell death and fibrosis. mPTP activation is responsible for the phenotype, as demonstrated by the fact that mating of these mice with cyclophilin D (CypD)-null mice rescues heart function and survival (Luongo et al., 2017).

7. Mitochondrial calcium uptake and ROS production

Reactive oxygen species (ROS) are molecules derived by the incomplete one-electron reduction of oxygen. ROS include free radicals such as superoxide ($O_2^-\cdot$), hydroxyl radical ($OH\cdot$), as well as non-radical species such as hydrogen peroxide (H_2O_2). Mitochondria are considered the major source of cellular ROS (Nickel et al., 2014). $O_2^-\cdot$ can be generated at different sites within mitochondria. Among them, the ubiquinone-binding sites in complex I (site IQ) and complex III (site IIIQo) of the respiratory chain, pyruvate and 2-oxoglutarate dehydrogenases have the highest capacity to generate superoxide. Interestingly, only site IIIQo (on complex III) and glycerol 3-phosphate dehydrogenase can release superoxide into the intermembrane spaces. $O_2^-\cdot$ quickly reacts with nearby molecules, dismutes into hydrogen peroxide (H_2O_2) and O_2, or is metabolized by the superoxide dismutase (SOD) enzymes. H_2O_2 is transformed in water by glutathione peroxidase (GPX), peroxiredoxin (PRX), and catalases (Nickel et al., 2014). In the presence of iron, the Fenton reaction transforms superoxide and H_2O_2 in highly reactive hydroxyl radicals, which can damage cellular proteins, RNA, DNA and lipids. Interaction of ROS with nitric oxide or fatty acids can lead to the formation of peroxynitrite or peroxyl radicals, respectively, that are also highly reactive. The regulation of the activity and the expression levels of these antioxidant enzymes are controlled by many mechanisms (Nickel et al., 2014) and tightly control the balance between ROS generation and ROS scavenging.

When ROS production is not compensated by ROS scavenging, ROS can exert several pathological effects by perturbing the redox state of signaling molecules and by triggering either hyper- or hypofunctionality of several signaling pathways (Sena and Chandel, 2012). Moreover, ROS can promote the opening of the mPTP. In this condition, the mitochondrial membrane becomes permeable to any molecule less than 1.5 kDa in size. Consequent dissipation of $\Delta\Psi m$ leads to a mitochondrial membrane depolarization, decreased ATP, and eventually cell apoptosis. In addition, mitochondrial membrane depolarization leads to cristae unfolding, uncoupling of oxidative phosphorylation, and the reverse electron transport (RET). RET, caused by electron flowing from ubiquinol back to respiratory complex I, generates a significant amount of ROS (Biasutto et al., 2016). Chronic ROS exposure promotes mitochondrial fragmentation and sublethal amounts of H_2O_2 or of other acute stressors induces hyperfusion and can be prevented using antioxidants (Liesa and Shirihai, 2013; Shutt et al., 2012).

Oxidative stress leads to severe cellular damage and cell death in many pathological disorders, such as myocardial infarct (Zorov et al., 2014), neurodegenerative disorders (De Mario et al., 2019) and the development of type 2 diabetes (Sánchez-Gómez et al., 2013). Aging and age-related diseases are influenced by ROS and disruption of mitochondrial redox signals contributes to the aging process (Balaban et al., 2005; Navarro and Boveris, 2007).

Although the pathological role of ROS has received considerable attention, ROS also can play a physiological role. Moreover, ROS are implicated in the activation of cellular signaling pathways and transcription factors, including phosphoinositide 3-kinase (PI3K)/Akt, mitogen-activated protein kinases (MAPK), nuclear factor (erythroid-derived 2)-like 2 (Nrf2)/ Kelch like-ECH-associated protein 1 (Keap1), nuclear factor-κB (NF-κB) and the tumor suppressor p53 (Covarrubias et al., 2008). In physiological conditions, where ROS production is induced in response to stress to facilitate cellular adaptation (Sena and Chandel, 2012), H_2O_2 is the primary mROS utilized for intracellular signaling. H_2O_2 can oxidize cysteine residues thereby modifying target protein functions. Cysteine residues exist as a thiolate anion (Cys–S-) at physiological pH and are more susceptible to oxidation compared to the protonated cysteine thiol (Cys–SH). During redox signaling, H_2O_2 oxidizes the thiolate anion to the sulfenic form (Cys–SOH), causing allosteric changes within the target protein. In turn, the sulfenic form can be reduced back to thiolate anion by the disulfide reductases thioredoxin (Trx) and glutaredoxin (Grx). Hence, first-degree oxidation of cysteine residues within proteins serves as a reversible signal transduction mechanism (Finkel, 2012).

Within mitochondria, several ROS-mediated redox reactions have been reported (Mailloux et al., 2014). These modifications include S-oxidation (sulfenylation and sulfinylation), S-glutathionylation and S-nitrosylation of mitochondrial thiolases, creatine kinase, homoaconitase and branched chain aminotransferase (Riemer et al., 2015). Moreover, ROS modulate the activity of several enzymes of the TCA cycle (isocitrate dehydrogenase, aconitase, succinyl-CoA synthetase and ketoglutarate dehydrogenase) (Riemer et al., 2015). ROS also influence mitochondrial function through sirtuins, a class of NAD^+-dependent deacetylases (Riemer et al., 2015). ROS-depletion of sirtuins results in an increased acetylation of mitochondrial proteins and decreased ATP production (Liu et al., 2017). Among the seven members of the sirtuin family, SIRT3, SIRT4 and SIRT5 are localized into mitochondria, where they govern mitochondrial processes. ROS can affect the activity of sirtuins by inducing or repressing their expression, by altering sirtuin-protein interactions, by changing NAD^+ levels or by inducing sirtuin post-translational modifications (Jones, 2006). ROS regulate also mitochondrial transport systems, including the carnitine/ acylcarnitine carrier responsible for the transport of fatty acids into mitochondria (Giangregorio et al., 2013). Another target of ROS is mitochondrial DNA (mtDNA), which is particularly susceptible to ROS-mediated damage due to its close proximity to the respiratory chain and the lack of protective histones. ROS-induced oxidative damage is probably a major cause of mitochondrial genomic instability and respiratory dysfunction (Richter, 1995).

The cellular redox state can also significantly modulate Ca^{2+} signaling (Görlach et al., 2015). For example, S-glutathionylation at a conserved cysteine in the NTD of the human MCU protein induces clustering of MCU channels and enhances MCU channel activity (Dong et al., 2017).

Viceversa, mitochondrial Ca^{2+} accumulation controls ROS production in different pathophysiological settings, including neurodegenerative disorders, diabetes, atherosclerosis, and ischemia/reperfusion (I/R) injury (Delierneux et al., 2020; Ryan et al., 2020; Wang and Wei, 2017). Decreased MCU expression reduces mROS production in different settings. In triple negative breast cancer, impaired MCU-dependent mROS generation hampers tumor growth and metastasis formation (Tosatto et al., 2016). On the other hand, mitochondrial Ca^{2+}-dependent mROS formation exerts a beneficial effect in promoting cell migration of non-transformed cells in wound healing. Skin wound repair in *C. elegans* requires local MCU-induced production of mROS, which act on a redox-sensitive motif of the Rho GTPase promoting actin-mediated wound closure (Xu and Chisholm, 2014). A similar MCU-mROS-Rho GTPase axis contributes to repair of injured skeletal muscle plasma membrane, an event that occurs during eccentric exercise (Horn et al., 2017).

8. Mitochondrial calcium accumulation and cell death

In addition to their role in aerobic metabolism, mitochondria exert a key role in several other processes, including cell death. In particular, apoptosis is a tightly regulated process involved in cell homeostasis and development. However, apoptosis also occurs during certain pathological conditions, including ischemia-reperfusion damage, infarction, neurodegenerative diseases and viral or chemical toxicity (Green and Reed, 1998).

Apoptosis in mammals is triggered either by extrinsic cues or by mitochondria-decoded signaling, in which mitochondria release a number of pro-apoptotic factors, such as cytochrome *c*, apoptosis inducing factor (AIF), procaspase-9, Smac/DIABLO as well as endonuclease G, that initiate the caspase cascade (Kroemer et al., 2007). The release of pro-apoptotic factors is preceded by permeabilization of the OMM. Several B-cell CLL/ lymphoma-2 (Bcl-2)-protein family members can affect the permeability of the OMM (Elmore, 2007). OMM rupture, and thus the release of apoptogenic factors, can occur subsequently to a long-lasting increase in IMM permeability, also known as mitochondrial permeability transition (MPT). Pore opening is a Ca^{2+}-dependent process, but it can be facilitated by other factors, such as inorganic phosphate, ATP depletion, low pH and oxidative stress (Rasola and Bernardi, 2007). mPTP opening is followed by osmotic swelling of the mitochondria and rupture of the mitochondrial membrane, leading to the release of pro-apoptotic factors into the cytosol. However, mitochondrial collapse might occur also in several forms of necrotic cell death, for example, that caused by oxidative stress, ischemia/reperfusion, and Ca^{2+} ionophores (Lemasters et al., 1999).

Before caspase activation, mitochondria undergo a macroscopic remodeling of their shape. Indeed, after apoptosis induction, they become fragmented, resulting in small, rounded and numerous organelles. Interestingly, if the equilibrium between fusion and fission rates is perturbed, cell death sensitivity is affected, too. Conditions in which mitochondrial fission is inhibited, such as DRP1 downregulation or mitofusins (MFN) overexpression, delay caspase activation and cell death. Similarly, stimulation of organelle fission (by DRP1 overexpression or MFN1/2 and OPA1 inhibition) promotes apoptosis by facilitating cytochrome c release (Youle and Karbowski, 2005). However, mitochondrial fragmentation is not necessarily related to apoptosis. Indeed, mitochondrial fission per se does not increase cell death, and DRP1 overexpression has been reported to protect cells from some apoptotic challenges, like those dependent on mitochondrial Ca^{2+} overload (Szabadkai et al., 2004).

ER Ca^{2+} release and hence mitochondrial Ca^{2+} uptake can sensitize mitochondria to release of cytochrome c, and of pro-apoptotic proteins such as Bax and Bak (Nutt et al., 2002). Moreover, overexpression of Mcl-1, an anti-apoptotic Bcl-2 family protein (Minagawa et al., 2005) or PPARgamma-coactivator-1alpha (PGC-1α) that triggers mitochondrial biogenesis (Bianchi et al., 2006) has been shown to lower the efficacy of mitochondrial Ca^{2+} uptake, the IP3R-mediated $[Ca^{2+}]m$ signal and the Ca^{2+}-dependent apoptosis.

While OMM permeabilization induced by Bid (and other BH3-only pro-apoptotic proteins) is insensitive to mitochondrial Ca^{2+}, in many other situations the apoptotic cascade (and OMM permeability) relies on mitochondrial Ca^{2+} overload, suggesting that mitochondrial Ca^{2+} plays a pivotal role in the control of the apoptotic process (Celsi et al., 2009). Mitochondrial Ca^{2+} can contribute to cell death also by stimulating mROS production, which in turn decreases the Ca^{2+} amount required to trigger mPTP opening. This may be important in ischemia/reperfusion, when an increase in intracellular Ca^{2+} level caused by hypoxia is followed by enhanced ROS generation upon re-oxygenation. Increased ROS production might also contribute to both mitochondria permeability transition and Bax/Bak-dependent mitochondrial membrane permeabilization and release of cytochrome *c* (Halestrap et al., 1997). In addition Bax-mediated permeabilization of the OMM might be triggered by ROS, since oxidation of conserved cys-62 on the Bax molecule has been linked to Bax activation and mitochondrial permeabilization (Nie et al., 2008).

9. Ca^{2+} in skeletal muscle: a bidirectional signaling route

Ca^{2+} is a key intracellular messenger in skeletal muscle fibers, where not only it binds troponin to trigger contraction but also it exerts additional functions, including modulation of gene transcription and activation of Ca^{2+}-dependent proteases. Excitation-contraction coupling (EC coupling), the process by which an action potential determines myofiber contraction, requires cytosolic Ca^{2+} rise. In skeletal muscle, this is possible thanks to the direct interaction between RyR, a Ca^{2+}-releasing channel located on the terminal cisternae of the SR, and dihydropyridine receptor (DHPR), a voltage-gated L-type Ca^{2+} channel, located on the sarcolemma of the transverse tubule (T-tubule) (Franzini-Armstrong and Protasi, 1997).

The adult skeletal muscle is characterized by a precise arrangement of RyRs and DHPRs. RyR1 proteins are positioned in ordered arrays within the SR terminal cisternae on both sides of the T-tubule. Each Ca^{2+} release channel is associated with a tetrad of DHPRs in the T-tubule membrane. In this setting, a Ca^{2+} release unit (CRU) is referred to as the association of a T-tubule containing DHPRs flanked on either side by two apposed SR terminals cisternae containing RyR1 channels (Franzini-Armstrong and Jorgensen, 1994).

Notably, during EC coupling the signal interaction between the two types of channels is bidirectional. The orthograde coupling is the mechanism by which the depolarization of the T-tubule membrane causes Ca^{2+} release from the SR through RyR1 opening. On the other hand, retrograde coupling is the influence of RyR1 on the Ca^{2+} gating properties of the DHPRs (Dirksen, 2002). Following RyR1 opening, Ca^{2+} is released from the SR into the cytosol generating a Ca^{2+} spark. The reversible binding of Ca^{2+}

to troponin C causes cross-bridge cycling and generates force. Troponin C, together with troponin T and tropomyosin form a regulatory unit that modulates the Ca^{2+}-dependency of muscle contraction (Schiaffino and Reggiani, 2011).

Ca^{2+} signaling not only couples excitation to contraction but also links excitation to gene transcription. Indeed, the signal transduction pathways typical of slow-oxidative fibers are activated by Ca^{2+} binding to calmodulin. Briefly, four members of the nuclear factor of activated T cell (NFAT) family are downstream targets of the Ca^{2+}/calmodulin-dependent phosphatase calcineurin. Upon nuclear translocation, they contribute to transcription of fiber type-specific genes. In particular, different combination of NFAT components are able to activate the transcription of specific myosin heavy chain (MyHC) isoforms (Calabria et al., 2009).

The increase of $[Ca^{2+}]c$ following RyR opening is sensed by mitochondria, that are strategically located close to sites of Ca^{2+} release allowing excitation-metabolism coupling during skeletal muscle activity (Ogata and Yamasaki, 1985). More recently, Protasi and coworkers described a bidirectional signaling route between mitochondria and SR based on molecular tethers that link the OMM to the SR. These robust contacts are achieved in the fully mature fibers as the result of postnatal remodeling consisting in a progressive shift of mitochondria from longitudinal rows to transverse orientation and the gradual maturation of the sarcotubular system (Boncompagni et al., 2009a).

The Ca^{2+} signaling from SR to mitochondria sustains aerobic metabolism by ensuring ATP production required to drive the actomyosin cross-bridge cycling. It has been well demonstrated that cytosolic Ca^{2+} peaks are transferred to mitochondria during muscle contraction (Brini et al., 1997; Rudolf et al., 2004). Mitochondria, in turn, activate Ca^{2+}-sensitive dehydrogenases of the TCA cycle (Denton, 2009).

10. The role of mitochondrial bioenergetics in the control of skeletal muscle function

Skeletal muscle metabolism largely depends on mitochondrial oxidative capacity. In this part of the chapter we will focus on the major role exerted by mitochondria to sustain aerobic metabolism, with an emphasis on how mitochondrial Ca^{2+} regulates skeletal muscle oxidative metabolism.

Skeletal muscle is characterized by the great capacity to rapidly boost the rate of energy production and substrate utilization in response to increased workload demand. On the other hand, during prolonged periods of low-intensity contractions, skeletal muscle is able to maintain a moderate increase of energy consumption. This ability is due to rapid changes in skeletal muscle energy metabolism. During contraction energy is consumed by the actin-myosin cross-bridge cycling and the prompt energy source is the hydrolysis of ATP. When a muscle is fully activated the small intracellular store of ATP is spent within 2 s. Therefore, both anaerobic and aerobic pathways must be activated to maintain the required rate of ATP synthesis. Anaerobic metabolism ensures the high intensity physical activity, whereas aerobic metabolism sustains prolonged sub-maximal exercise (Egan and Zierath, 2013). Carbohydrates and lipids are the major energetic substrates fueling skeletal muscle aerobic metabolism. Specifically, glycogen represent the predominant substrate, and glucose uptake increases during exercise through an insulin-independent pathway (Holloszy, 2003; Lee et al., 1995). Free fatty acids (FFAs) derived from triglyceride stores in skeletal muscle and adipose tissue are additional candidates for energy production. Finally, a less significant role is played by amino acids derived from muscle protein degradation (Lemon and Mullin, 1980).

Energy requirements of skeletal muscle can increase more that 100-fold during contraction. The creatine kinase shuttle together with the oxy-deoxy myoglobin shuttle promote the diffusion of ATP and O$_2$, respectively (Bessman and Geiger, 1981; Gros et al., 2010). However, in the absence of creatine, of creatine kinase or of myoglobin, skeletal muscle efficiency is still preserved with some adaptations (Garry et al., 1998; Lygate et al., 2013; van Deursen et al., 1993). These data suggest that the metabolite diffusion systems are not crucial for the normal muscle physiology. Balaban and coworkers proposed that the importance of the diffusion pathways becomes clear when muscular maximal performance is required, while in normal condition the membrane potential conduction ensures the control of energy production, thanks to the presence of a mitochondrial reticulum (Glancy et al., 2015), as detailed hereafter. The authors divided the mitochondrial network into four different categories, i.e., paravascular mitochondria (PVM), I-band mitochondria (IBM), fiber parallel mitochondria (FPM) and cross-fiber connection mitochondria (CFCM). They hypothesized that PVM are associated with the generation of the proton-motive force, whereas IBM use this force to produce ATP. The distribution of enzymes, which produce or utilize the proton-motive force is different between IBM and PVM, since complex IV, accounted for the generation of this force, is present abundantly in the PVM at the cell periphery. On the other hand, complex V, that catalyzes the phosphorylation of ADP in ATP applying the proton-motive force, is more abundant in the intra-fibrillar region. In line with this study Jaimovich and coworkers demonstrated that the propagation of the membrane potential from the periphery to the center of the fiber is Ca^{2+}-dependent (Díaz-Vegas et al., 2018). Further work is needed to clarify whether the mitochondrial Ca^{2+} uniporter components are differentially distributed among the different population of skeletal muscle mitochondria.

Mitochondrial Ca^{2+} uptake plays a central role in skeletal muscle physiology as shown by the disease phenotype of patients carrying mutations in the MICU1 gene. These patients are characterized by myopathy, extrapyramidal movement disorders and learning difficulties (Logan et al., 2014). In addition, a deletion of the exon 1 of MICU1 has been identified and associated with fatigue, lethargy, and weakness (Lewis-Smith et al., 2016). Interestingly, the phenotype of muscle-specific MICU1$^{-/-}$ mice resembles the patients' phenotype indicating that the muscle disease of the affected subjects is due to MICU1 impairment within the muscle, and it is not a consequence of a primary neuronal disorder (Debattisti et al., 2019). In particular, these mice show fatigue, weakness and increased susceptibility to demanding contractions.

Moreover, in line with the pivotal role exerted by mitochondrial Ca^{2+} uptake in the control of skeletal muscle function, it has been observed that MCU protein levels are decreased in a cellular model of Barth syndrome (BTHS), which is an X-linked disorder due to mutations in the TAZ gene (Ghosh et al., 2020). TAZ gene product is a phospholipid-lysophospholipid transacylase, an enzyme involved in the biosynthesis of cardiolipin, a fundamental component of mitochondrial membranes (Ghosh et al., 2019). Accordingly, muscles from patients with BTHS show abnormal mitochondria with aberrant cristae morphology and decrease respiratory chain function (Barth et al., 1983).

As reported above, in a constitutive MCU$^{-/-}$ mouse model, mitochondria display reduced resting matrix Ca^{2+} levels and increased Ca^{2+}-dependent PDH phosphorylation. Most importantly, these mice have a significant impairment in exercise capacity (Pan et al., 2013). These data highlight the fundamental role of mitochondrial Ca^{2+} uptake by enhancing the TCA cycle activity to increase ATP production necessary to maintain the regular muscle performance. However, it is worth to remind that in this model compensatory effects occur during embryonic development (Luongo et al., 2015; Murphy et al., 2014).

Additional studies, performed by modulating MCU expression exclusively in skeletal muscle, demonstrated that mitochondrial Ca^{2+} uptake is not only involved in the regulation of oxidative metabolism, but it also controls skeletal muscle mass by regulating the activity of two major skeletal muscle hypertrophic pathways, i.e., PGC-1α4 and IGF1-AKT/PKB. Specifically, MCU overexpression and downregulation lead to muscle hypertrophy and atrophy, respectively. Notably, these effects are independent of the control of aerobic metabolism, since PDH activity, although defective in MCU-silenced muscles, was unaffected in MCU overexpressing muscles. In addition, hypertrophy was comparable in both oxidative and glycolytic muscles and, finally, analyses of aerobic metabolism revealed no substantial changes (Mammucari et al., 2015).

A muscle-specific MCU knockout mouse (skMCU$^{-/-}$) was generated to unveil the contribution of mitochondrial Ca^{2+} uptake in the support of aerobic metabolism (Gherardi et al., 2018; Kwong et al., 2018). In accordance with previous reports (Pan et al., 2013), skMCU$^{-/-}$ mice showed loss of muscle force and impaired exercise performance (Gherardi et al., 2018). skMCU$^{-/-}$ myofibers were characterized by impaired glucose oxidation, which was evident by the decreased oxygen consumption rate (OCR) and increased lactate production. Skeletal muscle glucose uptake was increased as well as glycogen content, indicative of an inefficient glucose utilization. Nonetheless, skMCU$^{-/-}$ muscles were partially able to deal with work demand, thank to increased FA oxidation, which significantly contributed to basal OCR. Moreover, skMCU$^{-/-}$ animals displayed reduced running capacity in a sprinting protocol, whereas they showed enhanced muscle performance under condition of fatigue, where the FA metabolism is predominant (Kwong et al., 2018). These mice showed lipolysis with a consequent increase in circulating ketone bodies as a result of the failure in the muscle glucose utilization (Gherardi et al., 2018). The reduction in the PDH activity, which is tightly controlled by mitochondrial Ca^{2+}, is the main driver of this metabolic (Gherardi et al., 2018). In line with these findings, pyruvate entry into mitochondria is necessary for efficient skeletal muscle function. Indeed, a skeletal muscle-specific mitochondrial pyruvate carrier (MPC) knockout mouse strikingly resembles the phenotype of skMCU$^{-/-}$ mouse. In brief, MPC ablation decreases glucose oxidation despite enhanced glucose uptake and increased lactate production. This triggers enhanced FA oxidation and Cori cycling, with an overall increment in energy expenditure. Accordingly, skeletal muscle restricted MPC deletion accelerated fat mass loss when mice were fed again with a normal diet after high-fat diet-induced obesity (Sharma et al., 2019). All these data strengthen a model where altered mitochondrial Ca^{2+} accumulation and the block of pyruvate entry into mitochondria converge on a similar dysregulation of glucose oxidation, with the consequent coexistence of local and systemic metabolic defects.

Together with glucose, FAs are the major fuel source to sustain aerobic metabolism in skeletal muscle. Notably, even if the β-oxidation enzymes are known, the regulatory elements of this process are still under investigation. Recently, MOXI, a muscle-enriched micropeptide has been discovered and associated with the control of β-oxidation within skeletal muscle and heart (Makarewich et al., 2018). It has been demonstrated that MOXI is able to enhance β-oxidation through the interaction with the mitochondrial trifunctional protein (MTP). Indeed, its overexpression leads to an increase of FA oxidation capacity, whereas MOXI KO mouse shows impaired FA oxidation. Interestingly, the respiratory function is selectively impaired for β-oxidation, while the oxidation of other substrates remains constant. This discovery suggests that small molecules could regulated enzyme complexes involved in β-oxidation, paving the way to the identification of other regulatory elements.

11. Skeletal muscle mitochondrial biogenesis: A nuclear receptor network controls mitochondrial respiratory function and fuel selection

In skeletal muscle, mitochondrial content fits in response to changes in the metabolic demand, and even a single bout of exercise induces mitochondrial biogenesis. This process is a result of multiple molecular events that start with the activation of specific transcriptional factors. Newly formed proteins are transported into mitochondria where an orchestrated assembly of both nuclear and mitochondrial gene products takes place to expand the organelle net. In this scenario, peroxisome proliferator-activated receptor (PPAR) family, estrogen-related receptor (ERR) family and peroxisome proliferator-activated receptor gamma coactivator 1-α (PGC1α) constitute a nuclear receptor network responsible for the activation of genes essential for mitochondrial biogenesis. Here, we will focus on the regulation of fuel selection and respiratory capacity by these transcriptional factors.

PPAR family. This nuclear receptor family was originally identified as a regulator of peroxisome biogenesis and it is composed of three members: PPARα, PPARβ (also known as PPARδ) and PPARγ (Issemann and

Green, 1990). PPARα is the prototypical member of the family and it represents a key component in the regulation of substrate preference in skeletal muscle. Specifically, PPARα triggers gene expression of FA oxidation gene expression in skeletal muscle and in other highly metabolic-demanding tissues like heart and liver. In addition, its expression correlates with the different metabolic properties of skeletal muscle, showing the highest expression in type I myofibers (Brandt et al., 1998; Gulick et al., 1994; McMullen et al., 2014; van der Meer et al., 2010). The PPAR factors have large hydrophobic ligand binding domains that ensure the interaction of a variety of endogenous FAs or FA-derived ligands. Moreover, long chains of unsaturated FAs are able to bind and activate PPARα, which in turn stimulates the transcription of genes involved in the mitochondrial FA oxidation (Chakravarthy et al., 2009). Indeed, PPARα overexpression in mouse skeletal muscle increases the expression of genes involved in FA import and oxidation (Finck et al., 2005). Despite the pivotal role of PPARα in FA oxidation, the PPARα knockout mouse showed a decrease in the oxidative capacity only during starvation, indicating of a compensatory mechanisms exerted by PPARδ, which shares a large overlap of target genes with PPARα (Muoio et al., 2002).

ERR family. Three different orphan receptors are members of this superfamily of nuclear receptors expressed in tissues that preferentially metabolize fatty acids: ERRα, ERRβ, and ERRγ. It has been demonstrated that mitochondrial energy metabolism activities including glucose, FAs oxidation and oxidative phosphorylation are activated by ERR factors in cardiomyocytes. In addition, evidence supports the crucial role of this nuclear receptor family in the maintenance of the skeletal muscle metabolic homeostasis. Firstly, total ERRα KO mice are viable, fertile and display a reduced body weight with a decreased peripheral fat deposition (Luo et al., 2003). Regarding the skeletal muscle, these mice are hypoactive and show reduced energy tolerance with lactemia at exhaustion. A gene expression profiling analysis highlighted the role of ERRα in the control of genes involved in energy substrate transport, in TCA cycle activity and in oxidative metabolism. In addition, ERRα loss altered the expression of genes involved in the response to oxidative stress, in the maintenance of skeletal muscle integrity and in muscle plasticity. These results suggest that ERRα controls transcriptional programs crucial for mitochondrial metabolism and muscle performance (Perry et al., 2014). In addition, ERRα directly controls PPARα transcription. For this reason, ERRα and PPARα share a common signature of target genes (Huss et al., 2004). These data provide evidence of a feed-forward loop between ERRα and PPARα to activate oxidative metabolism in skeletal muscle. The role of ERRα in controlling energy metabolism impinges also on skeletal muscle repair after injury. Skeletal muscle-specific ERRα knockout mice showed impaired regeneration after intramuscular cardiotoxin injection, due to a reduced ATP content (LaBarge et al., 2014).

PGC1α. PGC1α is a key component of the system that matches the increased energy demand with the enhanced oxidative capacity to produce ATP. It was described as a cold-inducible transcriptional coactivator in the brown adipose tissue. PGC1α mRNA expression levels are increased upon cold exposure not only in brown adipose tissue but also in skeletal muscle, crucial organs involved in the thermogenesis (Puigserver et al., 1998). In particular, it has been showed that PGC1α stimulates mitochondrial biogenesis and respiration in muscle cells (Wu et al., 1999). PGC1α sustains oxidative metabolism and mitochondrial biogenesis interacting with multiple nuclear receptors such as NRF-1 (nuclear respiratory factor-1) and PPARγ. Importantly, for its transcriptional activity it requires the docking to PPARγ or NRF-1 to undergo a conformational change (Puigserver et al., 1999). Moreover, PGC1α enhances PPARα-mediated transcriptional activation of target elements, including FAO related genes. In skeletal muscle PGC1α is induced after an acute bout of exercise (Baar et al., 2002; Pilegaard et al., 2003) and its forced expression results in a huge induction of mitochondrial biogenesis (Lin et al., 2002). PGC1α is expressed preferentially in muscles enriched in slow-twitch fibers, which display a bigger mitochondria pool and rely on oxidative metabolism. Skeletal muscle-specific PGC1α transgenic mice are characterized by an enrichment in oxidative fibers expressing a protein asset similar to type I fibers, i.e., troponin I and myoglobin, and showing increase resistance to fatigue (Lin et al., 2002). On the other hand, skeletal muscle-specific PGC1α and β double knockout mice show a drastic reduction of exercise performance associated with a decrease in the total oxidative capacity with an impairment in the mitochondrial structure (Zechner et al., 2010). Surprisingly, loss of PGC1α and β does not altered the fiber type composition, suggesting that, while PGC1α pathway is required for mitochondrial biogenesis and to respond to acute stress, it is not necessary for chronic metabolic adaptations (Rowe et al., 2013; Zechner et al., 2010). In addition, it has been demonstrated that PGC1α positively regulates glucose uptake with a consequent accumulation of glycogen, mimicking the metabolic response to exercise recovery. Accordingly, the rate of post exercise glycogen repletion is decreased in PGC1α-deficient muscles (Wende et al., 2007). Furthermore, PGC1α is able to protect skeletal muscle from atrophy. In transgenic animals overexpressing PGC1α, denervation and fasting lead to a smaller decrease in fiber size and lower induction of atrophy-related ubiquitin ligases, atrogin-1 and MuRF-1, compared to control mice. This is explained by the fact that the overexpression of PGC1α in adult fibers reduces the ability of the atrophy-related transcription factor FoxO3 to bind atrogin-1 promoter and induce atrophy (Sandri et al., 2006). A muscle-specific splicing isoform of PGC1α exists, i.e., PGC1α4, which results from transcription driven by an alternative PGC1α promoter. This peculiar isoform is highly expressed in muscles subjected to exercise (i.e., resistance training or a combination of resistance and endurance training), but it does not participate to mito-chondrial biogenesis. In accordance, the comparison between the gene expression profiles of PGC1α and PGC1α4 overexpressing cells reveals a completely different pattern. Specifically, PGC1α4 induces IGF-1, represses myostatin and its overexpression in vitro and in vivo leads to hypertrophy. A direct consequence of this is the increase in exercise tolerance and the resistance to hindlimb suspension-induced muscle atrophy (Ruas et al., 2012). Different important signaling pathways converge in the regulation of PGC1α in skeletal muscle, such as β-adrenergic receptors (β-AR), CREB, AMPK and calcium. Clenbuterol, a β-2AR specific agonist, increased the mRNA levels of PGC1α in skeletal muscle. Conversely, treat-ment with propranolol, a β-AR antagonist, decrease its mRNA levels. In addition, the exercise-induced increase in PGC1α mRNA levels partially are reduced by a specific β-2-AR inhibitor, suggesting that the upregulation of PGC1α during exercise is mediated, at least in part, by the β-2AR (Miura et al., 2007). Moreover, it has been demonstrated that PGC1α expression is controlled by CREB (Herzig et al., 2001). CREB induces the expres-sion of hepatic gluconeogenic genes through PGC1α. Most importantly, PGC1α overexpression in CREB-deficient mice rescues the expression of gluconeogenic genes (Herzig et al., 2001). It is well known that the acti-vation of AMPK leads to an increase of glucose uptake by GLUT4 translo-cation to the plasma membrane (Kurth-Kraczek et al., 1999), and of FAO through the phosphorylation of ACC (Merrill et al., 1997; Vavvas et al., 1997). Interestingly, AMPK activity increases the PGC1α activation and different mutations on two AMPK phosphorylation sites in the PGC1α protein abolishes its activity. Subsequently, AMPK-dependent phosphory-lation on PGC1α modulates its ability to dock on transcription factors or to affect the binding of cofactors in the PGC1α coactivator complex

(Jäer et al., 2007). Also calcium signaling impinges on PGC1α activity in skeletal muscle. Indeed, several Ca^{2+}-regulated factors are involved in skeletal muscle adaptation to physical activity including CaMKIV (calcium/ calmodulin-dependent protein kinase IV) and calcineurin A (Berchtold et al., 2000). PGC1α promoter activity is under the control of both CaMKIV and calcineurin A. Briefly, CaMKIV induces PGC1α by activating CREB, which in turn, binds to a specific region of the PGC1α promoter. Moreover, calcineurin A acts additively with both CaMKIV and PGC1α (Handschin et al., 2003). Finally, PGC1α plays a role in shaping mitochondrial Ca^{2+} uptake in primary myotubes. PGC1α overexpression causes an increase in the mitochondrial volume of primary myotubes and reduces mitochondrial Ca^{2+} accumulation. This is due to PGC1α-dependent induction of UCPs expression and consequent activation of uncoupled respiration. By promoting protons entry into the mitochondrial matrix, UCPs decrease mitochondrial Ca^{2+} uptake. In addition, the increase in the mitochondrial volume leads to a decrease in Ca^{2+} "hot-spots" between mitochondria and SR, which are crucial for the Ca^{2+} transfer across these two organelles (Bianchi et al., 2006).

12. Impairment of Ca^{2+} homeostasis and mitochondrial Ca^{2+} signaling in muscular dystrophy

Muscular dystrophies are a heterogeneous group of diseases characterized by muscle wasting leading to an impairment in patient mobility and to premature death. Several pieces of evidence have highlighted the implication of Ca^{2+} deregulation in different muscular dystrophies, such as dystrophinopathies and myotonic dystrophy among others. In this section, we will discuss the actual knowledge regarding aberrant Ca^{2+} homeostasis in the different muscular dystrophies with particular focus on Duchenne muscular dystrophy (DMD).

DMD is an inherited X-linked neuromuscular disease affecting 1 in 3500 births that is characterized by severe and progressive skeletal muscle degeneration with premature death caused by cardiac dysfunction and respiratory failure. DMD is the result of lack of expression of dystrophin, a protein located at the cytoplasmic face of the sarcolemma where it maintains the integrity of the muscle fiber by linking the extracellular matrix to the actin cytoskeleton. Accordingly, dystrophin-deficient muscles show recurrent membrane damage due to sarcolemma fragility. Despite the great advancements toward the comprehension of the disease pathogenesis, the precise molecular mechanism causing fiber death is still under investigation.

In dystrophic muscles, intracellular Ca^{2+} levels increase, causing calpain activation and proteins degradation, and triggering mitochondrial Ca^{2+} overload with consequent PTP opening, eventually leading to fiber death (Burr and Molkentin, 2015) (Fig. 2). According to many studies conducted in patient biopsies and mouse models, the dysregulation of cytosolic Ca^{2+} homeostasis is an early event in this pathology. Many hypotheses have been proposed to explain this phenomenon. First, studies have demonstrated that the membrane ion channels TRPC are responsible for cytosolic Ca^{2+} elevation (Vandebrouck et al., 2002). Indeed, the block of TRPC channels in dystrophic mice reestablishes the normal Ca^{2+} influx. In addition, overexpression of TRPC3 in wildtype skeletal muscles increases Ca^{2+} influx and triggers a dystrophic phenotype (Millay et al., 2009). In addition to TRPC channels, also Orai1 and STIM1, both playing a fundamental role in the SOCE (store-operated Ca^{2+} entry) pathway, have been linked to DMD. In particular, it has been demonstrated that their expression is increased in mdx mice, indicating that the SOCE activation may contribute to the aberrant Ca^{2+} influx of this pathology (Edwards et al., 2010; Zhao et al., 2012).

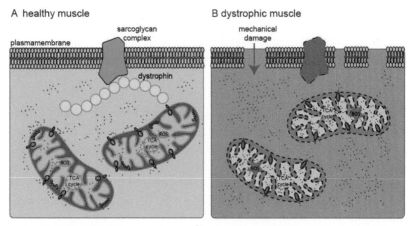

Fig. 2 Dysregulation of mitochondrial Ca^{2+} homeostasis in DMD. (A) In healthy muscles mitochondrial Ca^{2+} uptake is necessary to sustain multiple functions including aerobic metabolism. (B) In dystrophic muscles, elevated intracellular Ca^{2+} triggers mitochondrial Ca^{2+} overload with consequent PTP opening causing cell death. The several hypotheses underlying the molecular mechanism of cytosolic Ca^{2+} dysregulation are discussed in the text.

Also SR Ca^{2+} leakage through RyR and IP3R has been associated with aberrant Ca^{2+} homeostasis in DMD. Mdx mice show high hyper-nitrosylation of RyR1 cysteine residues that causes RyR leaking in resting condition due to depletion of calstabin-1 from the RyR1 complex. Interestingly, the prevention of calstabin-1 depletion is sufficient to inhibit the SR Ca^{2+} leak with a consequent improvement of muscle function in mdx mice (Bellinger et al., 2009). In addition to RyR, defects in IP3R conductivity are also involved in DMD pathology. An increase in IP3R-dependent Ca^{2+} transients, due to enhanced IP3 and IP3R protein levels, have been detected in mdx mice and myotubes from DMD patients (Basset et al., 2004; Cárdenas et al., 2010a; Liberona et al., 1998).

Moreover, myofibers from dystrophic mice and DMD patients exhibit impaired Ca^{2+} reuptake during relaxation (Divet and Huchet-Cadiou, 2002; Head, 1993; Nicolas-Metral et al., 2001). This defect can be rescued by the overexpression of SERCA1, ameliorating the muscle phenotype of mdx skeletal muscle (Goonasekera et al., 2011; Morine et al., 2010). These results indicate Ca^{2+} clearance potentiation during relaxation as a promising treatment to reduce myofiber necrosis. Accordingly, mdx mice treated with BGP-15, a drug that augments SERCA activity, show increased muscle force (Gehrig et al., 2012).

The contribution of mitochondrial Ca^{2+} signaling in DMD pathogenesis was clarified by the direct measurement of mitochondrial Ca^{2+} uptake in mdx myotubes that was augmented upon caffeine-induced Ca^{2+} release from the SR (Robert et al., 2001). Later, these data were confirmed in adult skeletal muscle fibers derived from mdx mice (Shkryl and Shirokova, 2006). In agreement with mitochondrial Ca^{2+} measurements, it has been demon-strated that MCU and EMRE protein levels are increased in skeletal muscles of mdx mice (Liu et al., 2020). Also in the skeletal muscle of the LAMA2 knockout mouse, a model of congenital merosin-deficient muscular dystro-phy where the gene encoding for the α2 subunit of laminin is deleted, the expression of MCU and EMRE is upregulated and the resting mito-chondria Ca^{2+} concentration is increased.

The first evidence that mitochondrial Ca^{2+} overload occurs in dystro-phic muscles in vivo was provided by the study of a mouse model carrying Col6a1 deficiency, mimicking Ullrich congenital muscular dystrophy (Angelin et al., 2007; Irwin et al., 2003). Muscle fibers of Col6a1 knockout mice showed mitochondrial dysfunction and signs of apoptosis. In addition, cell death could be inhibited by cyclosporine A, a cyclophilin D inhibitor

(Merlini et al., 2008), and by ablation of cyclophilin D (Palma et al., 2009). Finally, an amelioration of mitochondrial function and the reduction in cell death was observed in patients with Ullrich congenital MD treated with cyclosporine A (Merlini et al., 2011). Moreover, cyclophilin D reduced mitochondrial swelling also in mdx and Sgcd knockout mice (Millay et al., 2008). Even in this case, the deletion of cyclophilin D reduces the mitochondrial swelling suggesting that the opening of the mPTP, due to aberrant Ca^{2+} handling, is a common pathway of multiple muscle disease. The involvement of aberrant Ca^{2+} homeostasis in muscular dystrophy suggests multiple potential treatments. The preclinical efficacy of several molecules has been proven, such as Debio-025 and NHE1 (inhibitors of mPTP), S107 (ryanodine leaks inhibitor), BGP-15 (indirect SERCA activator) (Burr and Molkentin, 2015). In addition, gene therapy approaches are also emerging (Jaski et al., 2009).

Due to the role of mitochondrial Ca^{2+} overload in increasing ROS production in skeletal muscle (Brookes et al., 2004), ROS have been hypothesized as a possible cause of dystrophic muscle damage. It has been demonstrated that mdx myotubes are more susceptible to ROS–induced cell death with respect to healthy cells (Rando et al., 1998). Moreover, the oxidative damage in mdx muscles is evident even before any sign of muscle injury (Disatnik et al., 1998), suggesting that ROS production is not a secondary effect of muscle regeneration but rather it represents a primary feature of dystrophic muscle impairment. Furthermore, increased protein oxidation, triggering detrimental effects on muscle function, was observed both in mdx muscle and in DMD patients (Hauser et al., 1995; Haycock et al., 1997). Finally, treatment with the antioxidant N-acetylcysteine (NAC) has been shown to protect mdx skeletal muscles against damage by hampering the increased membrane permeability and by restoring muscle force (Whitehead et al., 2008).

Taken together these data demonstrate that the alteration in cytosolic Ca^{2+} signaling in skeletal muscle myopathies is a frequent event and it impinges on mitochondrial Ca^{2+} uptake with consequences on the organelle function. The molecular mechanisms are still under investigation. Nonetheless, the correction of Ca^{2+} homeostasis, that can be done at multiple levels (i.e., by restoration of cytosolic Ca^{2+} signaling or by inhibition of mPTP opening), have been proven to exert a beneficial effect on the different disease phenotypes.

13. Ryanodine receptor myopathies: Neuromuscular disorders affecting EC coupling and Ca^{2+} homeostasis

Congenital myopathies are a genetically heterogeneous group of neuromuscular disorders characterized by muscle weakness and peculiar structural abnormalities. Up to date the identified disorders are associated with defects in EC coupling, Ca^{2+} signaling or proteins involved in sarcomeric filament assembly and interaction. Congenital myopathies can be divided in four different disease entities based on the distinct pathological features: the dominantly inherited central core disease (CCD), the autosomal recessive multi-minicore disease (MmD), the centronuclear myopathy (CNM) and the nemaline myopathy (Platt and Griggs, 2009). In the next paragraphs, we will focus on CCD, the most common congenital myopathy.

In 1991, a mutation in RyR1 gene was discovered to be the cause of malignant hyperthermia (MH), a fatal hypermetabolic response to potent volatile anesthetic gases. Two years later, other RyR1 mutations were discovered and linked to CCD. With time, mutations in other 20 genes have been identified in patients with CCD (Jungbluth et al., 2018).

Despite the heterogeneity of the clinical aspects, the derangement of myofibrillar architecture is the main characteristic of CCD and results in the focal loss of mitochondria in a single central core (Sewry and Wallgren-Pettersson, 2017).

Since RyR1 is a key component of the EC coupling, many efforts have been made to understand the molecular consequences of RyR1 mutations. Two distinct pathogenic mechanisms have been proposed: (i) the leaky channel hypothesis suggests a constant Ca^{2+} leak from defective RyR1 channels with negative consequences on muscle contraction, (ii) the EC uncoupling theory proposes that mutated channels are functionally uncoupled from sarcolemma, and the response to action potential is a reduced Ca^{2+} release from the SR (Jungbluth et al., 2018) (Fig. 3).

An example that supported the first hypothesis is the Y522S mutation in RyR1 gene, which increases the channel release sensitivity promoting SR Ca^{2+} leak, and thus causing store depletion. As a consequence, resting intracellular [Ca^{2+}] is elevated and SR Ca^{2+} release during EC coupling is reduced (Dirksen and Avila, 2004; Tong et al., 1999). The specific molecular mechanism by which RyR1 mutations increase channel leak is still under investigation. It has been proposed that these mutations could

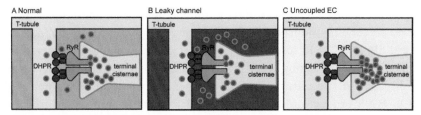

Fig. 3 Pathogenic mechanisms of RyR1 mutations in CCD. (A) Regular RyR1 function in healthy muscle. (B) Several RyR1 mutations increase the channel sensitivity to activation triggering SR Ca^{2+} leak and, eventually, store depletion. (C) Other RyR1 mutations affect the functional sarcolemma-SR coupling causing a reduced SR Ca^{2+} release.

impinge on the domain necessary to stabilize the channel in the close state. The disruption of this self-inhibition leads to increased basal RyR1 activity (Ikemoto and Yamamoto, 2000). In addition, an alternative mechanism was proposed by Chen and coworkers, where they suggested that RyR1 mutations enhance channel sensitivity due to an increased Ca^{2+} load in the SR; a similar mechanism was proposed for the increased RyR2-mediated spontaneous Ca^{2+} release occurring in the heart during catecholaminergic polymorphic ventricular tachycardia (CPVT) (Jiang et al., 2008).

Regarding the EC uncoupled hypothesis, while resting [Ca^{2+}] and SR Ca^{2+} content are unchanged, Ca^{2+} release from the SR is diminished. Most of the mutations causing this event are located in the selectivity filter region of the channel (Avila et al., 2003; Zhao et al., 1999). These mutations trigger drastic reduction in the RyR1 conductance, in the Ca^{2+} activation and in Ca^{2+} permeation (Gao et al., 2000). It has been proven that the mutation I4898T triggers a reduction in the SR Ca^{2+} release during muscle contraction affecting neither channel leak nor its sensitivity to activation by different stimuli (e.g., caffeine or halothane) (Avila et al., 2001). Notably, some controversy regarding the validity of the EC uncoupling mechanism comes from Ca^{2+} measurements performed in myotubes and lymphocytes derived from patients carrying the mutation I4898T. In these cells, RyR1 Ca^{2+} leak is actually increased (Ducreux et al., 2004; Tilgen et al., 2001).

During the last two decades, several RyR1 knockin mouse models have been generated to investigate in vivo the validity of the proposed hypotheses, i.e., the leak and the EC uncoupling mechanisms. The two models carrying homozygous mutations, i.e., Y522S and R163C, are embryonically lethal or die immediately after birth, while heterozygous mice are viable (Chelu et al., 2006; Yang et al., 2006). Y522S/+ and R163C/+ mice are MH susceptible as demonstrated by the fact that they die after isoflurane exposure or heat stress. Accordingly, muscle fibers from these mice show increased caffeine-induced SR Ca^{2+} release. In addition, in the Y522S/+ fibers, SR $[Ca^{2+}]$ and resting $[Ca^{2+}]c$ are unaffected at room temperature, while the levels of resting $[Ca^{2+}]c$ increases and SR Ca^{2+} content decreases at 30 °C. This peculiar temperature-dependency was due to increased ROS/RNS production leading to RyR1 S-nitrosylation which, in turn, triggers channel leak (Durham et al., 2008). This feed-forward mechanism is able to impair mitochondrial function in the long term. NAC treatment, a potent antioxidant, is beneficial both to prevent and to reduce mito-chondrial impairment in young, which carry a minimal damage, and adult Y522S/+ mice, which show evident structural damage, respectively (Michelucci et al., 2017a). In agreement with these data, treating Y522S/+ mice with azumolene, an analog of dantrolene, an FDA-approved drug used to treat MH crises, protects muscles from environmental heat stroke, a pathological condition similar to MH, by reducing myoplasmic Ca^{2+} concentration and nitro-oxidative stress (Michelucci et al., 2017b).

Moreover, electron microscopy analyses were conducted to investigate the molecular mechanisms involved in the development of central core. In myofibers of 2 month-old Y522S/+ mice mitochondria are swollen, disrupted and with alteration in the morphology (Boncompagni et al., 2009b). In addition, early core regions, which lack mitochondria oxidative enzyme activity and SR, are detected in contracted myofibrils in accordance with the role of SR and mitochondria to remove Ca^{2+}. At later stages (12 month-old mice) these early core regions evolve in larger unstructured cores which also lack contractile elements (Boncompagni et al., 2009b). It has been proposed that this dysfunctional progress of the cores is the results of RyR1 Ca^{2+} leak that causes mitochondrial Ca^{2+} overload with the conse-quent opening of the mitochondrial permeability transition pore. It is worth mentioning that CCD in humans is normally non-progressive, while cores in both Y522S and R163C knockin mice show a notable progression during the first year of age. These observations indicate that, although these mouse models are relevant to understand the genesis of cores, they do not fully recapitulate the human pathology, at least in terms of core progression.

In myotubes and human cell lines with RyR1 mutations that affect EC coupling IP3R expression is increased (Zhou et al., 2013) enhancing Ca^{2+} transfer between SR and mitochondria (Yi et al., 2012). A similar event occurs also in cardiomyocytes (Ibarra et al., 2013). Finally, it has been pro-posed that the increase in SR-mitochondria Ca^{2+} signaling stimulates mito-chondrial biogenesis and activity in myotubes from patients carrying RyR1 mutations (Suman et al., 2018).

Thus, despite the enormous advancements in elucidating CCD, further studies are needed to fully uncover the molecular mechanisms underlying this pathology.

References

Adams, V., Bosch, W., Schlegel, J., Wallimann, T., Brdiczka, D., 1989. Further characterization of contact sites from mitochondria of different tissues: topology of peripheral kinases. BBA-Biomembranes 981, 213–225. https://doi.org/10.1016/0005-2736(89)90031-X.

Angelin, A., Tiepolo, T., Sabatelli, P., Grumati, P., Bergamin, N., Golfieri, C., Mattioli, E., Gualandi, F., Ferlini, A., Merlini, L., Maraldi, N.M., Bonaldo, P., Bernardi, P., 2007. Mitochondrial dysfunction in the pathogenesis of Ullrich congenital muscular dystrophy and prospective therapy with cyclosporins. Proc. Natl. Acad. Sci. U. S. A. 104, 991–996. https://doi.org/10.1073/pnas.0610270104.

Antony, A.N., Paillard, M., Moffat, C., Juskeviciute, E., Correnti, J., Bolon, B., Rubin, E., Csordás, G., Seifert, E.L., Hoek, J.B., Hajnóczky, G., 2016. MICU1 regulation of mitochondrial Ca2+ uptake dictates survival and tissue regeneration. Nat. Commun. 7, 10955. https://doi.org/10.1038/ncomms10955.

Ashrafi, G., de Juan-Sanz, J., Farrell, R.J., Ryan, T.A., 2020. Molecular tuning of the axonal mitochondrial Ca2+ uniporter ensures metabolic flexibility of neurotransmission. Neuron 105, 678–687.e5. https://doi.org/10.1016/j.neuron.2019.11.020.

Avila, G., O'Brien, J.J., Dirksen, R.T., 2001. Excitation—contraction uncoupling by a human central core disease mutation in the ryanodine receptor. Proc. Natl. Acad. Sci. U. S. A. 98, 4215–4220. https://doi.org/10.1073/pnas.071048198.

Avila, G., O'Connell, K.M.S., Dirksen, R.T., 2003. The pore region of the skeletal muscle ryanodine receptor is a primary locus for excitation-contraction uncoupling in central core disease. J. Gen. Physiol. 121, 277–286. https://doi.org/10.1085/jgp.200308791.

Baar, K., Wende, A.R., Jones, T.E., Marison, M., Nolte, L.A., Chen, M., Kelly, D.P., Holloszy, J.O., 2002. Adaptations of skeletal muscle to exercise: rapid increase in the transcriptional coactivator PGC-1. FASEB J. 16, 1879–1886. https://doi.org/10.1096/fj.02-0367com.

Balaban, R.S., 2009. The role of Ca2+ signaling in the coordination of mitochondrial ATP production with cardiac work. Biochim. Biophys. Acta Bioenerg. 1787 (11), 1334–1341. https://doi.org/10.1016/j.bbabio.2009.05.011.

Balaban, R.S., Nemoto, S., Finkel, T., 2005. Mitochondria, oxidants, and aging. Cell 120, 483–495. https://doi.org/10.1016/j.cell.2005.02.001.

Baradaran, R., Wang, C., Siliciano, A.F., Long, S.B., 2018. Cryo-EM structures of fungal and metazoan mitochondrial calcium uniporters. Nature 559, 580–584. https://doi.org/10.1038/s41586-018-0331-8.

Barth, P.G., Scholte, H.R., Berden, J.A., Van Der Klei-Van Moorsel, J.M., Luyt-Houwen, I.E.M., Van'T Veer-Korthof, E.T., Van Der Harten, J.J., Sobotka-Plojhar, M.A., 1983. An X-linked mitochondrial disease affecting cardiac muscle, skeletal muscle and neutrophil leucocytes. J. Neurol. Sci. 62, 327–355. https://doi.org/10.1016/0022-510X(83)90209-5.

Basset, O., Boittin, F.X., Dorchies, O.M., Chatton, J.Y., Van Breemen, C., Ruegg, U.T., 2004. Involvement of inositol 1,4,5-trisphosphate in nicotinic calcium responses in dystrophic myotubes assessed by near-plasma membrane calcium measurement. J. Biol. Chem. 279, 47092–47100. https://doi.org/10.1074/jbc.M405054200.

Baughman, J.M., Perocchi, F., Girgis, H.S., Plovanich, M., Belcher-Timme, C.A., Sancak, Y., Bao, X.R., Strittmatter, L., Goldberger, O., Bogorad, R.L., Koteliansky, V.,

Mootha, V.K., 2011. Integrative genomics identifies MCU as an essential component of the mitochondrial calcium uniporter. Nature 476, 341–345. https://doi.org/10.1038/nature10234.

Bellinger, A.M., Reiken, S., Carlson, C., Mongillo, M., Liu, X., Rothman, L., Matecki, S., Lacampagne, A., Marks, A.R., 2009. Hypernitrosylated ryanodine receptor calcium release channels are leaky in dystrophic muscle. Nat. Med. 15, 325–330. https://doi.org/10.1038/nm.1916.

Berchtold, M.W., Brinkmeier, H., Müntener, M., 2000. Calcium ion in skeletal muscle: its crucial role for muscle function, plasticity, and disease. Physiol. Rev. 80 (3), 1215–1265. https://doi.org/10.1152/physrev.2000.80.3.1215.

Bernardi, P., Vassanelli, S., Veronese, P., Colonna, R., Szabo, I., Zoratti, M., 1992. Modulation of the mitochondrial permeability transition pore. Effect of protons and divalent cations. J. Biol. Chem. 267, 2934–2939.

Berridge, M.J., Lipp, P., Bootman, M.D., 2000. 2000—Review-calcium signalling. Nat. Rev. Mol. Cell Biol. 1, 11–21.

Bessman, S.P., Geiger, P.J., 1981. Transport of energy in muscle: the phosphorylcreatine shuttle. Science 211, 448–452. https://doi.org/10.1126/science.6450446.

Bezprozvanny, I., Watras, J., Ehrlich, B.E., 1991. Bell-shaped calcium-response curves of Ins(1,4,5)P3- and calcium-gated channels from endoplasmic reticulum of cerebellum. Nature 351, 751–754. https://doi.org/10.1038/351751a0.

Bianchi, K., Vandecasteele, G., Carli, C., Romagnoli, A., Szabadkai, G., Rizzuto, R., 2006. Regulation of Ca2+ signalling and Ca2+-mediated cell death by the transcriptional coactivator PGC-1α. Cell Death Differ. 13, 586–596. https://doi.org/10.1038/sj.cdd.4401784.

Biasutto, L., Azzolini, M., Szabò, I., Zoratti, M., 2016. The mitochondrial permeability transition pore in AD 2016: an update. Biochim. Biophys. Acta Mol. Cell Res. 1863, 2515–2530. https://doi.org/10.1016/j.bbamcr.2016.02.012.

Boitier, E., Rea, R., Duchen, M.R., 1999. Mitochondria exert a negative feedback on the propagation of intracellular Ca2+ waves in rat cortical astrocytes. J. Cell Biol. 145, 795–808. https://doi.org/10.1083/jcb.145.4.795.

Boncompagni, S., Rossi, A.E., Micaroni, M., Beznoussenko, G.V., Polishchuk, R.S., Dirksen, R.T., Protasi, F., 2009a. Mitochondria are linked to calcium stores in striated muscle by developmentally regulated tethering structures. Mol. Biol. Cell 20, 1058–1067. https://doi.org/10.1091/mbc.e08-07-0783.

Boncompagni, S., Rossi, A.E., Micaroni, M., Hamilton, S.L., Dirksen, R.T., Franzini-Armstrong, C., Protasi, F., 2009b. Characterization and temporal development of cores in a mouse model of malignant hyperthermia. Proc. Natl. Acad. Sci. U. S. A. 106, 21996–22001. https://doi.org/10.1073/pnas.0911496106.

Brandt, J.M., Djouadi, F., Kelly, D.P., 1998. Fatty acids activate transcription of the muscle carnitine palmitoyltransferase I gene in cardiac myocytes via the peroxisome proliferator-activated receptor α. J. Biol. Chem. 273, 23786–23792. https://doi.org/10.1074/jbc.273.37.23786.

Bricker, D.K., Taylor, E.B., Schell, J.C., Orsak, T., Boutron, A., Chen, Y.C., Cox, J.E., Cardon, C.M., Van Vranken, J.G., Dephoure, N., Redin, C., Boudina, S., Gygi, S.P., Brivet, M., Thummel, C.S., Rutter, J., 2012. A mitochondrial pyruvate carrier required for pyruvate uptake in yeast, Drosophila, and humans. Science 336, 96–100. https://doi.org/10.1126/science.1218099.

Brini, M., De Giorgi, F., Murgia, M., Marsault, R., Massimino, M.L., Cantini, M., Rizzuto, R., Pozzan, T., 1997. Subcellular analysis of Ca2+ homeostasis in primary cultures of skeletal muscle myotubes. Mol. Biol. Cell 8, 129–143. https://doi.org/10.1091/mbc.8.1.129.

Brookes, P.S., Yoon, Y., Robotham, J.L., Anders, M.W., Sheu, S.S., 2004. Calcium, ATP, and ROS: a mitochondrial love-hate triangle. Am. J. Physiol. Cell Physiol. 287. https://doi.org/10.1152/ajpcell.00139.2004.

Burr, A.R., Molkentin, J.D., 2015. Genetic evidence in the mouse solidifies the calcium hypothesis of myofiber death in muscular dystrophy. Cell Death Differ. 22, 1402–1412. https://doi.org/10.1038/cdd.2015.65.

Calabria, E., Ciciliot, S., Moretti, I., Garcia, M., Picard, A., Dyar, K.A., Pallafacchina, G., Tothova, J., Schiaffino, S., Murgia, M., 2009. NFAT isoforms control activity-dependent muscle fiber type specification. Proc. Natl. Acad. Sci. U. S. A. 106, 13335–13340. https://doi.org/10.1073/pnas.0812911106.

Cárdenas, C., Juretić, N., Bevilacqua, J.A., García, I.E., Figueroa, R., Hartley, R., Taratuto, A.L., Gejman, R., Riveros, N., Molgó, J., Jaimovich, E., 2010a. Abnormal distribution of inositol 1,4,5-trisphosphate receptors in human muscle can be related to altered calcium signals and gene expression in Duchenne dystrophy-derived cells. FASEB J. 24, 3210–3221. https://doi.org/10.1096/fj.09-152017.

Cárdenas, C., Miller, R.A., Smith, I., Bui, T., Molgó, J., Müller, M., Vais, H., Cheung, K.H., Yang, J., Parker, I., Thompson, C.B., Birnbaum, M.J., Hallows, K.R., Foskett, J.K., 2010b. Essential regulation of cell bioenergetics by constitutive InsP3 receptor Ca2+ transfer to mitochondria. Cell 142, 270–283. https://doi.org/10.1016/j.cell.2010.06.007.

Celsi, F., Pizzo, P., Brini, M., Leo, S., Fotino, C., Pinton, P., Rizzuto, R., 2009. Mitochondria, calcium and cell death: a deadly triad in neurodegeneration. Biochim. Biophys. Acta Bioenerg. 1787 (5), 335–344. https://doi.org/10.1016/j.bbabio.2009.02.021.

Chakravarthy, M.V., Lodhi, I.J., Yin, L., Malapaka, R.R.V., Xu, H.E., Turk, J., Semenkovich, C.F., 2009. Identification of a physiologically relevant endogenous ligand for PPARα in liver. Cell 138, 476–488. https://doi.org/10.1016/j.cell.2009.05.036.

Chelu, M.G., Goonasekera, S.A., Durham, W.J., Tang, W., Lueck, J.D., Riehl, J., Pessah, I.N., Zhang, P., Bhattacharjee, M.B., Dirksen, R.T., Hamilton, S.L., 2006. Heat- and anesthesia-induced malignant hyperthermia in an RyR1 knock-in mouse. FASEB J. 20, 329–330. https://doi.org/10.1096/fj.05-4497fje.

Colombini, M., 2012. VDAC structure, selectivity, and dynamics. Biochim. Biophys. Acta Biomembr. 1818, 1457–1465. https://doi.org/10.1016/j.bbamem.2011.12.026.

Covarrubias, L., Hernández-García, D., Schnabel, D., Salas-Vidal, E., Castro-Obregón, S., 2008. Function of reactive oxygen species during animal development: passive or active? Dev. Biol. 320, 1–11. https://doi.org/10.1016/j.ydbio.2008.04.041.

Csordás, G., Thomas, A.P., Hajnóczky, G., 1999. Quasi-synaptic calcium signal transmission between endoplasmic reticulum and mitochondria. EMBO J. 18, 96–108. https://doi.org/10.1093/emboj/18.1.96.

Csordás, G., Golenár, T., Seifert, E.L., Kamer, K.J., Sancak, Y., Perocchi, F., Moffat, C., Weaver, D., de la Fuente Perez, S., Bogorad, R., Koteliansky, V., Adijanto, J., Mootha, V.K., Hajnóczky, G., 2013. MICU1 controls both the threshold and cooperative activation of the mitochondrial Ca^{2+} uniporter. Cell Metab. 17, 976–987. https://doi.org/10.1016/j.cmet.2013.04.020.

De Mario, A., Peggion, C., Massimino, M.L., Viviani, F., Castellani, A., Giacomello, M., Lim, D., Bertoli, A., Sorgato, M.C., 2017. The prion protein regulates glutamate-mediated Ca2+ entry and mitochondrial Ca2+ accumulation in neurons. J. Cell Sci. 130, 2736–2746. https://doi.org/10.1242/jcs.196972.

De Mario, A., Peggion, C., Massimino, M.L., Norante, R.P., Zulian, A., Bertoli, A., Sorgato, M.C., 2019. The link of the prion protein with Ca2+ metabolism and ROS production, and the possible implication in Aβ toxicity. Int. J. Mol. Sci. 20 (18), 4640. https://doi.org/10.3390/ijms20184640.

De Stefani, D., Raffaello, A., Teardo, E., Szabò, I., Rizzuto, R., 2011. A forty-kilodalton protein of the inner membrane is the mitochondrial calcium uniporter. Nature 476, 336–340. https://doi.org/10.1038/nature10230.

De Stefani, D., Bononi, A., Romagnoli, A., Messina, A., De Pinto, V., Pinton, P., Rizzuto, R., 2012. VDAC1 selectively transfers apoptotic Ca2 signals to mitochondria. Cell Death Differ. 19, 267–273. https://doi.org/10.1038/cdd.2011.92.

Debattisti, V., Horn, A., Singh, R., Seifert, E.L., Hogarth, M.W., Mazala, D.A., Huang, K.T., Horvath, R., Jaiswal, J.K., Hajnóczky, G., 2019. Dysregulation of mitochondrial Ca2+ uptake and sarcolemma repair underlie muscle weakness and wasting in patients and mice lacking MICU1. Cell Rep. 29, 1274–1286.e6. https://doi.org/10.1016/j.celrep.2019.09.063.

Delierneux, C., Kouba, S., Shanmughapriya, S., Potier-Cartereau, M., Trebak, M., Hempel, N., 2020. Mitochondrial calcium regulation of redox signaling in cancer. Cell 9, 432. https://doi.org/10.3390/cells9020432.

Denton, R.M., 2009. Regulation of mitochondrial dehydrogenases by calcium ions. Biochim. Biophys. Acta Bioenerg. 1787, 1309–1316. https://doi.org/10.1016/J.BBABIO.2009.01.005.

Denton, R.M., Randle, P.J., Martin, B.R., 1972. Stimulation by calcium ions of pyruvate dehydrogenase phosphate phosphatase. Biochem. J. 128, 161–163.

Denton, R.M., Richards, D.A., Chin, J.G., 1978. Calcium ions and the regulation of NAD+-linked isocitrate dehydrogenase from the mitochondria of rat heart and other tissues. Biochem. J. 176, 899–906. https://doi.org/10.1042/bj1760899.

Di Benedetto, G., Scalzotto, E., Mongillo, M., Pozzan, T., 2013. Mitochondrial Ca2+ uptake induces cyclic AMP generation in the matrix and modulates organelle ATP levels. Cell Metab. 17, 965–975. https://doi.org/10.1016/j.cmet.2013.05.003.

Díaz-Vegas, A.R., Cordova, A., Valladares, D., Llanos, P., Hidalgo, C., Gherardi, G., De Stefani, D., Mammucari, C., Rizzuto, R., Contreras-Ferrat, A., Jaimovich, E., 2018. Mitochondrial calcium increase induced by RyR1 and IP$_3$R channel activation after membrane depolarization regulates skeletal muscle metabolism. Front. Physiol. 9. https://doi.org/10.3389/fphys.2018.00791.

Dirksen, R.T., 2002. Bi-directional coupling between dihydropyridine receptors and ryanodine receptors. Front. Biosci. 7, d659–d670.

Dirksen, R.T., Avila, G., 2004. Distinct effects on Ca2+ handling caused by malignant hyperthermia and central core disease mutations in RyR1. Biophys. J. 87, 3193–3204. https://doi.org/10.1529/biophysj.104.048447.

Disatnik, M.H., Dhawan, J., Yu, Y., Beal, M.F., Whirl, M.M., Franco, A.A., Rando, T.A., 1998. Evidence of oxidative stress in mdx mouse muscle: studies of the pre-necrotic state. J. Neurol. Sci. 161, 77–84. https://doi.org/10.1016/S0022-510X(98)00258-5.

Divet, A., Huchet-Cadiou, C., 2002. Sarcoplasmic reticulum function in slow- and fast-twitch skeletal muscles from mdx mice. Pflügers Arch. Eur. J. Physiol. 444, 634–643. https://doi.org/10.1007/s00424-002-0854-5.

Dong, Z., Shanmughapriya, S., Tomar, D., Siddiqui, N., Lynch, S., Nemani, N., Breves, S.L., Zhang, X., Tripathi, A., Palaniappan, P., Riitano, M.F., Worth, A.M., Seelam, A., Carvalho, E., Subbiah, R., Jaña, F., Soboloff, J., Peng, Y., Cheung, J.Y., Joseph, S.K., Caplan, J., Rajan, S., Stathopulos, P.B., Madesh, M., 2017. Mitochondrial Ca2+ uniporter is a mitochondrial luminal redox sensor that augments MCU channel activity. Mol. Cell 65, 1014–1028.e7. https://doi.org/10.1016/j.molcel.2017.01.032.

Drago, I., Davis, R.L., 2016. Inhibiting the mitochondrial calcium uniporter during development impairs memory in adult Drosophila. Cell Rep. 16, 2763–2776. https://doi.org/10.1016/j.celrep.2016.08.017.

Ducreux, S., Zorzato, F., Müller, C., Sewry, C., Muntoni, F., Quinlivan, R., Restagno, G., Girard, T., Treves, S., 2004. Effect of ryanodine receptor mutations on interleukin-6 release and intracellular calcium homeostasis in human myotubes from malignant hyperthermia-susceptible individuals and patients affected by central core disease. J. Biol. Chem. 279, 43838–43846. https://doi.org/10.1074/jbc.M403612200.

Durham, W.J., Aracena-Parks, P., Long, C., Rossi, A.E., Goonasekera, S.A., Boncompagni, S., Galvan, D.L., Gilman, C.P., Baker, M.R., Shirokova, N.,

Protasi, F., Dirksen, R., Hamilton, S.L., 2008. RyR1 S-nitrosylation underlies environmental heat stroke and sudden death in Y522S RyR1 knockin mice. Cell 133, 53–65. https://doi.org/10.1016/j.cell.2008.02.042.

Edwards, J.N., Friedrich, O., Cully, T.R., Von Wegner, F., Murphy, R.M., Launikonis, B.S., 2010. Upregulation of store-operated Ca2+ entry in dystrophic mdx mouse muscle. Am. J. Physiol. Cell Physiol. 299, C42–C50. https://doi.org/10.1152/ajpcell.00524.2009.

Egan, B., Zierath, J.R., 2013. Exercise metabolism and the molecular regulation of skeletal muscle adaptation. Cell Metab. 17, 162–184. https://doi.org/10.1016/j.cmet.2012.12.012.

Elmore, S., 2007. Apoptosis: a review of programmed cell death. Toxicol. Pathol. 35, 495–516. https://doi.org/10.1080/01926230701320337.

Fan, C., Fan, M., Orlando, B.J., Fastman, N.M., Zhang, J., Xu, Y., Chambers, M.G., Xu, X., Perry, K., Liao, M., Feng, L., 2018. X-ray and cryo-EM structures of the mitochondrial calcium uniporter. Nature 559, 575–579. https://doi.org/10.1038/s41586-018-0330-9.

Fan, M., Zhang, J., Tsai, C.-W., Orlando, B.J., Rodriguez, M., Xu, Y., Liao, M., Tsai, M.-F., Feng, L., 2020. Structure and mechanism of the mitochondrial Ca2+ uniporter holocomplex. Nature 582, 1–5. https://doi.org/10.1038/s41586-020-2309-6.

Fieni, F., Lee, S.B., Jan, Y.N., Kirichok, Y., 2012. Activity of the mitochondrial calcium uniporter varies greatly between tissues. Nat. Commun. 3, 1317. https://doi.org/10.1038/ncomms2325.

Finck, B.N., Bernal-Mizrachi, C., Han, D.H., Coleman, T., Sambandam, N., LaRiviere, L.L., Holloszy, J.O., Semenkovich, C.F., Kelly, D.P., 2005. A potential link between muscle peroxisome proliferator- activated receptor-α signaling and obesity-related diabetes. Cell Metab. 1, 133–144. https://doi.org/10.1016/j.cmet.2005.01.006.

Finkel, T., 2012. From sulfenylation to sulfhydration: what a thiolate needs to tolerate. Sci. Signal. 5, 1–4. https://doi.org/10.1126/scisignal.2002943.

Franzini-Armstrong, C., Jorgensen, A.O., 1994. Structure and development of E-C coupling units in skeletal muscle. Annu. Rev. Physiol. 56, 509–534. https://doi.org/10.1146/annurev.ph.56.030194.002453.

Franzini-Armstrong, C., Protasi, F., 1997. Ryanodine receptors of striated muscles: a complex channel capable of multiple interactions. Physiol. Rev. 77 (3), 699–729. https://doi.org/10.1152/physrev.1997.77.3.699.

Gao, L., Balshaw, D., Xu, L., Tripathy, A., Xin, C., Meissner, G., 2000. Evidence for a role of the lumenal M3-M4 loop in skeletal muscle Ca2+ release channel (ryanodine receptor) activity and conductance. Biophys. J. 79, 828–840. https://doi.org/10.1016/S0006-3495(00)76339-9.

Garry, D.J., Ordway, G.A., Lorenz, J.N., Radford, N.B., Chin, E.R., Grange, R.W., Bassel-Duby, R., Williams, R.S., 1998. Mice without myoglobin. Nature 395, 905–908. https://doi.org/10.1038/27681.

Gehrig, S.M., Van Der Poel, C., Sayer, T.A., Schertzer, J.D., Henstridge, D.C., Church, J.E., Lamon, S., Russell, A.P., Davies, K.E., Febbraio, M.A., Lynch, G.S., 2012. Hsp72 preserves muscle function and slows progression of severe muscular dystrophy. Nature 484, 394–398. https://doi.org/10.1038/nature10980.

Gherardi, G., Nogara, L., Ciciliot, S., Fadini, G.P., Blaauw, B., Braghetta, P., Bonaldo, P., De Stefani, D., Rizzuto, R., Mammucari, C., 2018. Loss of mitochondrial calcium uniporter rewires skeletal muscle metabolism and substrate preference. Cell Death Differ. 1, 362–381. https://doi.org/10.1038/s41418-018-0191-7.

Ghosh, S., Iadarola, D.M., Ball, W.B., Gohil, V.M., 2019. Mitochondrial dysfunctions in Barth syndrome. IUBMB Life 71 (7), 791–801. https://doi.org/10.1002/iub.2018.

Ghosh, S., Ball, W.B., Madaris, T.R., Srikantan, S., Madesh, M., Mootha, V.K., Gohil, V.M., 2020. An essential role for cardiolipin in the stability and function of the mitochondrial calcium uniporter. Proc. Natl. Acad. Sci. U. S. A. 117, 16383–16390. https://doi.org/10.1073/pnas.2000640117.

Giangregorio, N., Palmieri, F., Indiveri, C., 2013. Glutathione controls the redox state of the mitochondrial carnitine/acylcarnitine carrier Cys residues by glutathionylation. Biochim. Biophys. Acta Gen. Subj. 1830, 5299–5304. https://doi.org/10.1016/j.bbagen.2013.08.003.

Glancy, B., Willis, W.T., Chess, D.J., Balaban, R.S., 2013. Effect of calcium on the oxidative phosphorylation cascade in skeletal muscle mitochondria. Biochemistry 52, 2793–2809. https://doi.org/10.1021/bi3015983.

Glancy, B., Hartnell, L.M., Malide, D., Yu, Z.-X., Combs, C.A., Connelly, P.S., Subramaniam, S., Balaban, R.S., 2015. Mitochondrial reticulum for cellular energy distribution in muscle. Nature 523, 617–620. https://doi.org/10.1038/nature14614.

Goonasekera, S.A., Lam, C.K., Millay, D.P., Sargent, M.A., Hajjar, R.J., Kranias, E.G., Molkentin, J.D., 2011. Mitigation of muscular dystrophy in mice by SERCA over-expression in skeletal muscle. J. Clin. Invest. 121, 1044–1052. https://doi.org/10.1172/JCI43844.

Görlach, A., Bertram, K., Hudecova, S., Krizanova, O., 2015. Calcium and ROS: a mutual interplay. Redox Biol. 6, 260–271. https://doi.org/10.1016/j.redox.2015.08.010.

Granatiero, V., Giorgio, V., Calì, T., Patron, M., Brini, M., Bernardi, P., Tiranti, V., Zeviani, M., Pallafacchina, G., De Stefani, D., Rizzuto, R., 2016. Reduced mitochondrial Ca2+ transients stimulate autophagy in human fibroblasts carrying the 13514A > G mutation of the ND5 subunit of NADH dehydrogenase. Cell Death Differ. 23, 231–241. https://doi.org/10.1038/cdd.2015.84.

Green, D.R., Reed, J.C., 1998. Mitochondria and apoptosis. Science 281, 1309–1312. https://doi.org/10.1126/science.281.5381.1309.

Gros, G., Wittenberg, B.A., Jue, T., 2010. Myoglobin's old and new clothes: from molecular structure to function in living cells. J. Exp. Biol. 213, 2713–2725. https://doi.org/10.1242/jeb.043075.

Gulick, T., Cresci, S., Caira, T., Moore, D.D., Kelly, D.P., 1994. The peroxisome proliferator-activated receptor regulates mitochondrial fatty acid oxidative enzyme gene expression. Proc. Natl. Acad. Sci. U. S. A. 91, 11012–11016. https://doi.org/10.1073/pnas.91.23.11012.

Gunter, K.K., Zuscik, M.J., Gunter, T.E. (Eds.), 1991. The Na+-independent Ca2+ efflux mechanism of liver mitochondria is not a passive Ca2+/2H+ exchanger. J. Biol. Chem. 266, 21640–21648.

Hajnóczky, G., Robb-Gaspers, L.D., Seitz, M.B., Thomas, A.P., 1995. Decoding of cytosolic calcium oscillations in the mitochondria. Cell 82, 415–424.

Hajnóczky, G., Hager, R., Thomas, A.P., 1999. Mitochondria suppress local feedback activation of inositol 1,4,5- trisphosphate receptors by Ca2+. J. Biol. Chem. 274, 14157–14162. https://doi.org/10.1074/jbc.274.20.14157.

Halestrap, A.P., Woodfield, K.Y., Connern, C.P., 1997. Oxidative stress, thiol reagents, and membrane potential modulate the mitochondrial permeability transition by affecting nucleotide binding to the adenine nucleotide translocase. J. Biol. Chem. 272, 3346–3354. https://doi.org/10.1074/jbc.272.6.3346.

Handschin, C., Rhee, J., Lin, J., Tarr, P.T., Spiegelman, B.M., 2003. An autoregulatory loop controls peroxisome proliferator-activated receptor γ coactivator 1α expression in muscle. Proc. Natl. Acad. Sci. U. S. A. 100, 7111–7116. https://doi.org/10.1073/pnas.1232352100.

Hauser, E., Hoger, H., Bittner, R., Widhalm, K., Herkner, K., Lubec, G., 1995. Oxyradical damage and mitochondrial enzyme activities in the mdx mouse. Neuropediatrics 26, 260–262. https://doi.org/10.1055/s-2007-979768.

Haycock, J.W., Mac Neil, S., Jones, P., Harris, J.B., Mantle, D., 1997. Oxidative damage to muscle protein in Duchenne muscular dystrophy. Neuroreport 8, 357–361. https://doi. org/10.1097/00001756-199612200-00070.

Head, S.I., 1993. Membrane potential, resting calcium and calcium transients in isolated muscle fibres from normal and dystrophic mice. J. Physiol. 469, 11–19.

Herzig, S., Long, F., Jhala, U.S., Hedrick, S., Quinn, R., Bauer, A., Rudolph, D., Schutz, G., Yoon, C., Puigserver, P., Spiegelman, B., Montminy, M., 2001. CREB regulates hepatic gluconeogenesis through the coactivator PGC-1. Nature 413, 179–183. https://doi.org/10.1038/35093131.

Herzig, S., Raemy, E., Montessuit, S., Veuthey, J.L., Zamboni, N., Westermann, B., Kunji, E.R.S., Martinou, J.C., 2012. Identification and functional expression of the mitochondrial pyruvate carrier. Science 336, 93–96. https://doi.org/10.1126/science.1218530.

Holloszy, J.O., 2003. A forty-year memoir of research on the regulation of glucose transport into muscle. Am. J. Physiol. Endocrinol. Metab. 284 (3), E453–E467. https://doi.org/10.1152/ajpendo.00463.2002.

Holmström, K.M., Finkel, T., 2014. Cellular mechanisms and physiological consequences of redox-dependent signalling. Nat. Rev. Mol. Cell Biol. 15, 411–421. https://doi.org/10.1038/nrm3801.

Holmström, K.M., Pan, X., Liu, J.C., Menazza, S., Liu, J., Nguyen, T.T., Pan, H., Parks, R.J., Anderson, S., Noguchi, A., Springer, D., Murphy, E., Finkel, T., 2015. Assessment of cardiac function in mice lacking the mitochondrial calcium uniporter. J. Mol. Cell. Cardiol. 85, 178–182. https://doi.org/10.1016/j.yjmcc.2015.05.022.

Horn, A., Van Der Meulen, J.H., Defour, A., Hogarth, M., Sreetama, S.C., Reed, A., Scheffer, L., Chandel, N.S., Jaiswal, J.K., 2017. Mitochondrial redox signaling enables repair of injured skeletal muscle cells. Sci. Signal. 10 (495), eaaj1978. https://doi.org/10.1126/scisignal.aaj1978.

Huang, G., Vercesi, A.E., Docampo, R., 2013. Essential regulation of cell bioenergetics in Trypanosoma brucei by the mitochondrial calcium uniporter. Nat. Commun. 4, 2865. https://doi.org/10.1038/ncomms3865.

Huo, J., Lu, S., Kwong, J.Q., Bround, M.J., Grimes, K.M., Sargent, M.A., Brown, M.E., Davis, M.E., Bers, D.M., Molkentin, J.D., 2020. MCUb induction protects the heart from postischemic remodeling. Circ. Res. 127, 379–390. https://doi.org/10.1161/CIRCRESAHA.119.316369.

Huss, J.M., Torra, I.P., Staels, B., Giguère, V., Kelly, D.P., 2004. Estrogen-related receptor alpha directs peroxisome proliferator-activated receptor alpha signaling in the transcriptional control of energy metabolism in cardiac and skeletal muscle. Mol. Cell. Biol. 24, 9079–9091. https://doi.org/10.1128/MCB.24.20.9079-9091.2004.

Ibarra, C., Vicencio, J.M., Estrada, M., Lin, Y., Rocco, P., Rebellato, P., Munoz, J.P., Garcia-Prieto, J., Quest, A.F.G., Chiong, M., Davidson, S.M., Bulatovic, I., Grinnemo, K.H., Larsson, O., Szabadkai, G., Uhlén, P., Jaimovich, E., Lavandero, S., 2013. Local control of nuclear calcium signaling in cardiac myocytes by perinuclear microdomains of sarcolemmal insulin-like growth factor 1 receptors. Circ. Res. 112, 236–245. https://doi.org/10.1161/CIRCRESAHA.112.273839.

Ikemoto, N., Yamamoto, T., 2000. Postulated role of Inter-domain interaction within the ryanodine receptor in Ca2+ channel regulation. Trends Cardiovasc. Med. 10 (7), 310–316. https://doi.org/10.1016/S1050-1738(01)00067-6.

Irwin, W.A., Bergamin, N., Sabatelli, P., Reggiani, C., Megighian, A., Merlini, L., Braghetta, P., Columbaro, M., Volpin, D., Bressan, G.M., Bernardi, P., Bonaldo, P., 2003. Mitochondrial dysfunction and apoptosis in myopathic mice with collagen VI deficiency. Nat. Genet. 35, 367–371. https://doi.org/10.1038/ng1270.

Ishii, K., Hirose, K., Iino, M., 2006. Ca2+ shuttling between endoplasmic reticulum and mitochondria underlying Ca2+ oscillations. EMBO Rep. 7, 390–396. https://doi.org/10.1038/sj.embor.7400620.

Issemann, I., Green, S., 1990. Activation of a member of the steroid hormone receptor superfamily by peroxisome proliferators. Nature 347, 645–650. https://doi.org/10.1038/347645a0.

Jäer, S., Handschin, C., St-Pierre, J., Spiegelman, B.M., 2007. AMP-activated protein kinase (AMPK) action in skeletal muscle via direct phosphorylation of PGC-1α. Proc. Natl. Acad. Sci. U. S. A. 104, 12017–12022. https://doi.org/10.1073/pnas.0705070104.

Jaski, B.E., Jessup, M.L., Mancini, D.M., Cappola, T.P., Pauly, D.F., Greenberg, B., Borow, K., Dittrich, H., Zsebo, K.M., Hajjar, R.J., 2009. Calcium upregulation by percutaneous administration of gene therapy in cardiac disease (CUPID Trial), a first-in-human phase 1/2 clinical trial. J. Card. Fail. 15, 171–181. https://doi.org/10.1016/j.cardfail.2009.01.013.

Jiang, D., Chen, W., Xiao, J., Wang, R., Kong, H., Jones, P.P., Zhang, L., Fruen, B., Chen, S.R.W., 2008. Reduced threshold for luminal Ca2+ activation of RyR1 underlies a causal mechanism of porcine malignant hyperthermia. J. Biol. Chem. 283, 20813–20820. https://doi.org/10.1074/jbc.M801944200.

Jones, D.P., 2006. Redefining oxidative stress. Antioxid. Redox Signal. 8, 1865–1879. https://doi.org/10.1089/ars.2006.8.1865.

Jouaville, L.S., Pinton, P., Bastianutto, C., Rutter, G.A., Rizzuto, R., 1999. Regulation of mitochondrial ATP synthesis by calcium: evidence for a long-term metabolic priming. Proc. Natl. Acad. Sci. U. S. A. 96, 13807–13812. https://doi.org/10.1073/pnas.96.24.13807.

Jung, D.W., Baysal, K., Brierley, G.P., 1995. The sodium-calcium antiport of heart mitochondria is not electroneutral. J. Biol. Chem. 270, 672–678. https://doi.org/10.1074/jbc.270.2.672.

Jungbluth, H., Treves, S., Zorzato, F., Sarkozy, A., Ochala, J., Sewry, C., Phadke, R., Gautel, M., Muntoni, F., 2018. Congenital myopathies: disorders of excitation-contraction coupling and muscle contraction. Nat. Rev. Neurol. 14 (3), 151–167. https://doi.org/10.1038/nrneurol.2017.191.

Kamer, K.J., Mootha, V.K., 2014. MICU$_1$ and MICU$_2$ play nonredundant roles in the regulation of the mitochondrial calcium uniporter. EMBO Rep. 15, 299–307. https://doi.org/10.1002/embr.201337946.

Kovács-Bogdán, E., Sancak, Y., Kamer, K.J., Plovanich, M., Jambhekar, A., Huber, R.J., Myre, M.A., Blower, M.D., Mootha, V.K., 2014. Reconstitution of the mitochondrial calcium uniporter in yeast. Proc. Natl. Acad. Sci. U. S. A. 111, 8985–8990. https://doi.org/10.1073/pnas.1400514111.

Kroemer, G., Galluzzi, L., Brenner, C., 2007. Mitochondrial membrane permeabilization in cell death. Physiol. Rev. 87, 99–163. https://doi.org/10.1152/physrev.00013.2006.

Kurth-Kraczek, E.J., Hirshman, M.F., Goodyear, L.J., Winder, W.W., 1999. 5' AMP-activated protein kinase activation causes GLUT4 translocation in skeletal muscle. Diabetes 48, 1667–1671. https://doi.org/10.2337/diabetes.48.8.1667.

Kwong, J.Q., Lu, X., Correll, R.N., Schwanekamp, J.A., Vagnozzi, R.J., Sargent, M.A., York, A.J., Zhang, J., Bers, D.M., Molkentin, J.D., 2015. The mitochondrial calcium uniporter selectively matches metabolic output to acute contractile stress in the heart. Cell Rep. 12 (1), 15–22. https://doi.org/10.1016/j.celrep.2015.06.002.

Kwong, J.Q., Huo, J., Bround, M.J., Boyer, J.G., Schwanekamp, J.A., Ghazal, N., Maxwell, J.T., Jang, Y.C., Khuchua, Z., Shi, K., Bers, D.M., Davis, J., Molkentin, J.D., 2018. The mitochondrial calcium uniporter underlies metabolic fuel preference in skeletal muscle. JCI Insight 3. https://doi.org/10.1172/jci.insight.121689.

LaBarge, S., McDonald, M., Smith-Powell, L., Auwerx, J., Huss, J.M., 2014. Estrogen-related receptor-α (ERRα) deficiency in skeletal muscle impairs regeneration in response to injury. FASEB J. 28, 1082–1097. https://doi.org/10.1096/fj.13-229211.

Lambert, J.P., Luongo, T.S., Tomar, D., Jadiya, P., Gao, E., Zhang, X., Lucchese, A.M., Kolmetzky, D.W., Shah, N.S., Elrod, J.W., 2019. MCUB regulates the molecular composition of the mitochondrial calcium uniporter channel to limit mitochondrial calcium overload during stress. Circulation 140, 1720–1733. https://doi.org/10.1161/CIRCULATIONAHA.118.037968.

Lee, A.D., Hansen, P.A., Holloszy, J.O., 1995. Wortmannin inhibits insulin-stimulated but not contraction-stimulated glucose transport activity in skeletal muscle. FEBS Lett. 361, 51–54. https://doi.org/10.1016/0014-5793(95)00147-2.

Lee, Y., Min, C.K., Kim, T.G., Song, H.K., Lim, Y., Kim, D., Shin, K., Kang, M., Kang, J.Y., Youn, H., Lee, J., An, J.Y., Park, K.R., Lim, J.J., Kim, J.H., Kim, J.H., Park, Z.Y., Kim, Y., Wang, J., Kim, D.H., Eom, S.H., 2015. Structure and function of the N-terminal domain of the human mitochondrial calcium uniporter. EMBO Rep. 16, 1318–1333. https://doi.org/10.15252/embr.201540436.

Lemasters, J.J., Qian, T., Bradham, C.A., Brenner, D.A., Cascio, W.E., Trost, L.C., Nishimura, Y., Nieminen, A.L., Herman, B., 1999. Mitochondrial dysfunction in the pathogenesis of necrotic and apoptotic cell death. J. Bioenerg. Biomembr. 31, 305–319. https://doi.org/10.1023/A:1005419617371.

Lemon, P.W.R., Mullin, J.P., 1980. Effect of initial muscle glycogen levels on protein catabolism during exercise. J. Appl. Physiol. Respir. Environ. Exerc. Physiol. 48, 624–629. https://doi.org/10.1152/jappl.1980.48.4.624.

Lewis-Smith, D., Kamer, K.J., Griffin, H., Childs, A.-M., Pysden, K., Titov, D., Duff, J., Pyle, A., Taylor, R.W., Yu-Wai-Man, P., Ramesh, V., Horvath, R., Mootha, V.K., Chinnery, P.F., 2016. Homozygous deletion in MICU1 presenting with fatigue and lethargy in childhood. Neurol. Genet. 2, e59. https://doi.org/10.1212/NXG.0000000000000059.

Liberona, J.L., Powell, J.A., Shenoi, S., Petherbridge, L., Caviedes, R., Jaimovich, E., 1998. Differences in both inositol 1,4,5-triphosphate mass and inositol 1,4,5- triphosphate receptors between normal and dystrophic between normal and dystrophic skeletal muscle cell lines. Muscle Nerve 21, 902–909. https://doi.org/10.1002/(SICI)1097-4598(199807)21:7 < 902::AID-MUS8 > 3.0.CO;2-A.

Liesa, M., Shirihai, O.S., 2013. Mitochondrial dynamics in the regulation of nutrient utilization and energy expenditure. Cell Metab. 17, 491–506. https://doi.org/10.1016/j.cmet.2013.03.002.

Lin, J., Wu, H., Tarr, P.T., Zhang, C.Y., Wu, Z., Boss, O., Michael, L.F., Puigserver, P., Isotani, E., Olson, E.N., Lowell, B.B., Bassel-Duby, R., Spiegelman, B.M., 2002. Transcriptional co-activator PGC-1α drives the formation of slow-twitch muscle fibres. Nature 418, 797–801. https://doi.org/10.1038/nature00904.

Liu, G., Park, S.H., Imbesi, M., Nathan, W.J., Zou, X., Zhu, Y., Jiang, H., Parisiadou, L., Gius, D., 2017. Loss of NAD-dependent protein deacetylase sirtuin-2 alters mitochondrial protein acetylation and dysregulates mitophagy. Antioxid. Redox Signal. 26, 846–863. https://doi.org/10.1089/ars.2016.6662.

Liu, J.C., Syder, N.C., Ghorashi, N.S., Willingham, T.B., Parks, R.J., Sun, J., Fergusson, M.M., Liu, J., Holmström, K.M., Menazza, S., Springer, D.A., Liu, C., Glancy, B., Finkel, T., Murphy, E., 2020. EMRE is essential for mitochondrial calcium uniporter activity in a mouse model. JCI Insight 5, e134063. https://doi.org/10.1172/jci.insight.134063.

Logan, C.V., Szabadkai, G., Sharpe, J.A., Parry, D.A., Torelli, S., Childs, A.-M., Kriek, M., Phadke, R., Johnson, C.A., Roberts, N.Y., Bonthron, D.T., Pysden, K.A., Whyte, T., Munteanu, I., Foley, A.R., Wheway, G., Szymanska, K., Natarajan, S., Abdelhamed, Z.A., Morgan, J.E., Roper, H., Santen, G.W.E., Niks, E.H., van der Pol, W.L., Lindhout, D., Raffaello, A., De Stefani, D., den Dunnen, J.T., Sun, Y.,

Ginjaar, I., Sewry, C.A., Hurles, M., Rizzuto, R., Duchen, M.R., Muntoni, F., Sheridan, E., 2014. Loss-of-function mutations in MICU1 cause a brain and muscle disorder linked to primary alterations in mitochondrial calcium signaling. Nat. Genet. 46, 188–193. https://doi.org/10.1038/ng.2851.

Luo, J., Sladek, R., Carrier, J., Bader, J.-A., Richard, D., Giguère, V., 2003. Reduced fat mass in mice lacking orphan nuclear receptor estrogen-related receptor α. Mol. Cell. Biol. 23, 7947–7956. https://doi.org/10.1128/mcb.23.22.7947-7956.2003.

Luongo, T.S., Lambert, J.P., Yuan, A., Zhang, X., Gross, P., Song, J., Shanmughapriya, S., Gao, E., Jain, M., Houser, S.R., Koch, W.J., Cheung, J.Y., Madesh, M., Elrod, J.W., 2015. The mitochondrial calcium uniporter matches energetic supply with cardiac workload during stress and modulates permeability transition. Cell Rep. 12, 23–34. https://doi.org/10.1016/j.celrep.2015.06.017.

Luongo, T.S., Lambert, J.P., Gross, P., Nwokedi, M., Lombardi, A.A., Shanmughapriya, S., Carpenter, A.C., Kolmetzky, D., Gao, E., van Berlo, J.H., Tsai, E.J., Molkentin, J.D., Chen, X., Madesh, M., Houser, S.R., Elrod, J.W., 2017. The mitochondrial Na+/Ca2+ exchanger is essential for Ca2+ homeostasis and viability. Nature 545, 93–97. https://doi.org/10.1038/nature22082.

Lygate, C.A., Aksentijevic, D., Dawson, D., Ten Hove, M., Phillips, D., De Bono, J.P., Medway, D.J., Sebag-Montefiore, L., Hunyor, I., Channon, K.M., Clarke, K., Zervou, S., Watkins, H., Balaban, R.S., Neubauer, S., 2013. Living without creatine: unchanged exercise capacity and response to chronic myocardial infarction in creatine-deficient mice. Circ. Res. 112, 945–955. https://doi.org/10.1161/CIRCRESAHA.112.300725.

Mailloux, R.J., Jin, X., Willmore, W.G., 2014. Redox regulation of mitochondrial function with emphasis on cysteine oxidation reactions. Redox Biol. 2, 123–139. https://doi.org/10.1016/j.redox.2013.12.011.

Makarewich, C.A., Baskin, K.K., Munir, A.Z., Bezprozvannaya, S., Sharma, G., Khemtong, C., Shah, A.M., McAnally, J.R., Malloy, C.R., Szweda, L.I., Bassel-Duby, R., Olson, E.N., 2018. MOXI is a mitochondrial micropeptide that enhances fatty acid β-oxidation. Cell Rep. 23, 3701–3709. https://doi.org/10.1016/j.celrep.2018.05.058.

Mallilankaraman, K., Doonan, P., Cárdenas, C., Chandramoorthy, H.C., Müller, M., Miller, R., Hoffman, N.E., Gandhirajan, R.K., Molgó, J., Birnbaum, M.J., Rothberg, B.S., Mak, D.-O.D., Foskett, J.K., Madesh, M., 2012. MICU1 Is an essential gatekeeper for MCU-mediated mitochondrial Ca2+ uptake that regulates cell survival. Cell 151, 630–644. https://doi.org/10.1016/j.cell.2012.10.011.

Mammucari, C., Gherardi, G., Zamparo, I., Raffaello, A., Boncompagni, S., Chemello, F., Cagnin, S., Braga, A., Zanin, S., Pallafacchina, G., Zentilin, L., Sandri, M., De Stefani, D., Protasi, F., Lanfranchi, G., Rizzuto, R., 2015. The mitochondrial calcium uniporter controls skeletal muscle trophism in vivo. Cell Rep. 10 (8), 1269–1279. https://doi.org/10.1016/j.celrep.2015.01.056.

McCormack, J.G., Denton, R.M., 1984. Role of Ca2+ ions in the regulation of intramitochondrial metabolism in rat heart. Evidence from studies with isolated mitochondria that adrenaline activates the pyruvate dehydrogenase and 2-oxoglutarate dehydrogenase complexes by increasing the intramito. Biochem. J. 218, 235–247. https://doi.org/10.1042/bj2180235.

McMullen, P.D., Bhattacharya, S., Woods, C.G., Sun, B., Yarborough, K., Ross, S.M., Miller, M.E., McBride, M.T., Lecluyse, E.L., Clewell, R.A., Andersen, M.E., 2014. A map of the PPARα transcription regulatory network for primary human hepatocytes. Chem. Biol. Interact. 209, 14–24. https://doi.org/10.1016/j.cbi.2013.11.006.

Merlini, L., Angelin, A., Tiepolo, T., Braghetta, P., Sabatelli, P., Zamparelli, A., Ferlini, A., Maraldi, N.M., Bonaldo, P., Bernardi, P., 2008. Cyclosporin A corrects mitochondrial dysfunction and muscle apoptosis in patients with collagen VI myopathies. Proc. Natl. Acad. Sci. U. S. A. 105, 5225–5229. https://doi.org/10.1073/pnas.0800962105.

Merlini, L., Sabatelli, P., Armaroli, A., Gnudi, S., Angelin, A., Grumati, P., Michelini, M.E., Franchella, A., Gualandi, F., Bertini, E., Maraldi, N.M., Ferlini, A., Bonaldo, P., Bernardi, P., 2011. Cyclosporine a in Ullrich congenital muscular dystrophy: long-term results. Oxid. Med. Cell. Longev. 2011. https://doi.org/10.1155/2011/139194.

Merrill, G.F., Kurth, E.J., Hardie, D.G., Winder, W.W., 1997. AICA riboside increases AMP-activated protein kinase, fatty acid oxidation, and glucose uptake in rat muscle. Am. J. Physiol. Endocrinol. Metab. 273, E1107–E1112. https://doi.org/10.1152/ajpendo.1997.273.6.e1107.

Messina, A., Reina, S., Guarino, F., De Pinto, V., 2012. VDAC isoforms in mammals. Biochim. Biophys. Acta Biomembr. 1818, 1466–1476. https://doi.org/10.1016/j.bbamem.2011.10.005.

Michelucci, A., De Marco, A., Guarnier, F.A., Protasi, F., Boncompagni, S., 2017a. Antioxidant treatment reduces formation of structural cores and improves muscle function in RYR1Y522S/WT mice. Oxid. Med. Cell. Longev. 2017, 3649–3662. https://doi.org/10.1155/2017/6792694.

Michelucci, A., Paolini, C., Boncompagni, S., Canato, M., Reggiani, C., Protasi, F., 2017b. Strenuous exercise triggers a life-threatening response in mice susceptible to malignant hyperthermia. FASEB J. 31, 3649–3662. https://doi.org/10.1096/fj.201601292R.

Millay, D.P., Sargent, M.A., Osinska, H., Baines, C.P., Barton, E.R., Vuagniaux, G., Sweeney, H.L., Robbins, J., Molkentin, J.D., 2008. Genetic and pharmacologic inhibition of mitochondrial-dependent necrosis attenuates muscular dystrophy. Nat. Med. 14, 442–447. https://doi.org/10.1038/nm1736.

Millay, D.P., Goonasekera, S.A., Sargent, M.A., Maillet, M., Aronow, B.J., Molkentin, J.D., 2009. Calcium influx is sufficient to induce muscular dystrophy through a TRPC-dependent mechanism. Proc. Natl. Acad. Sci. U. S. A. 106, 19023–19028. https://doi.org/10.1073/pnas.0906591106.

Minagawa, N., Kruglov, E.A., Dranoff, J.A., Robert, M.E., Gores, G.J., Nathanson, M.H., 2005. The anti-apoptotic protein Mcl-1 inhibits mitochondrial Ca2+ signals. J. Biol. Chem. 280, 33637–33644. https://doi.org/10.1074/jbc.M503210200.

Miura, S., Kawanaka, K., Kai, Y., Tamura, M., Goto, M., Shiuchi, T., Minokoshi, Y., Ezaki, O., 2007. An increase in murine skeletal muscle peroxisome proliferator-activated receptor-γ coactivator-1α (PGC-1α) mRNA in response to exercise is mediated by β-adrenergic receptor activation. Endocrinology 148, 3441–3448. https://doi.org/10.1210/en.2006-1646.

Montero, M., Alonso, M.T., Carnicero, E., Cuchillo-Ibá~ez, I., Albillos, A., García, A.G., García-Sancho, J., Alvarez, J., 2000. Chromaffin-cell stimulation triggers fast millimolar mitochondrial Ca2+ transients that modulate secretion. Nat. Cell Biol. 2, 57–61. https://doi.org/10.1038/35000001.

Morine, K.J., Sleeper, M.M., Barton, E.R., Sweeney, H.L., 2010. Overexpression of SERCA1a in the mdx diaphragm reduces susceptibility to contraction-induced damage. Hum. Gene Ther. 21, 1735–1739. https://doi.org/10.1089/hum.2010.077.

Muoio, D.M., MacLean, P.S., Lang, D.B., Li, S., Houmard, J.A., Way, J.M., Winegar, D.A., Christopher Corton, J., Lynis Dohm, G., Kraus, W.E., 2002. Fatty acid homeostasis and induction of lipid regulatory genes in skeletal muscles of peroxisome proliferator-activated receptor (PPAR) α knock-out mice. Evidence for compensatory regulation by PPARδ. J. Biol. Chem. 277, 26089–26097. https://doi.org/10.1074/jbc.M203997200.

Murphy, E., Pan, X., Nguyen, T., Liu, J., Holmström, K.M., Finkel, T., 2014. Unresolved questions from the analysis of mice lacking MCU expression. Biochem. Biophys. Res. Commun. 449, 384–385. https://doi.org/10.1016/j.bbrc.2014.04.144.

Navarro, A., Boveris, A., 2007. The mitochondrial energy transduction system and the aging process. Am. J. Physiol. Cell Physiol. 292, 670–686. https://doi.org/10.1152/ajpcell.00213.2006.

Nemani, N., Dong, Z., Daw, C.C., Madaris, T.R., Ramachandran, K., Enslow, B.T., Rubannelsonkumar, C.S., Shanmughapriya, S., Mallireddigari, V., Maity, S., SinghMalla, P., Natarajanseenivasan, K., Hooper, R., Shannon, C.E., Tourtellotte, W.G., Singh, B.B., Reeves, W.B., Sharma, K., Norton, L., Srikantan, S., Soboloff, J., Madesh, M., 2020. Mitochondrial pyruvate and fatty acid flux modulate MICU1-dependent control of MCU activity. Sci. Signal. 13, eaaz6206. https://doi.org/10.1126/scisignal.aaz6206.

Nguyen, N.X., Armache, J.-P., Lee, C., Yang, Y., Zeng, W., Mootha, V.K., Cheng, Y., Bai, X., Jiang, Y., 2018. Cryo-EM structure of a fungal mitochondrial calcium uniporter. Nature 559, 570–574. https://doi.org/10.1038/s41586-018-0333-6.

Nichols, M., Elustondo, P.A., Warford, J., Thirumaran, A., Pavlov, E.V., Robertson, G.S., 2017. Global ablation of the mitochondrial calcium uniporter increases glycolysis in cortical neurons subjected to energetic stressors. J. Cereb. Blood Flow Metab. 37, 3027–3041. https://doi.org/10.1177/0271678X16682250.

Nichols, M., Pavlov, E.V., Robertson, G.S., 2018. Tamoxifen-induced knockdown of the mitochondrial calcium uniporter in Thy1-expressing neurons protects mice from hypoxic/ischemic brain injury article. Cell Death Dis. 9, 606. https://doi.org/10.1038/s41419-018-0607-9.

Nickel, A., Kohlhaas, M., Maack, C., 2014. Mitochondrial reactive oxygen species production and elimination. J. Mol. Cell. Cardiol. 73, 26–33. https://doi.org/10.1016/j.yjmcc.2014.03.011.

Nicolas-Metral, V., Raddatz, E., Kucera, P., Ruegg, U.T., 2001. Mdx myotubes have normal excitability but show reduced contraction-relaxation dynamics. J. Muscle Res. Cell Motil. 22, 69–75. https://doi.org/10.1023/A:1010384625954.

Nie, C., Tian, C., Zhao, L., Petit, P.X., Mehrpour, M., Chen, Q., 2008. Cysteine 62 of Bax is critical for its conformational activation and its proapoptotic activity in response to H2O2-induced apoptosis. J. Biol. Chem. 283, 15359–15369. https://doi.org/10.1074/jbc.M800847200.

Nutt, L.K., Pataer, A., Pahler, J., Fang, B., Roth, J., McConkey, D.J., Swisher, S.G., 2002. Bax and Bak promote apoptosis by modulating endoplasmic reticular and mitochondrial Ca2+ stores. J. Biol. Chem. 277, 9219–9225. https://doi.org/10.1074/jbc.M106817200.

Ogata, T., Yamasaki, Y., 1985. Scanning electron-microscopic studies on the three-dimensional structure of mitochondria in the mammalian red, white and intermediate muscle fibers. Cell Tissue Res. 241, 251–256. https://doi.org/10.1007/BF00217168.

Pacher, P., Thomas, A.P., Hajnóczky, G., 2002. Ca2+ marks: miniature calcium signals in single mitochondria driven by ryanodine receptors. Proc. Natl. Acad. Sci. U. S. A. 99, 2380–2385. https://doi.org/10.1073/pnas.032423699.

Paillard, M., Csordás, G., Szanda, G., Golenár, T., Debattisti, V., Bartok, A., Wang, N., Moffat, C., Seifert, E.L., Spät, A., Hajnóczky, G., 2017. Tissue-specific mitochondrial decoding of cytoplasmic Ca 2+ signals is controlled by the stoichiometry of MICU1/2 and MCU. Cell Rep. 18, 2291–2300. https://doi.org/10.1016/j.celrep.2017.02.032.

Palma, E., Tiepolo, T., Angelin, A., Sabatelli, P., Maraldi, N.M., Basso, E., Forte, M.A., Bernardip, P., Bonaldo, P., 2009. Genetic ablation of cyclophilin D rescues mitochondrial defects and prevents muscle apoptosis in collagen VI myopathic mice. Hum. Mol. Genet. 18, 2024–2031. https://doi.org/10.1093/hmg/ddp126.

Palmieri, L., Pardo, B., Lasorsa, F.M., Del Arco, A., Kobayashi, K., Iijima, M., Runswick, M.J., Walker, J.E., Saheki, T., Satrústegui, J., Palmieri, F., 2001. Citrin and aralar1 are Ca2+-stimulated aspartate/glutamate transporters in mitochondria. EMBO J. 20, 5060–5069. https://doi.org/10.1093/emboj/20.18.5060.

Palty, R., Silverman, W.F., Hershfinkel, M., Caporale, T., Sensi, S.L., Parnis, J., Nolte, C., Fishman, D., Shoshan-Barmatz, V., Herrmann, S., Khananshvili, D., Sekler, I., 2010. NCLX is an essential component of mitochondrial Na+/Ca 2+ exchange. Proc. Natl. Acad. Sci. U. S. A. 107, 436–441. https://doi.org/10.1073/pnas.0908099107.

Pan, X., Liu, J., Nguyen, T., Liu, C., Sun, J., Teng, Y., Fergusson, M.M., Rovira, I.I., Allen, M., Springer, D.A., Aponte, A.M., Gucek, M., Balaban, R.S., Murphy, E., Finkel, T., 2013. The physiological role of mitochondrial calcium revealed by mice lacking the mitochondrial calcium uniporter. Nat. Cell Biol. 15, 1464–1472. https://doi.org/10.1038/ncb2868.

Park, M.K., Ashby, M.C., Erdemli, G., Petersen, O.H., Tepikin, A.V., 2001. Perinuclear, perigranular and sub-plasmalemmal mitochondria have distinct functions in the regulation of cellular calcium transport. EMBO J. 20, 1863–1874. https://doi.org/10.1093/emboj/20.8.1863.

Patron, M., Checchetto, V., Raffaello, A., Teardo, E., VecellioReane, D., Mantoan, M., Granatiero, V., Szabò, I., DeStefani, D., Rizzuto, R., 2014. MICU1 and MICU2 finely tune the mitochondrial Ca2+ uniporter by exerting opposite effects on MCU activity. Mol. Cell 53, 726–737. https://doi.org/10.1016/j.molcel.2014.01.013.

Patron, M., Granatiero, V., Espino, J., Rizzuto, R., De Stefani, D., 2019. MICU3 is a tissue-specific enhancer of mitochondrial calcium uptake. Cell Death Differ. 26, 179–195. https://doi.org/10.1038/s41418-018-0113-8.

Payne, R., Hoff, H., Roskowski, A., Foskett, J.K., 2017. MICU2 restricts spatial crosstalk between InsP3R and MCU channels by regulating threshold and gain of MICU1-mediated inhibition and activation of MCU. Cell Rep. 21, 3141–3154. https://doi.org/10.1016/j.celrep.2017.11.064.

Perocchi, F., Gohil, V.M., Girgis, H.S., Bao, X.R., McCombs, J.E., Palmer, A.E., Mootha, V.K., 2010. MICU1 encodes a mitochondrial EF hand protein required for Ca2+ uptake. Nature 467, 291–296. https://doi.org/10.1038/nature09358.

Perry, M.-C., Dufour, C.R., Tam, I.S., B'chir, W., Giguère, V., 2014. Estrogen-related receptor-α coordinates transcriptional programs essential for exercise tolerance and muscle fitness. Mol. Endocrinol. 28, 2060–2071. https://doi.org/10.1210/me.2014-1281.

Pilegaard, H., Saltin, B., Neufer, D.P., 2003. Exercise induces transient transcriptional activation of the PGC-1α gene in human skeletal muscle. J. Physiol. 546 (3), 851–858. https://doi.org/10.1113/jphysiol.2002.034850.

Platt, D., Griggs, R., 2009. Skeletal muscle channelopathies: new insights into the periodic paralyses and nondystrophic myotonias. Curr. Opin. Neurol. 22 (5), 524–531. https://doi.org/10.1097/WCO.0b013e32832efa8f.

Plovanich, M., Bogorad, R.L., Sancak, Y., Kamer, K.J., Strittmatter, L., Li, A.A., Girgis, H.S., Kuchimanchi, S., De Groot, J., Speciner, L., Taneja, N., OShea, J., Koteliansky, V., Mootha, V.K., 2013. MICU2, a Paralog of MICU1, resides within the mitochondrial uniporter complex to regulate calcium handling. PLoS One 8, e55785. https://doi.org/10.1371/journal.pone.0055785.

Pozzan, T., Rizzuto, R., 2000. The renaissance of mitochondrial calcium transport. Eur. J. Biochem. 267, 5269–5273. https://doi.org/10.1046/j.1432-1327.2000.01567.x.

Prudent, J., Popgeorgiev, N., Bonneau, B., Thibaut, J., Gadet, R., Lopez, J., Gonzalo, P., Rimokh, R., Manon, S., Houart, C., Herbomel, P., Aouacheria, A., Gillet, G., 2013. Bcl-wav and the mitochondrial calcium uniporter drive gastrula morphogenesis in zebrafish. Nat. Commun. 4, 2330. https://doi.org/10.1038/ncomms3330.

Puigserver, P., Wu, Z., Park, C.W., Graves, R., Wright, M., Spiegelman, B.M., 1998. A cold-inducible coactivator of nuclear receptors linked to adaptive thermogenesis. Cell 92, 829–839. https://doi.org/10.1016/S0092-8674(00)81410-5.

Puigserver, P., Adelmant, G., Wu, Z., Fan, M., Xu, J., O'Malley, B., Spiegelman, B.M., 1999. Activation of PPARγ coactivator-1 through transcription factor docking. Science 286, 1368–1371. https://doi.org/10.1126/science.286.5443.1368.

Raffaello, A., De Stefani, D., Sabbadin, D., Teardo, E., Merli, G., Picard, A., Checchetto, V., Moro, S., Szabò, I., Rizzuto, R., 2013. The mitochondrial calcium uniporter is a multimer that can include a dominant-negative pore-forming subunit. EMBO J. 32, 2362–2376. https://doi.org/10.1038/emboj.2013.157.

Rando, T.A., Disatnik, M.H., Yu, Y., Franco, A., 1998. Muscle cells from mdx mice have an increased susceptibility to oxidative stress. Neuromuscul. Disord. 8, 14–21. https://doi.org/10.1016/S0960-8966(97)00124-7.

Rapizzi, E., Pinton, P., Szabadkai, G., Wieckowski, M.R., Vandecasteele, G., Baird, G., Tuft, R.A., Fogarty, K.E., Rizzuto, R., 2002. Recombinant expression of the voltage-dependent anion channel enhances the transfer of Ca2 + microdomains to mitochondria. J. Cell Biol. 159, 613–624. https://doi.org/10.1083/jcb.200205091.

Rasola, A., Bernardi, P., 2007. The mitochondrial permeability transition pore and its involvement in cell death and in disease pathogenesis. Apoptosis 12, 815–833. https://doi.org/10.1007/s10495-007-0723-y.

Richter, C., 1995. Oxidative damage to mitochondrial DNA and its relationship to ageing. Int. J. Biochem. Cell Biol. 27, 647–653. https://doi.org/10.1016/1357-2725(95)00025-K.

Riemer, J., Schwarzländer, M., Conrad, M., Herrmann, J.M., 2015. Thiol switches in mitochondria: operation and physiological relevance. Biol. Chem. 396, 465–482. https://doi.org/10.1016/j.beem.

Rizzuto, R., Simpson, A.W.M., Brini, M., Pozzan, T., 1992. Rapid changes of mitochondrial Ca2 + revealed by specifically targeted recombinant aequorin. Nature 358, 325–327. https://doi.org/10.1038/358325a0.

Rizzuto, R., Pinton, P., Carrington, W., Fay, F.S., Fogarty, K.E., Lifshitz, L.M., Tuft, R.A., Pozzan, T., 1998. Close contacts with the endoplasmic reticulum as determinants of mitochondrial Ca2 + responses. Science 280, 1763–1766. https://doi.org/10.1126/science.280.5370.1763.

Rizzuto, R., De Stefani, D., Raffaello, A., Mammucari, C., 2012. Mitochondria as sensors and regulators of calcium signalling. Nat. Rev. Mol. Cell Biol. 13, 566–578. https://doi.org/10.1038/nrm3412.

Robert, V., Massimino, M.L., Tosello, V., Marsault, R., Cantini, M., Sorrentino, V., Pozzan, T., 2001. Alteration in calcium handling at the subcellular level in mdx myotubes. J. Biol. Chem. 276, 4647–4651. https://doi.org/10.1074/jbc.M006337200.

Rowe, G.C., Patten, I.S., Zsengeller, Z.K., El-Khoury, R., Okutsu, M., Bampoh, S., Koulisis, N., Farrell, C., Hirshman, M.F., Yan, Z., Goodyear, L.J., Rustin, P., Arany, Z., 2013. Disconnecting mitochondrial content from respiratory chain capacity in PGC-1-deficient skeletal muscle. Cell Rep. 3, 1449–1456. https://doi.org/10.1016/j.celrep.2013.04.023.

Ruas, J.L., White, J.P., Rao, R.R., Kleiner, S., Brannan, K.T., Harrison, B.C., Greene, N.P., Wu, J., Estall, J.L., Irving, B.A., Lanza, I.R., Rasbach, K.A., Okutsu, M., Nair, K.S., Yan, Z., Leinwand, L.A., Spiegelman, B.M., 2012. A PGC-1α isoform induced by resistance training regulates skeletal muscle hypertrophy. Cell 151, 1319–1331. https://doi.org/10.1016/j.cell.2012.10.050.

Rudolf, R., Mongillo, M., Magalhães, P.J., Pozzan, T., 2004. In vivo monitoring of Ca(2 +) uptake into mitochondria of mouse skeletal muscle during contraction. J. Cell Biol. 166, 527–536. https://doi.org/10.1083/jcb.200403102.

Ryan, K.C., Ashkavand, Z., Norman, K.R., 2020. The role of mitochondrial calcium homeostasis in Alzheimer's and related diseases. Int. J. Mol. Sci. 21, 1–17. https://doi.org/10.3390/ijms21239153.

Sancak, Y., Markhard, A.L., Kitami, T., Kovács-Bogdán, E., Kamer, K.J., Udeshi, N.D., Carr, S.A., Chaudhuri, D., Clapham, D.E., Li, A.A., Calvo, S.E., Goldberger, O., Mootha, V.K., 2013. EMRE is an essential component of the mitochondrial calcium uniporter complex. Science 342, 1379–1382. https://doi.org/10.1126/science.1242993.

Sánchez-Gómez, F.J., Espinosa-Díez, C., Dubey, M., Dikshit, M., Lamas, S., 2013. S-glutathionylation: relevance in diabetes and potential role as a biomarker. Biol. Chem. 394, 1263–1280. https://doi.org/10.1515/hsz-2013-0150.

Sandri, M., Lin, J., Handschin, C., Yang, W., Arany, Z.P., Lecker, S.H., Goldberg, A.L., Spiegelman, B.M., 2006. PGC-1α protects skeletal muscle from atrophy by suppressing FoxO3 action and atrophy-specific gene transcription. Proc. Natl. Acad. Sci. U. S. A. 103, 16260–16265. https://doi.org/10.1073/pnas.0607795103.

Satrústegui, J., Pardo, B., Del Arco, A., 2007. Mitochondrial transporters as novel targets for intracellular calcium signaling. Physiol. Rev. 87, 29–67. https://doi.org/10.1152/physrev.00005.2006.

Schiaffino, S., Reggiani, C., 2011. Fiber types in mammalian skeletal muscles. Physiol. Rev. 91, 1447–1531. https://doi.org/10.1152/physrev.00031.2010.

Sena, L.A., Chandel, N.S., 2012. Physiological roles of mitochondrial reactive oxygen species. Mol. Cell 48, 158–167. https://doi.org/10.1016/j.molcel.2012.09.025.

Sewry, C.A., Wallgren-Pettersson, C., 2017. Myopathology in congenital myopathies. Neuropathol. Appl. Neurobiol. 43 (1), 5–23. https://doi.org/10.1111/nan.12369.

Sharma, A., Oonthonpan, L., Sheldon, R.D., Rauckhorst, A.J., Zhu, Z., Tompkins, S.C., Cho, K., Grzesik, W.J., Gray, L.R., Scerbo, D.A., Pewa, A.D., Cushing, E.M., Dyle, M.C., Cox, J.E., Adams, C., Davies, B.S., Shields, R.K., Norris, A.W., Patti, G., Zingman, L.V., Taylor, E.B., 2019. Impaired skeletal muscle mitochondrial pyruvate uptake rewires glucose metabolism to drive whole-body leanness. Elife 8, e45873. https://doi.org/10.7554/eLife.45873.

Shkryl, V.M., Shirokova, N., 2006. Transfer and tunneling of Ca2 + from sarcoplasmic reticulum to mitochondria in skeletal muscle. J. Biol. Chem. 281, 1547–1554. https://doi.org/10.1074/jbc.M505024200.

Shutt, T., Geoffrion, M., Milne, R., McBride, H.M., 2012. The intracellular redox state is a core determinant of mitochondrial fusion. EMBO Rep. 13, 909–915. https://doi.org/10.1038/embor.2012.128.

Stock, D., Leslie, A.G.W., Walker, J.E., 1999. Molecular architecture of the rotary motor in ATP synthase. Science 286, 1700–1705. https://doi.org/10.1126/science.286.5445.1700.

Suman, M., Sharpe, J.A., Bentham, R.B., Kotiadis, V.N., Menegollo, M., Pignataro, V., Molgó, J., Muntoni, F., Duchen, M.R., Pegoraro, E., Szabadkai, G., 2018. Inositol trisphosphate receptor-mediated Ca2 + signalling stimulates mitochondrial function and gene expression in core myopathy patients. Hum. Mol. Genet. 27, 2367–2382. https://doi.org/10.1093/hmg/ddy149.

Szabadkai, G., Simoni, A.M., Chami, M., Wieckowski, M.R., Youle, R.J., Rizzuto, R., 2004. Drp-1-dependent division of the mitochondrial network blocks intraorganellar Ca2 + waves and protects against Ca 2 +-mediated apoptosis. Mol. Cell 16, 59–68. https://doi.org/10.1016/j.molcel.2004.09.026.

Szibor, M., Gizatullina, Z., Gainutdinov, T., Endres, T., Debska-Vielhaber, G., Kunz, M., Karavasili, N., Hallmann, K., Schreiber, F., Bamberger, A., Schwarzer, M., Doenst, T., Heinze, H.J., Lessmann, V., Vielhaber, S., Kunz, W.S., Gellerich, F.N., Colbran, R.J., 2020. Cytosolic, but not matrix, calcium is essential for adjustment of mitochondrial pyruvate supply. J. Biol. Chem. 295, 4383–4397. https://doi.org/10.1074/jbc.RA119.011902.

Territo, P.R., Mootha, V.K., French, S.A., Balaban, R.S., 2000. Ca^{2+} activation of heart mitochondrial oxidative phosphorylation: role of the F_0/F_1-ATPase. Am. J. Physiol. Physiol. 278, C423–C435. https://doi.org/10.1152/ajpcell.2000.278.2.C423.

Tilgen, N., Zorzato, F., Halliger-Keller, B., Muntoni, F., Sewry, C., Palmucci, L.M., Schneider, C., Hauser, E., Lehmann-Horn, F., Müller, C.R., Treves, S., 2001. Identification of four novel mutations in the C-terminal membrane spanning domain of the ryanodine receptor 1: association with central core disease and alteration of calcium homeostasis. Hum. Mol. Genet. 10, 2879–2887. https://doi.org/10.1093/hmg/10.25.2879.

Tong, J., McCarthy, T.V., MacLennan, D.H., 1999. Measurement of resting cytosolic Ca2 +
    concentrations and Ca2 + store size in HEK-293 cells transfected with malignant hyper-
    thermia or central core disease mutant Ca2 + release channels. J. Biol. Chem. 274,
    693–702. https://doi.org/10.1074/jbc.274.2.693.
Tosatto, A., Sommaggio, R., Kummerow, C., Bentham, R.B., Blacker, T.S., Berecz, T.,
    Duchen, M.R., Rosato, A., Bogeski, I., Szabadkai, G., Rizzuto, R., Mammucari, C.,
    2016. The mitochondrial calcium uniporter regulates breast cancer progression via
    HIF-1α. EMBO Mol. Med. 8, 569–585. https://doi.org/10.15252/emmm.201606255.
Vais, H., Mallilankaraman, K., Mak, D.-O.D., Hoff, H., Payne, R., Tanis, J.E., Foskett, J.K.,
    2016. EMRE is a matrix Ca 2 + sensor that governs gatekeeping of the mitochondrial Ca
    2 + uniporter. Cell Rep. 14, 403–410. https://doi.org/10.1016/j.celrep.2015.12.054.
van der Meer, D.L.M., Degenhardt, T., Väisänen, S., de Groot, P.J., Heinäniemi, M., de
    Vries, S.C., Müller, M., Carlberg, C., Kersten, S., 2010. Profiling of promoter occu-
    pancy by PPARalpha in human hepatoma cells via ChIP-chip analysis. Nucleic Acids
    Res. 38, 2839–2850. https://doi.org/10.1093/nar/gkq012.
van Deursen, J., Heerschap, A., Oerlemans, F., Rultenbeek, W., Jap, P., ter Laak, H.,
    Wieringa, B., 1993. Skeletal muscles of mice deficient in muscle creatine kinase lack
    burst activity. Cell 74, 621–631. https://doi.org/10.1016/0092-8674(93)90510-W.
Vandebrouck, C., Martin, D., Van Schoor, M.C., Debaix, H., Gailly, P., 2002. Involvement
    of TRPC in the abnormal calcium influx observed in dystrophic (mdx) mouse skeletal
    muscle fibers. J. Cell Biol. 158, 1089–1096. https://doi.org/10.1083/jcb.200203091.
Vavvas, D., Apazidis, A., Saha, A.K., Gamble, J., Patel, A., Kemp, B.E., Witters, L.A.,
    Ruderman, N.B., 1997. Contraction-induced changes in acetyl-CoA carboxylase and
    5'-AMP- activated kinase in skeletal muscle. J. Biol. Chem. 272, 13255–13261.
    https://doi.org/10.1074/jbc.272.20.13255.
Vecellio Reane, D., Vallese, F., Checchetto, V., Acquasaliente, L., Butera, G., De
    Filippis, V., Szabò, I., Zanotti, G., Rizzuto, R., Raffaello, A., 2016. A MICU1 splice
    variant confers high sensitivity to the mitochondrial Ca 2 + uptake machinery of skeletal
    muscle. Mol. Cell 64, 760–773. https://doi.org/10.1016/j.molcel.2016.10.001.
Voronina, S.G., Barrow, S.L., Simpson, A.W.M., Gerasimenko, O.V., da Silva Xavier, G.,
    Rutter, G.A., Petersen, O.H., Tepikin, A.V., 2010. Dynamic changes in cytosolic and
    mitochondrial ATP levels in pancreatic acinar cells. Gastroenterology 138, 1976–1987.
    https://doi.org/10.1053/j.gastro.2010.01.037.
Wang, C.H., Wei, Y.H., 2017. Role of mitochondrial dysfunction and dysregulation of
    Ca2 + homeostasis in the pathophysiology of insulin resistance and type 2 diabetes.
    J. Biomed. Sci. 24, 1–11. https://doi.org/10.1186/s12929-017-0375-3.
Wende, A.R., Schaeffer, P.J., Parker, G.J., Zechner, C., Han, D.H., Chen, M.M.,
    Hancock, C.R., Lehman, J.J., Huss, J.M., McClain, D.A., Holloszy, J.O., Kelly, D.P.,
    2007. A role for the transcriptional coactivator PGC-1α in muscle refueling. J. Biol.
    Chem. 282, 36642–36651. https://doi.org/10.1074/jbc.M707006200.
Whitehead, N.P., Pham, C., Gervasio, O.L., Allen, D.G., 2008. N-acetylcysteine amelio-
    rates skeletal muscle pathophysiology in mdx mice. J. Physiol. 586, 2003–2014.
    https://doi.org/10.1113/jphysiol.2007.148338.
Wu, Z., Puigserver, P., Andersson, U., Zhang, C., Adelmant, G., Mootha, V., Troy, A.,
    Cinti, S., Lowell, B., Scarpulla, R.C., Spiegelman, B.M., 1999. Mechanisms controlling
    mitochondrial biogenesis and respiration through the thermogenic coactivator PGC-1.
    Cell 98, 115–124. https://doi.org/10.1016/S0092-8674(00)80611-X.
Wu, Y., Rasmussen, T.P., Koval, O.M., Joiner, M.-L.A., Hall, D.D., Chen, B., Luczak, E.D.,
    Wang, Q., Rokita, A.G., Wehrens, X.H.T., Song, L.-S., Anderson, M.E., 2015. The
    mitochondrial uniporter controls fight or flight heart rate increases. Nat. Commun. 6,
    6081. https://doi.org/10.1038/ncomms7081.

Wu, W., Shen, Q., Zhang, R., Qiu, Z., Wang, Y., Zheng, J., Jia, Z., 2020. The structure of the MICU$_1$-MICU$_2$ complex unveils the regulation of the mitochondrial calcium uniporter. EMBO J. 39, e104285. https://doi.org/10.15252/embj.2019104285.

Xu, S., Chisholm, A.D., 2014. C. elegans epidermal wounding induces a mitochondrial ROS burst that promotes wound repair. Dev. Cell 31, 48–60. https://doi.org/10.1016/j.devcel.2014.08.002.

Yang, T., Riehl, J., Esteve, E., Matthaei, K.I., Goth, S., Allen, P.D., Pessah, I.N., Lopez, J.R., 2006. Pharmacologic and functional characterization of malignant hyperthermia in the R163C RyR1 knock-in mouse. Anesthesiology 105, 1164–1175. https://doi.org/10.1097/00000542-200612000-00016.

Yi, M., Weaver, D., Eisner, V., Várnai, P., Hunyady, L., Ma, J., Csordás, G., Hajnóczky, G., 2012. Switch from ER-mitochondrial to SR-mitochondrial calcium coupling during muscle differentiation. Cell Calcium 52, 355–365. https://doi.org/10.1016/j.ceca.2012.05.012.

Yoo, J., Wu, M., Yin, Y., Herzik, M.A., Lander, G.C., Lee, S.-Y., 2018. Cryo-EM structure of a mitochondrial calcium uniporter. Science 361, 506–511. https://doi.org/10.1126/science.aar4056.

Youle, R.J., Karbowski, M., 2005. Mitochondrial fission in apoptosis. Nat. Rev. Mol. Cell Biol. 6, 657–663. https://doi.org/10.1038/nrm1697.

Zechner, C., Lai, L., Zechner, J.F., Geng, T., Yan, Z., Rumsey, J.W., Collia, D., Chen, Z., Wozniak, D.F., Leone, T.C., Kelly, D.P., 2010. Total skeletal muscle PGC-1 deficiency uncouples mitochondrial derangements from fiber type determination and insulin sensitivity. Cell Metab. 12, 633–642. https://doi.org/10.1016/j.cmet.2010.11.008.

Zhao, M., Li, P., Li, X., Zhang, L., Winkfein, R.J., Chen, S.R.W., 1999. Molecular identification of the ryanodine receptor pore-forming segment. J. Biol. Chem. 274, 25971–25974. https://doi.org/10.1074/jbc.274.37.25971.

Zhao, X., Moloughney, J.G., Zhang, S., Komazaki, S., Weisleder, N., 2012. Orai1 mediates exacerbated Ca2+ entry in dystrophic skeletal muscle. PLoS One 7, e49862. https://doi.org/10.1371/journal.pone.0049862.

Zhou, H., Rokach, O., Feng, L., Munteanu, I., Mamchaoui, K., Wilmshurst, J.M., Sewry, C., Manzur, A.Y., Pillay, K., Mouly, V., Duchen, M., Jungbluth, H., Treves, S., Muntoni, F., 2013. RyR1 deficiency in congenital myopathies disrupts excitation-contraction coupling. Hum. Mutat. 34, 986–996. https://doi.org/10.1002/humu.22326.

Zhuo, W., Zhou, H., Guo, R., Yi, J., Zhang, L., Yu, L., Sui, Y., Zeng, W., Wang, P., Yang, M., 2020. Structure of intact human MCU supercomplex with the auxiliary MICU subunits. Protein Cell 34 (7), 986–996. https://doi.org/10.1007/s13238-020-00776-w.

Zorov, D.B., Juhaszova, M., Sollott, S.J., 2014. Mitochondrial reactive oxygen species (ROS) and ROS-induced ROS release. Physiol. Rev. 94, 909–950. https://doi.org/10.1152/physrev.00026.2013.

CHAPTER SEVEN

Mitochondrial Ca^{2+} homeostasis in trypanosomes

Roberto Docampo[a,*], Anibal E. Vercesi[b], Guozhong Huang[a], Noelia Lander[a], Miguel A. Chiurillo[a], and Mayara Bertolini[a]

[a]Center for Tropical and Emerging Global Diseases and Department of Cellular Biology, University of Georgia, Athens, GA, United States
[b]Departamento de Patologia Clinica, Universidade Estadual de Campinas, São Paulo, Brazil
[*]Corresponding author: e-mail address: rdocampo@uga.edu

Contents

Abstract

Mitochondrial calcium ion (Ca^{2+}) uptake is important for buffering cytosolic Ca^{2+} levels, for regulating cell bioenergetics, and for cell death and autophagy. Ca^{2+} uptake is mediated by a mitochondrial Ca^{2+} uniporter (MCU) and the discovery of this channel in trypanosomes has been critical for the identification of the molecular nature of the channel in all eukaryotes. However, the trypanosome uniporter, which has been studied in detail in *Trypanosoma cruzi*, the agent of Chagas disease, and *T. brucei*, the agent of human and animal African trypanosomiasis, has lineage-specific adaptations which include the lack of some homologues to mammalian subunits, and the presence of unique subunits. Here, we review newly emerging insights into the role of mitochondrial Ca^{2+} homeostasis in trypanosomes, the composition of the uniporter, its functional characterization, and its role in general physiology.

International Review of Cell and Molecular Biology, Volume 362
ISSN 1937-6448
https://doi.org/10.1016/bs.ircmb.2021.01.002

1. Introduction

Trypanosomatids are the etiologic agents of several neglected tropical diseases that cause significant morbidity and mortality in millions of people and animals in the world. *Trypanosoma cruzi* is the agent of Chagas disease, which is endemic in the Americas. The *Trypanosoma brucei* group causes human African trypanosomiasis or sleeping sickness, and nagana in cattle, and is endemic in Sub-Saharan Africa. *Leishmania* spp. cause cutaneous, mucocutaneous and visceral leishmaniasis in several continents. *T. cruzi* has four main life cycle stages, epimastigotes and metacyclic trypomastigotes in the triatomine vector, and bloodstream trypomastigote and intracellular amastigote in the mammalian host. *T. brucei* has two best studied stages, procyclic trypomastigotes in the *tsetse* fly vector midgut and bloodstream trypomastigotes in the mammalian host. *Leishmania* spp. has two well-studied forms, the promastigote in the sand fly vector and the intracellular amastigote in the mammalian host.

Trypanosomatids belong to the eukaryotic supergroup Discoba, which is distantly related to the supergroup Opisthokonta, which includes animals and fungi (Adl et al., 2019). Like animals and fungi, trypanosomatids possess mitochondria, although with lineage-specific adaptations. They have only one mitochondrion per cell that is characterized for the presence of the kinetoplast. The kinetoplast is formed by thousands of concatenated DNA minicircles and a few DNA maxicircles encoding a few mitochondrial proteins and rRNA (Shlomai, 2004). Several mitochondrial mRNA are edited by a complex mechanism, first discovered in these cells (Benne et al., 1986). As the mitochondrial genome of trypanosomes does not possess tRNA genes, these molecules have to be imported into the mitochondrion (Seidman et al., 2012). Some respiratory complexes are incomplete (Surve et al., 2012; van Hellemond et al., 2005), or absent in some stages. While the insect stages of *T. brucei* (procyclic forms), as well those of *T. cruzi* (epimastigotes), have fully developed mitochondria with respiratory chain and oxidative phosphorylation, the *T. brucei* bloodstream forms have more rudimentary mitochondria, do not possess a functional respiratory chain or oxidative phosphorylation, possesses an alternative oxidase (Clarkson et al., 1989), lacks the tricarboxylic acid cycle and depend on glycolysis as energy source (Verner et al., 2015).

Despite these distinct characteristics, mitochondrial Ca^{2+} homeostasis mechanisms are conserved in trypanosomatids and will be the subject of this review. For more detailed information on this topic in animal cells there are excellent reviews on what is known about the mitochondrial Ca^{2+} transport and functions (Kamer and Mootha, 2015; Rizzuto et al., 2012) and more general aspects of Ca^{2+} homeostasis and signaling (Cali et al., 2017; Clapham, 2007).

2. The mitochondrial Ca^{2+} uniporter: Discovery

The mitochondrial Ca^{2+} uniporter (MCU) activity was discovered almost 60 years ago in rat kidney mitochondria (Deluca and Engstrom, 1961; Vasington and Murphy, 1962). Ca^{2+} transport was found to be energized by coupled respiration, blocked by respiratory chain inhibitors and oxidative phosphorylation uncouplers, and resulted in large amounts of Ca^{2+} taken up (Vasington and Murphy, 1962). Further work revealed that the process does not require ATP, except when the respiratory chain is inhibited and, in this case, it is sensitive to oligomycin because it is driven by the ATP synthase working in reverse as an ATPase (Lehninger et al., 1963). Phosphate is also needed for Ca^{2+} uptake (Lehninger et al., 1963) and the uniporter is inhibited by ruthenium red (Moore, 1971), or its derivative, Ru360 (Ying et al., 1991). The uniporter is a Ca^{2+}-selective channel (Kirichok et al., 2004), although other cations, such as Mn^{2+} (Bartley and Amoore, 1958) and Sr^{2+} (Greenawalt and Carafoli, 1966) can also be taken up.

The finding that MCU was absent in *Saccharomyces cerevisiae* mitochondria (Carafoli et al., 1970) led to the proposal that this process was absent in non-animal species (Carafoli and Lehninger, 1971; McCormack and Denton, 1986). However, functional evidence of a mitochondrial Ca^{2+} uniporter with similar properties to the animal MCU was found in *T. cruzi* (Docampo and Vercesi, 1989a, 1989b). *T. cruzi* mitochondrial Ca^{2+} transport has all the characteristics of the animal MCU: it is electrogenic, has high capacity and low affinity for Ca^{2+}, and is inhibited by ruthenium red (Docampo and Vercesi, 1989a, 1989b).

Other trypanosomatids were later shown to possess an MCU, like several *Leishmania* spp. (Benaim et al., 1990; Vercesi and Docampo, 1992; Vercesi et al., 1990), and *T. brucei* (Moreno et al., 1992; Vercesi et al., 1992., 1993; Xiong et al., 1997). Even the mitochondria of the bloodstream form of *T. brucei*, which lacks a respiratory chain and oxidative phosphorylation is able to transport Ca^{2+}, but in this case using the electrochemical gradient generated by the ATP synthase working in reverse, as an ATPase. This Ca^{2+} transport is inhibited by oligomycin (Vercesi et al., 1992).

Interestingly, comparison of the mitochondrial proteins expressed in mammals with those expressed in trypanosomes, which were known to possess the MCU (Docampo and Vercesi, 1989a), and those expressed in yeast, which were known to lack the uniporter (Carafoli et al., 1970), led to the identification, first of the gene encoding a modulator of MCU in animals, the mitochondrial calcium uptake protein 1 (MICU1) (Perocchi et al., 2010), and then of the gene encoding the MCU pore subunit (Baughman et al., 2011; De Stefani et al., 2011; Docampo and Lukes, 2012).

3. The mitochondrial Ca^{2+} uniporter of trypanosomes: The pore subunits

Following the discovery of the molecular nature of the MCU pore subunit, other components of the MCU complex (known as uniplex or holocomplex) were described in mammals, such as MCU regulator 1 (MCUR1) (Mallilankaraman et al., 2012a), MICU2 and MICU3 (Plovanich et al., 2013), MCUb (Raffaello et al., 2013), and essential MCU regulator or EMRE (Sancak et al., 2013).

The trypanosomatid MCU complex differs from the mammalian one (Fig. 1A) in several aspects. First, some subunits, like MCUR1, MICU3, and EMRE, are absent (Docampo et al., 2014). Second, additional subunits, named MCUc, and MCUd form part of the complex and together with the MCU and MCUb subunits, have Ca^{2+}-transporting roles (Chiurillo et al., 2019; Huang and Docampo, 2018) (Fig. 1B).

The MCU subunit was the first MCU complex component described in trypanosomes (Huang et al., 2013b). The gene is single copy in trypanosomes and the encoded protein in *T. brucei* (TbMCU) has only 20% identity and 33% similarity with the human MCU. In contrast, the orthologues in *T. cruzi* (TcMCU) and *Leishmania major* (LmMCU) share 49 and 41% identity, respectively, with the *T. brucei* protein. All these MCU subunits have two transmembrane domains and a mitochondrial targeting signal and localize to the inner mitochondrial membrane. The processed proteins have an apparent molecular weight of ∼30 kDa.

Knockdown of *TbMCU* by RNAi significantly affects the growth of procyclic (insect form) and bloodstream (mammalian form) forms, reduces mitochondrial Ca^{2+} uptake, and the ability of their mitochondria to accumulate

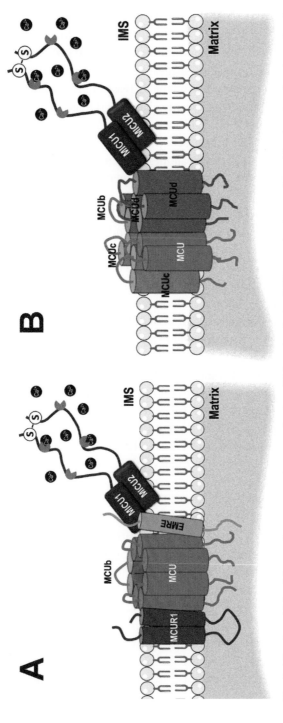

Fig. 1 Mitochondrial calcium uniporter complex organization. (A) In animals the MCU complex is constituted by the pore-forming subunit MCU, and regulator proteins MCUb, MICU1, MICU2, EMRE, and MCUR1. (B) The MCU complex in trypanosomes is composed by the pore forming subunits MCU, MCUb, MCUc, and MCUd, probably forming a hetero-hexamer, and Ca^{2+}-sensing subunits MICU1 and MICU2. IMS, inter membrane space; black balls, Ca^{2+} ions; dark orange circular sectors: EF hand domains, S letters enclosed in circle are cysteine residues that can form a disulfide bridge.

large Ca^{2+} quantities (Huang et al., 2013b). The mitochondrial membrane potential ($\Delta\Psi_m$) is not affected, indicating specific MCU inhibition. Procyclic forms in which $TbMCU$ is downregulated by RNAi have a higher AMP/ATP ratio and increased autophagy, as revealed by the increase in the number of autophagosomes per cell and the increase in the autophagy marker Atg8.2-II, orthologue to LC3-II in mammalian cells (Huang et al., 2013b). In contrast, overexpression of $TbMCU$ in procyclic forms leads to increased mitochondrial Ca^{2+} uptake and mitochondrial Ca^{2+} overload, changes that make the cells more sensitive to proapoptotic agents like C2-ceramide and H_2O_2, increases production of reactive oxygen species (ROS), and cell death (Huang et al., 2013b).

Conditional knockdown of $TbMCU$ in the bloodstream stage greatly affects its growth, mitochondrial Ca^{2+} uptake, and virulence in mice, demonstrating its essentiality (Huang et al., 2013b). The bloodstream form of *T. brucei* uses glucose as the main source of energy, as it does not have an active tricarboxylic acid cycle or respiratory chain. The end product of glycolysis is pyruvate, that is mostly excreted with protons to maintain their intracellular pH (Vanderheyden et al., 2000). However, some pyruvate is needed in the mitochondria where a pyruvate dehydrogenase (PDH) catalyzes its conversion into acetyl-CoA (Huang et al., 2013b; Zhuo et al., 2017). Acetyl-CoA is used for intramitochondrial fatty acid synthesis (FAS II) to generate lipoic acid and myristic acid (Stephens et al., 2007). Alternatively, acetyl-CoA is used to generate acetate (Van Hellemond et al., 1998) that is transferred to the cytosol (Mazet et al., 2013), where it is converted back to acetyl-CoA by acetyl-CoA synthetase (Millerioux et al., 2012) and can be used for fatty acid synthesis. Acetyl-CoA is also used to produce succinyl-CoA by the acetate: succinate CoA transferase (ASCT), which is coupled to ATP production with succinyl-CoA synthetase (SCS) in a process called the ASCT/SCS cycle (Mochizuki et al., 2020). Pyruvate dehydrogenase is one of the mitochondrial dehydrogenases that has been demonstrated to be regulated by Ca^{2+} (McCormack and Denton, 1986). Ca^{2+} stimulates a PDH phosphatase (PDP) that dephosphorylates the E1α subunit of PDH stimulating its activity, which explains the partial rescue of the lethal effect of $TbMCU$ downregulation by addition of threonine to the culture medium (Huang et al., 2013b). Bloodstream forms have a threonine dehydrogenase (Linstead et al., 1977) able to bypass the need of a Ca^{2+}-stimulated step for generation of acetyl-CoA. More recent work demonstrated that Ca^{2+} directly

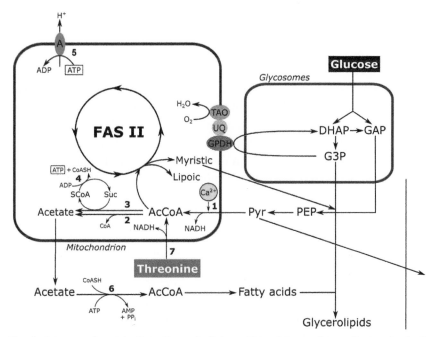

Fig. 2 Scheme of metabolic pathway in *T. brucei* bloodstream forms. Rectangles indicate steps of glucose and threonine metabolism. A, ATPase; AcCoA, acetyl-CoA; FAS II, type II fatty-acid biosynthesis pathway; DHAP, dihydroxyacetone phosphate; GAP, glyceraldehyde 3-phosphate; G3P, glycerol 3-phosphate; GPDH, glycerol 3-phosphate dehydrogenase; PEP, phosphoenolpyruvate; Pyr, pyruvate; UQ, ubiquinone; TAO, trypanosome alternative oxidase. Enzymes are: 1. Pyruvate dehydrogenase; 2. Acetyl-CoA hydrolase; 3. Acetate: succinate CoA transferase (ASCT); 4. Succinyl CoA synthetase (SCS); 5. ATPase (ATP synthase in reverse); 6. Acetyl CoA synthetase; 7, threonine dehydrogenase. Activity stimulated by Ca^{2+} is in yellow. *Modified with permission from Huang, G., Vercesi, A.E., Docampo, R., 2013b. Essential regulation of cell bioenergetics in Trypanosoma brucei by the mitochondrial calcium uniporter. Nat. Commun. 4, 2865.*

stimulate the PDH phosphatase of *T. brucei* (Lander et al., 2018). Fig. 2 shows a scheme of the reactions that occur in these cells.

Screening for genes encoding TbMCU orthologs led to the identification of three putative proteins, each with two transmembrane helices (TMH) that were designated TbMCUb, TbMCUc, and TbMCUd. The open reading frames predict proteins of 254, 249, and 214 amino acids, with apparent molecular weights of 28.4, 27.8, and 24.7 kDa, respectively. TbMCUb, TbMCUc, and TbMCUd have 16%–19% identity and 28%–34% similarity with TbMCU and, in addition to two TMH, they have a modified putative Ca^{2+}selectivity filter (WDXXEPXXY, the section of the pore responsible for selective passage of Ca^{2+} through the channel) and belong to the MCU family (Pfam: PF04678) (Huang and Docampo, 2018).

The *T. cruzi* ortholog *TcMCUb* was studied first, together with *TcMCU* (Chiurillo et al., 2017). Both genes were knocked out using the CRISPR/ Cas9 system (Lander et al., 2015) and the mutants lost their mitochondrial capacity to take up Ca^{2+} without any alteration in their mitochondrial membrane potential. Complementation of *TcMCU*-KO cells with an exogenous *TcMCU* gene, but not with the human orthologue (*HsMCU*) or with a *TcMCU* gene mutated in the critical aspartate (D) and glutamate (E) amino acids of the putative pore region, was able to restore mitochondrial Ca^{2+} uptake.

Overexpression of both genes (*TcMCU*-OE and *TcMCUb*-OE) led to increased mitochondrial Ca^{2+} uptake, also without alterations in $\Delta\Psi_m$. The results suggest that both proteins are Ca^{2+} transporting subunits. This is in contrast with the mammalian *MCUb*, which encodes a dominant negative subunit that inhibits mitochondrial Ca^{2+} transport when over-expressed (Raffaello et al., 2013). Both proteins, TcMCU and TcMCUb, co-immunoprecipitate indicating that they form oligomers. Growth of both mutants was affected by varying degrees. Growth was slightly slower in *TcMCU*-KO epimastigotes (vector form), especially in a glucose-deficient medium, but recovered during stationary phase, suggesting an alternative source of energy. In agreement with these results, *TcMCU*-KO epimastigotes have more lipid droplets, which suggests that they might use fatty acids as alternative source of energy in the stationary phase. These mutants have higher ability to differentiate into the infective forms (metacyclic tryp-omastigotes) and trypomastigotes were able to infect host cells and replicate normally as intracellular amastigotes. In contrast, *TcMCUb*-KO epimastigotes were difficult to differentiate to infective metacyclic trypomastigotes and it was not possible to obtain infected host cells, indicating the relevance of this subunit for infection. Concurrently with the higher mitochondrial Ca^{2+} uptake and overload, *TcMCU*-OE and *TcMCUb*-OE also caused oxidative stress. On the other hand, *TcMCUb*-KO, but not *TcMCU*-KO epimastigotes, had decreased respiratory rate, lower mitochondrial mass, and increased autophagy, suggesting a better adaptation of *TcMCU*-KO parasites to the lower mitochondrial ability to take up Ca^{2+} (Chiurillo et al., 2017).

The two other subunits identified in the trypanosome proteome, MCUc and MCUd, have only orthologues in trypanosomatids (Chiurillo et al., 2019; Huang and Docampo, 2018). Interestingly, Fig. 3 shows that human

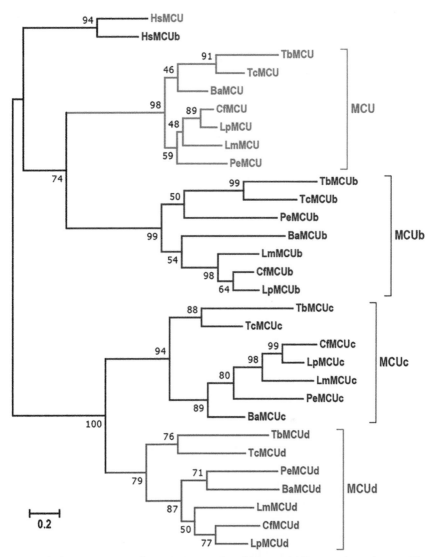

Fig. 3 Phylogenetic tree of trypanosomatid and human MCU complex subunits. The TriTrypDB and GenBank accession numbers for 30 MCUC subunits were described in Huang and Docampo (2018). The scale bar corresponds to a distance of 20 changes per 100 amino acid positions. Numbers indicate bootstrap support. The MCU complex subunits from *Homo sapiens* (Hs), *T. brucei* (Tb), *T. cruzi* (Tc) *Blechomonas ayalai* (Ba), *Crithidia fasciculata* (Cf), *Leptomonas pyrrhocoris* (Lp), *Leishmania major* (Lm), and *Phytomonas* sp. isolate EM1 (Pe) are shown. *Reproduced with permission from Huang, G., Docampo, R., 2018. The mitochondrial Ca²⁺ uniporter complex (MCUC) of Trypanosoma brucei is a hetero-oligomer that contains novel subunits essential for Ca²⁺ uptake. MBio 9, e1700–18.*

MCUb (HsMCUb) forms a well-supported monophyletic clade while TbMCUb forms a monophyletic clade with orthologs from other trypanosomes suggesting that although humans encode a gene for a dominant negative MCU subunit (MCUb), which is expressed in specific tissues, trypanosomes do not. The various MCU subunits of trypanosomes are likely products of gene duplication.

Knockdown of all T. brucei proteins (TbMCU, TbMCUb, TbMCUc, and TbMCUd) by RNAi decreased mitochondrial Ca^{2+} uptake without affecting $\Delta\Psi_m$ (Huang and Docampo, 2018) and their overexpression enhanced Ca^{2+} uptake. Therefore, TbMCUb is not a dominant negative subunit as it occurs with the animal MCUb (Raffaello et al., 2013). In addition, knockout of each of the T. cruzi subunits by CRISPR/Cas9 suppressed, while their overexpression increased, mitochondrial Ca^{2+} uptake (Chiurillo et al., 2019). Taken together the results indicate that MCU, MCUb, MCUc, and MCUd are all Ca^{2+}-transporting subunits. All the T. brucei subunits co-immunoprecipitate and exist in a large protein complex with a net molecular weight of $\sim$380 kDa, suggesting that the complex is a hetero-oligomer (Huang and Docampo, 2018). Further evidence of its hetero-oligomeric structure was provided using the split-ubiquitin membrane-based yeast two hybrid (MYTH) and by co-immunoprecipitation assays. Combining mutagenesis analysis with MYTH assays determined that the transmembrane helices (TMH) of the subunits are involved in these interactions and that the subunits form a hetero-hexamer (Fig. 4A). Mutagenesis of TM helix 1 (TMH1) and especially 2 (TMH2) of each of the four subunits of the complex showed that they are required for their interactions (Huang and Docampo, 2018). These results are apparently different from those reported in animals (Baradaran et al., 2018) and fungi (Baradaran et al., 2018; Fan et al., 2018; Nguyen et al., 2018; Yoo et al., 2018) in which structural studies proposed that the MCU subunit forms homo-tetramers. However, these studies were done using recombinant MCU and how MCU interacts with its membrane partners, like MCUb, and which is their oligomeric structure in vivo remains to be investigated.

Knockdown of TbMCUc and TbMCUd reduced procyclic trypomastigotes growth in glucose-deficient media (Huang and Docampo, 2018), and increased the AMP/ATP ratio (Huang and Docampo, 2020), in agreement with the importance of mitochondrial metabolism for these stages (Lamour et al., 2005). Similar results were observed after ablation of TcMCUc or TcMCUd by CRISPR/Cas9 in epimastigotes (Chiurillo et al.,

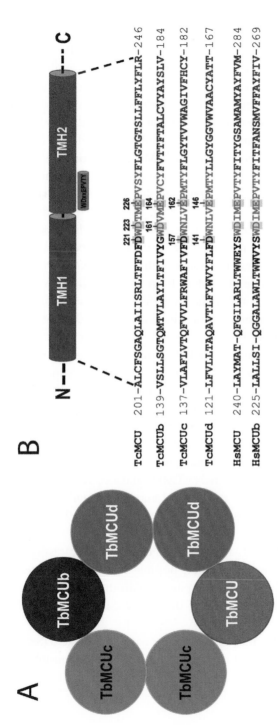

Fig. 4 (A) Scheme depicting the putative organization and composition of a hetero-hexameric TbMCU complex. (B) Conserved WDXXEPXTY motif in TcMCU complex subunits. Alignment of the C-terminal fragment of the second transmembrane helix TMH2 (*green*), including conserved WDXXEPXTY motif (*in red*), of TcMCU complex subunits with human MCU and MCUb. Conserved putative critical acidic residues in or near the WDXXEPXTY selectivity filter are indicated in light gray. TcMCU, TcMCUb, TcMCUc, and TcMCUd exhibit a substitution of the Ru360-sensitive residue S to D/G in the pore region. Residues in each TcMCU complex subunit that were subjected to substitutions are indicated with the corresponding number in their sequence. *Panel (A): Reproduced with permission from Huang, G., Docampo, R., 2018. The mitochondrial Ca^{2+} uniporter complex (MCUC) of Trypanosoma brucei is a hetero-oligomer that contains novel subunits essential for Ca^{2+} uptake. MBio 9, e1700–18. Panel (B): Reproduced with permission from Chiurillo, M.A., Lander, N., Bertolini, M.S., Vercesi, A.E., Docampo, R., 2019. Functional analysis and importance for host cell infection of the Ca^{2+}-conducting subunits of the mitochondrial calcium uniporter of Trypanosoma cruzi. Mol. Biol. Cell 30, 1676–1690.*

2019). *TcMCUc*-KO and *TcMCUd*-KO epimastigotes have also alterations in their respiratory rate, citrate synthase activity, and AMP/ATP ratio, but they normally differentiate into metacyclic trypomastigotes, while trypomastigotes are less infective and amastigotes have a reduced replication rate. All these results are in agreement with the relevance of mitochondrial metabolism for *T. cruzi* invasion of host cells (Schenkman et al., 1991) and for replication of intracellular amastigotes (Dumoulin and Burleigh, 2018).

Complementation of the knockout of each MCU subunit of *T. cruzi* with mutant genes in which the glutamate (E) and aspartate (D) of the pore motif (WDXXEPXXY) were mutated revealed that the E is essential and the D is important for Ca^{2+} uptake (Chiurillo et al., 2019) (Fig. 4B, indicated in light gray). The results are in agreement with structural studies that have shown that the selectivity filter is formed by symmetrical arrangement of WDXXEPXXY sequences in transmembrane helix 2 (TMH2) from each monomer around the pore. The D residues form an acid mouth (Site 1), and the E residues form a second acidic ring (Site 2) (Baradaran et al., 2018). Site 2 is a high-affinity binding site for Ca^{2+} in MCU. Mutations of the D residues located in the filter region of TcMCU (D^{223}) or TcMCUb (D^{161}), or the D residues located N-terminal to the selectivity filter of TcMCU (D^{221}), TcMCUc (D^{157}) or TcMCUd (D^{141}), reduced but did not completely abolish mitochondrial Ca^{2+} transport (Chiurillo et al., 2019). On the other hand, mutations of the E residues of each monomer suppressed mitochondrial Ca^{2+} transport. Since mitochondrial Ca^{2+} uptake in *TcMCU*-KO cells can be restored with TcMCU$^{R214W/D219V}$ mutant, these other residues are not important for Ca^{2+} transport in *T. cruzi* (Chiurillo et al., 2017) (Fig. 4B). These residues are located in TMH1 and loop region near the WDXXEPXXY motif but outside the channel entrance (Baradaran et al., 2018; Fan et al., 2018; Nguyen et al., 2018; Oxenoid et al., 2016; Yoo et al., 2018).

Complementation of *TcMCU*-KO epimastigotes by co-expression of *HsMCU* and *HsEMRE* were unsuccessful, possibly because the proteins did not insert properly in the inner membrane or with the right topology, failed to interact, or did not form part of the MCU complex (Chiurillo et al., 2019).

All *T. cruzi* MCU complex monomers lack the serine amino acid (Ser^{259} in the human MCU) (Fig. 4B) that has been proposed to be responsible for the sensitivity of the uniporter to the ruthenium red derivative Ru360 (Baughman et al., 2011). As mitochondrial Ca^{2+} transport in *T. cruzi* is sensitive to Ru360, this suggests that there is another target for Ru360 besides Ser^{259} (Chiurillo et al., 2019).

In conclusion, four pore subunits MCU, MCUb, MCUc, and MCUd form a hetero-oligomer, probably a hetero-hexamer, required for mitochondrial Ca^{2+} uptake in trypanosomes (Chiurillo et al., 2019; Huang and Docampo, 2018) (Fig. 1B). The selectivity filter is formed by symmetrical arrangement of WDXXEPXXY sequences in TMH2 from each subunit around the pore. The lack of any of the subunits abolishes Ca^{2+} uptake. MCUb is not a dominant negative subunit of the uniporter as the protein of the same denomination in animal cells (Chiurillo et al., 2017; Huang and Docampo, 2018). MCUc and MCUd are subunits present only in trypanosomatids (Chiurillo et al., 2019; Huang and Docampo, 2018). All subunits are required for growth under glucose-limited conditions, revealing the importance of the uniporter in mitochondrial metabolism (Chiurillo et al., 2019; Huang and Docampo, 2018). Complementation studies of all the pore subunits with mutant subunits found that the glutamate and aspartate of the selectivity filter are essential, and important, respectively, for Ca^{2+} uptake (Chiurillo et al., 2019). Complementation of *MCU*-KO cells with human *MCU* (Chiurillo et al., 2017) or human *MCU* and *EMRE* (Chiurillo et al., 2019) or attempts to reconstitute the uniporter in yeast transforming them with *TcMCU* (Chiurillo et al., 2017) failed, suggesting that exogenous subunits could not form part of the trypanosome uniporter and that probably all the endogenous subunits of the pore would be required for reconstitution of its function in yeast. The presence of two additional Ca^{2+}-transporting subunits (MCUc and MCUd) in trypanosomes reflect the parallel evolution of the uniporter in different supergroups of eukaryotes (Pittis et al., 2020).

4. The mitochondrial Ca^{2+} uniporter of trypanosomes: The gatekeeper subunits

MICU1 was reported to act as a gatekeeper of the uniporter by inhibiting Ca^{2+} uptake at low cytosolic Ca^{2+} concentration ([Ca^{2+}]$_{cyt}$), thus preventing mitochondrial Ca^{2+} overload under physiological ([Ca^{2+}]$_{cyt}$) (Csordas et al., 2013; Mallilankaraman et al., 2012b). Later work suggested that MICU2, which binds covalently to MICU1, was the most important inhibitor (Patron et al., 2014). Subsequently either MICU1 or MICU2 or both were considered more relevant gatekeepers in a variety of cells

(Kamer et al., 2017; Liu et al., 2016; Matesanz-Isabel et al., 2016; Paillard et al., 2017; Payne et al., 2017).

T. cruzi and *T. brucei* possess orthologs to MICU1 and MICU2, but orthologs to MICU2 are absent in *Leishmania* spp. (Docampo et al., 2014). The *T. cruzi* proteins have been studied in more detail (Bertolini et al., 2019). TcMICU1 and TcMICU2 have estimated molecular masses of 46.7 and 53.2 kDa, respectively, have 20% identity and 38% similarity between them, possess mitochondrial targeting signals, and two canonical and two noncanonical EF-hand domains. TcMICU1 and TcMICU2 have only 22% and 23.9% overall sequence identity (44.4% and 40% of similarity), respectively, to their human orthologs (Bertolini et al., 2019).

Ablation of *TcMICU1* or *TcMICU2* reduces mitochondrial Ca^{2+} uptake and increases the Ca^{2+} concentration needed for opening the uniporter without affecting the $\Delta\Psi_m$. Interestingly, these mitochondria are less efficient in taking up Ca^{2+} across a wide range of Ca^{2+} concentrations and the threshold for Ca^{2+} uptake is elevated (Bertolini et al., 2019). These results indicate that although these proteins have a role in Ca^{2+} sensing in the intermembrane space, they have individually no role as gatekeepers at low Ca^{2+} concentrations, as it occurs with the mammalian proteins. Ca^{2+} transport at low Ca^{2+} concentrations is already inhibited in the absence of either protein. Both proteins are required for normal growth and respiration of epimastigotes, for normal metacyclogenesis (differentiation of epimastigotes to metacyclic trypomastigotes), for trypomastigote invasion of host cells, and for intracellular replication of amastigotes, indicating their relevance for parasite survival (Bertolini et al., 2019). However, these cells do not show changes in mitochondrial mass, AMP/ATP ratio, or autophagy. The lower Ca^{2+} transport correlates with the increase in phosphorylation of the PDH, reflecting the lower stimulation of the *T. cruzi* PDH phosphatase (TcPDP) by Ca^{2+} (Bertolini et al., 2019).

Overexpression of MICU1 or MICU2 in HeLa cells increases or decreases, respectively, mitochondrial Ca^{2+} accumulation (Patron et al., 2014). However, this does not occur in *T. cruzi* where no changes are observed when TcMICU1 or TcMICU2 are overexpressed. Overexpression does not affect epimastigote growth, and these overexpressed proteins do not form covalently bound oligomeric complexes (Bertolini et al., 2019).

In conclusion, trypanosomes appear to also differ from mammalian cells in the ability of either MICU1 or MICU2 to individually act as gatekeepers of the uniporter at low Ca^{2+} concentrations, in their formation of covalently-bound dimers, and in changes in mitochondrial Ca^{2+} accumulation by their overexpression, although they are still able to sense changes in Ca^{2+} concentration in the mitochondrial intermembrane space and are important for parasite survival and infectivity.

5. The mitochondrial Ca^{2+} uniporter of trypanosomes: Interaction with the ATP synthase

Recent work has indicated that the MCU complex of trypanosomes physically interacts with the subunit c of the ATP synthase (Huang and Docampo, 2020). Tandem affinity purification using overexpressed TbMCU, combined with mass spectrometry (MS), resulted in the identification of 19 subunits of the ATP synthase together with the voltage dependent anion channel (VDAC), the adenine nucleotide translocator (ANT), the phosphate carrier (PiC), and the MCU complex components MCU, MCUb and MCUc. The ATP synthase, the ANT, and the PiC constitute what is known as ATP synthasome in mammals (Ko et al., 2003). Similar results were obtained by immunoprecipitation of tagged TbMCU combined with MS. When two subunits of the ATP synthase (TbATPβ, TbATPp18) were in situ tagged they were also able to immunoprecipitate TbMCU. In situ tagged ANT and PiC were not able to do the same, suggesting that TbMCU is closely associated with the ATP synthase but only loosely associated with ANT and PiC.

The split-ubiquitin membrane-based yeast two-hybrid (MYTH) assay was used to validate the interaction of TbMCU with 10 of the ATP synthase subunits but only subunit c (TbATPc) was shown to specifically physically interact with TbMCU. Each TbMCU subunit, except TbMCUb, (TbMCU, TbMCUc, and TbMCUd), as well as *T. cruzi* MCU, and human MCU, were also able to physically interact with the corresponding subunit c of the same species. Expression of truncated or substituted forms of the different TbMCU subunits (TbMCU, TbMCUc, and TbMCUd) or of the TbATPc subunit (without the C- or N-terminal regions, with mutations in conserved residues of the TMHs or with substitutions of TMHs with artificial helices) indicated that TMH1s of each TbMCU complex subunit (except for TbMCUb) specifically interacts with TMH1 of TbATPc via the conserved motifs of the TMH1s. TbMCU subunits (TbMCU, TbMCUc, and TbMCUd) also co-immunoprecipitated with TbATPc.

Blue native PAGE and immunodetection analyses showed that the TbMCU complex physically interacts with TbATPc in a large protein complex of ~900 kDa. Further evidence of this interaction was the co-immunoprecipitation of TbMCU and TbMCUc with TbATPb and TbATPp18.

Interestingly, MYTH assays using the human MCU (HsMCU) and ATPc (HsATPc) validated that their specific physical interaction is also through their respective TMH1s. Reciprocal co-immunoprecipitations of HsMCU and HsATPc using HEK-293T and HeLa cells confirmed this interaction in vivo, which was not detected when HsMCU-KO HEK-293T cells were used as control.

In conclusion, three of the *T. brucei* MCU subunits (TbMCU, TbMCUc, and TbMCUd) physically interact with mitochondrial ATP synthase subunit c (TbATPc) when expressed in yeast membranes. These interactions are also observed with the *T. cruzi* (TcMCU) or human MCU (HsMCU) and the corresponding ATP synthase subunit c (TcATPc, HsATPc), and in all cases these results were confirmed by co-immunoprecipitations, and by their co-localizations, as studied by immunofluorescence microscopy. These interactions were also confirmed in trypanosomes or human cells in vivo by their co-immunoprecipitations from cell lysates and in the case of *T. brucei* by blue native PAGE and immunodetection analysis. As a result of these interactions it was possible to pull down the ATP synthase complex together with the ANT and PiC by TbMCU, suggesting the presence of an ATP synthasome megacomplex that includes the TbMCU complex (Huang and Docampo, 2020). In this regard, the association of the ATP synthase with the adenine nucleotide carrier was also reported in *Leishmania mexicana* mitochondria (Detke and Elsabrouty, 2008) but not in *T. brucei* PCF mitochondria (Gnipova et al., 2015).

These results suggest several intriguing possibilities: (1) As these interactions involve the TMH1s of both the TbMCU subunits and the TbATPc, the results suggest that, if this interaction occurs in situ, the TbMCU complex would be within the c-ring of the *T. brucei* ATP synthase (Fig. 5) (Huang and Docampo, 2020). In this context, the presence of a protein within the c-ring of the porcine ATP synthase, as studied in situ by cryo-electron microscopy, has been reported (Gu et al., 2019), and further work is needed to investigate whether it corresponds to a MCU component. (2) The MCU complex could have a potential role in the formation of the mitochondrial permeability transition pore (mPTP) (Huang and Docampo, 2020). This is because the c-ring of the

ATP synthase (Alavian et al., 2014; Bonora et al., 2013), a channel inside the ATP synthase dimers (Bonora et al., 2013; Bonora et al., 2017), or the purified ATP synthase itself (Urbani et al., 2019), have been proposed to form the pore of this channel. In this regard, overexpression of a mitochondrial cyclophilin D homolog (TcPyD22) in *T. cruzi* epimastigotes enhances loss of mitochondrial membrane potential and cell viability when H_2O_2 is present, via a mechanism sensitive to cyclosporin A (Bustos et al., 2017), suggesting that a mPTP may be formed in these parasites (Bustos et al., 2015, 2017). A mPTP opening was also induced by some lectins in *T. cruzi* (Fernandes et al., 2014). (3) the formation of a megacomplex including the ATP synthase, the MCU complex, the ANT, and the PiC suggest the coupling between ADP and Pi transport with ATP synthesis, which is stimulated by Ca^{2+} (Huang and Docampo, 2020). In this regard several reports indicate that the ATP synthase binds or is stimulated by Ca^{2+} (Giorgio et al., 2017; Hubbard and McHugh, 1996; Territo et al., 2000, 2001; Zakharov et al., 1993).

6. Mechanism of mitochondrial Ca^{2+} efflux in trypanosomes

In mammalian cells mitochondrial Ca^{2+} is extruded by either a H^+- or a Na^+-coupled exchanger. While the activity of the Na^+/Ca^{2+} exchanger is found in most cell types and is especially important in excitable cells, a Ca^{2+}/H^+ exchanger is mainly found in non-excitable cells (Palty et al., 2010). Early work described the activity of the Na^+/Ca^{2+} exchanger (Carafoli et al., 1974) but the molecular identity of the exchanger was discovered more recently (Palty et al., 2010). Trypanosomes lack orthologues to this gene. In contrast, evidence of the presence of a Ca^{2+}/H^+ exchanger was found as judged by the response of *T. cruzi* mitochondria to the additions of Ca^{2+} and EGTA (Docampo and Vercesi, 1989b). A mitochondrial Ca^{2+}/H^+ exchanger named Letm1 was identified in mammalian cells (Jiang et al., 2009; Tsai et al., 2014), and trypanosomes do have orthologues to this gene. Studies in *T. brucei* concluded that TbLetm1 is involved in maintaining mitochondrial volume via K^+/H^+ exchange across the inner membrane (Hashimi et al., 2013) but its function in Ca^{2+} transport was not investigated. Letm1 was found to be active also in budding yeast, which does not have MCU and thus does not accumulate Ca^{2+}, and this has been proposed as evidence of Letm1 being primarily involved

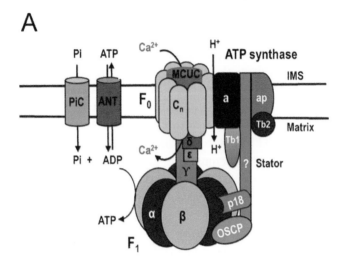

MCUC-ATP synthasome "megacomplex"

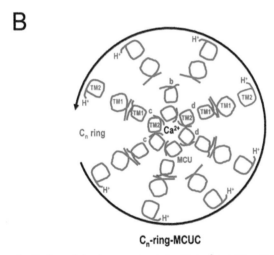

C_n-ring-MCUC

Fig. 5 Hypothetical models showing organization of putative MCU complex-ATP synthasome megacomplex in trypanosomes. (A) The MCU complex physically interacts with the ATP synthasome (ATP synthase, ANT, and PiC) via the c-ring of the F_0 ATP synthase. In trypanosomes, ATP synthase consists of F_1 region with the central stalk ($\alpha3\beta3$, γ, δ and ε) for ATP synthesis, F_0 region with the putative stator (c_n, a, p18, Tb1, Tb2, and OSCP) for proton (H^+) translocation, and trypanosome-specific associated proteins (ap), while the molecular identity of the peripheral stalk is unknown. OSCP, oligomycin sensitivity-conferring protein; ANT, adenine nucleotide translocator; PiC, phosphate carrier. (B) Cross section model to hypothetical *T. brucei* c_n-ring-MCU complex. The *T. brucei* heterohexameric MCU complex consisting of four different subunits

in potassium efflux (Nowikovsky et al., 2004). Further work is needed to investigate the role of trypanosome Letm1 in Ca^{2+} influx and/or efflux.

7. Role of mitochondrial Ca^{2+} uptake

Mitochondrial Ca^{2+} uptake in mammalian cells is important as cytosolic Ca^{2+}-buffering system for regulation of spatially confined cytosolic Ca^{2+} rises, for regulation of mitochondrial metabolism, and for cell survival (Rizzuto et al., 2012). Some of these functions are conserved in trypanosomes.

7.1 Cytosolic Ca^{2+} buffering system

Experiments with *T. brucei* expressing the genetically-encoded Ca^{2+} indicator aequorin targeted to the mitochondria found that intramitochondrial Ca^{2+} concentrations can reach values 10-fold higher than cytosolic Ca^{2+} levels when Ca^{2+} influx through the plasma membrane or Ca^{2+} release from acidocalcisomes are stimulated (Xiong et al., 1997). Mitochondrial Ca^{2+} uptake can be induced at nanomolar Ca^{2+} concentrations (Xiong and Ruben, 1998), suggesting a very close proximity of these organelles and the presence of microdomains of high Ca^{2+} concentration in the vicinity of the plasma membrane or acidocalcisomes (Xiong et al., 1997). Such membrane contact sites between acidocalcisomes and mitochondria were reported in both *T. brucei* (Ramakrishnan et al., 2018) and *T. cruzi* (Miranda et al., 2000). The inositol 1,4,5-trisphosphate receptor is located in acidocalcisomes of both trypanosome species (Huang et al., 2013a; Lander et al., 2016) and close contacts between these organelles facilitate the mitochondrial Ca^{2+} transfer to the mitochondria when the IP$_3$ receptor is activated (Chiurillo et al., 2020).

(MCU, MCUb, MCUc, and MCUd), with a molecular weight of approximately 145 kDa, is within the c-ring of ATP synthase. TMH1 (TM1 in the figure) of each MCU complex subunit (excluding MCUb) interacts with TMH1 (TM1 in the figure) of ATPc. The c-ring rotates in counterclockwise direction and translocates H$^+$ from the intermembrane space to matrix during ATP synthesis. *Reproduced with permission from Huang, G., Docampo, R., 2020. The mitochondrial calcium uniporter interacts with subunit c of the ATP synthase of trypanosomes and humans. MBio 11, e00268–e00320.*

7.2 Regulation of mitochondrial metabolism

Mitochondrial Ca^{2+} in vertebrate cells is important for regulation of the activity of several mitochondrial dehydrogenases (McCormack and Denton, 1986) and the ATP synthase (Territo et al., 2001). Only one of the dehydrogenases stimulated by Ca^{2+} in vertebrates has been studied in detail in trypanosomatids (Lander et al., 2018). The pyruvate dehydrogenase is activated by dephosphorylation of the E1α subunit catalyzed by a PDH phosphatase (PDP), which is stimulated by Ca^{2+}. Both *T. brucei* and *T. cruzi* recombinant PDPs catalyze the dephosphorylation of a synthetic phosphopeptide from either species containing the phosphorylated sites that regulate PDH activity (Lander et al., 2018). TcPDP and TbPDP exhibit maximal activity at 100 nM and 1 μM Ca^{2+}, respectively, suggesting a physiological response. Interestingly, although the binding site for mammalian PDP is formed in the presence of PDH E2 subunit (Turkan et al., 2004), the parasite enzymes are able to directly dephosphorylate E1α phosphopeptides (Lander et al., 2018), in agreement with earlier studies with the mammalian enzyme that identified an E1α binding site (Teague et al., 1982). Knockout of TcPDP results in reduced growth of epimastigotes, defective metacyclogenesis, and reduced host cell invasion by trypomastigotes (Lander et al., 2018). The epimastigotes have a respiratory deficiency, lower citrate synthase activity, higher AMP/ATP ratio and increased autophagy. These cells have an increase in amino acid metabolism, as revealed by the increased ammonia production, probably needed to compensate the lower conversion of pyruvate to acetyl-CoA (Lander et al., 2018).

The stimulation by Ca^{2+} of two other dehydrogenases that are stimulated by Ca^{2+} in animals, isocitrate dehydrogenase and the α-ketoglutarate dehydrogenase, has not been studied in trypanosomes. However, in contrast to the Ca^{2+}-regulated mammalian NAD-dependent enzyme, the mitochondrial isocitrate dehydrogenase present in trypanosomatids is NADP-dependent (Leroux et al., 2011). Another dehydrogenase regulated by Ca^{2+}, the glycerol phosphate dehydrogenase (Denton, 2009) is devoid of the Ca^{2+}-binding EF-hands domains in trypanosomes and presumably insensitive to Ca^{2+}. The aspartate–glutamate carrier (AGC) and the ATP-Mg-Pi carriers (SCaMCs) are known to be regulated by Ca^{2+} in mammalian cells (Satrustegui et al., 2007). However, the trypanosomatid homologues lack EF-hand domains and are therefore potentially Ca^{2+} insensitive. Finally, regulation of the ATP synthase by Ca^{2+} has not been studied in trypanosomes.

7.3 Regulation of cell survival

Mitochondrial Ca^{2+} is important for programmed cell death (PCD), or apoptosis, in trypanosomatids. Early work on the effects of naphthoquinones on *T. cruzi* demonstrated changes in their morphology that can be attributed to PCD, such as shrinking, membrane blebbing, mitochondrial alterations and chromatin condensation (Docampo et al., 1977). However, trypanosomes lack some key regulatory or effector molecules involved in apoptosis in mammalian cells, such as the tumor necrosis factor (TNF)-related family of receptors, Bcl-2 family members, and caspases (Kaczanowski et al., 2011; Smirlis et al., 2010). Mitochondrial Ca^{2+} overload affects the mitochondrial membrane potential, induces the generation of reactive oxygen species (ROS) generation and releases cytochrome *c* in trypanosomatids (Smirlis and Soteriadou, 2011). The production of ROS in *T. brucei* impairs mitochondrial Ca^{2+} uptake, and leads to its accumulation in the nucleus, resulting in cell death (Ridgley et al., 1999). A mitochondrial endonuclease G is released and translocated to the nucleus in *Leishmania* spp. (Gannavaram et al., 2008) and this change stimulates a caspase-independent, apoptosis-like cell death (reviewed in Smirlis and Soteriadou, 2011). *T. cruzi* is highly resistant to mitochondrial permeability transition (Docampo and Vercesi, 1989b), and apoptosis-like death upon mitochondrial Ca^{2+} overload is dependent on superoxide anion generation (Irigoin et al., 2009).

In conclusion, mitochondrial Ca^{2+} uptake in trypanosomatids has a role in shaping the amplitude of cytosolic Ca^{2+} increases after influx through the plasma membrane or release from acidocalcisomes, in the regulation of ATP production, and in apoptosis-like death.

8. Conclusions and open questions

The mitochondria of trypanosomes possess a Ca^{2+} uniporter for Ca^{2+} uptake and a putative Ca^{2+}/H$^+$ exchanger for Ca^{2+} release. The finding of a Ca^{2+} transporting mechanism in trypanosomes with similar characteristics to those of the mammalian uniporter and its absence in yeast, together with the elucidation of the genomes of these and mammalian organisms, led to the discovery of the molecular nature of MICU1 and MCU subunits. However, the MCU complex of trypanosomes has lineage-specific adaptations not seen in the vertebrate uniporter complex. For example, some subunits present in vertebrate cells, such as MCUR1, MICU3, and EMRE, are absent in trypanosomatids. Trypanosomes possess four Ca^{2+}-transporting subunits (MCU, MCUb, MCUc, and MCUd) that form hetero-oligomers where each subunit contributes to the formation of the pore of the channel, interacting through their TMHs. Interestingly, the MCUc and MCUd subunits are exclusive components of the MCU complex of trypanosomatids. In addition, MICU1 and MICU2 do not form covalently-bound dimers and do not act individually as gatekeepers of the channel at low Ca^{2+} concentrations. Concerning trypanosome biology, the MCU complex is essential for normal growth in vitro and in vivo. Downregulation of the uniporter expression leads to increase in the AMP/ATP ratio and autophagy while overexpression leads to Ca^{2+} overload, reactive oxygen species (ROS) generation, and cell death. The MCU complex interacts with the subunit c of the ATP synthase and contributes to the formation of a megacomplex including the phosphate carrier (PiC) and the adenine nucleotide translocator (ANT), which couples ADP and Pi uptake with ATP synthesis stimulated by Ca^{2+}. Finally, Ca^{2+} has a direct role in the stimulation of the mitochondrial PDH activity by dephosphorylation of the E1α subunit catalyzed by a Ca^{2+}-sensitive PDH phosphatase (PDP).

It has not been possible to reconstitute the trypanosome uniporter in yeast, or to complement the lack of the MCU subunit in trypanosomes by either *HsMCU* alone or together with *HsEMRE*, suggesting that the four Ca^{2+}-transporting subunits would be needed for reconstitution and that the human MCU subunit is not compatible with the trypanosome-specific subunits. Further work is needed to identify whether Letm1 is the Ca^{2+}/H^+ exchanger in trypanosomes and to determine how MICU1/MICU2 interact with the pore of the channel, given the apparent absence of an EMRE ortholog. It will also be important to confirm, by structural studies, whether the Ca^{2+}-transporting subunits form hetero-hexamers in situ, and the nature of the megacomplex involving the ATP synthase in situ. The regulation of the activity of the MCU complex in vivo also needs to be investigated.

Acknowledgments

This work was funded by the U.S. National Institutes of Health (grants AI140421 and AI108222) and the São Paulo Research Foundation (FAPESP, Brazil, grant 2013/50624-0). N.L. was a postdoctoral trainee supported by the U.S. National Institutes of Health under award number K99AI137322.

References

Adl, S.M., Bass, D., Lane, C.E., Lukes, J., Schoch, C.L., Smirnov, A., Agatha, S., Berney, C., Brown, M.W., Burki, F., Cardenas, P., Cepicka, I., Chistyakova, L., Del Campo, J., Dunthorn, M., Edvardsen, B., Eglit, Y., Guillou, L., Hampl, V.,

Heiss, A.A., Hoppenrath, M., James, T.Y., Karnkowska, A., Karpov, S., Kim, E., Kolisko, M., Kudryavtsev, A., Lahr, D.J.G., Lara, E., Le Gall, L., Lynn, D.H., Mann, D.G., Massana, R., Mitchell, E.A.D., Morrow, C., Park, J.S., Pawlowski, J.W., Powell, M.J., Richter, D.J., Rueckert, S., Shadwick, L., Shimano, S., Spiegel, F.W., Torruella, G., Youssef, N., Zlatogursky, V., Zhang, Q., 2019. Revisions to the classification, nomenclature, and diversity of eukaryotes. J. Eukaryot. Microbiol. 66, 4–119.

Alavian, K.N., Beutner, G., Lazrove, E., Sacchetti, S., Park, H.A., Licznerski, P., Li, H., Nabili, P., Hockensmith, K., Graham, M., Porter Jr., G.A., Jonas, E.A., 2014. An uncoupling channel within the c-subunit ring of the F1FO ATP synthase is the mitochondrial permeability transition pore. Proc. Natl. Acad. Sci. U. S. A. 111, 10580–10585.

Baradaran, R., Wang, C., Siliciano, A.F., Long, S.B., 2018. Cryo-EM structures of fungal and metazoan mitochondrial calcium uniporters. Nature 559, 580–584.

Bartley, W., Amoore, J.E., 1958. The effects of manganese on the solute content of rat-liver mitochondria. Biochem. J. 69, 348–360.

Baughman, J.M., Perocchi, F., Girgis, H.S., Plovanich, M., Belcher-Timme, C.A., Sancak, Y., Bao, X.R., Strittmatter, L., Goldberger, O., Bogorad, R.L., Koteliansky, V., Mootha, V.K., 2011. Integrative genomics identifies MCU as an essential component of the mitochondrial calcium uniporter. Nature 476, 341–345.

Benaim, G., Bermudez, R., Urbina, J.A., 1990. Ca^{2+} transport in isolated mitochondrial vesicles from Leishmania braziliensis promastigotes. Mol. Biochem. Parasitol. 39, 61–68.

Benne, R., Van den Burg, J., Brakenhoff, J.P., Sloof, P., Van Boom, J.H., Tromp, M.C., 1986. Major transcript of the frameshifted coxII gene from trypanosome mitochondria contains four nucleotides that are not encoded in the DNA. Cell 46, 819–826.

Bertolini, M.S., Chiurillo, M.A., Lander, N., Vercesi, A.E., Docampo, R., 2019. MICU1 and MICU2 play an essential role in mitochondrial Ca^{2+} uptake, growth, and infectivity of the human pathogen Trypanosoma cruzi. MBio 10. e00348-19.

Bonora, M., Bononi, A., De Marchi, E., Giorgi, C., Lebiedzinska, M., Marchi, S., Patergnani, S., Rimessi, A., Suski, J.M., Wojtala, A., Wieckowski, M.R., Kroemer, G., Galluzzi, L., Pinton, P., 2013. Role of the c subunit of the F$_0$ ATP synthase in mitochondrial permeability transition. Cell Cycle 12, 674–683.

Bonora, M., Morganti, C., Morciano, G., Pedriali, G., Lebiedzinska-Arciszewska, M., Aquila, G., Giorgi, C., Rizzo, P., Campo, G., Ferrari, R., Kroemer, G., Wieckowski, M.R., Galluzzi, L., Pinton, P., 2017. Mitochondrial permeability transition involves dissociation of F$_1$F$_0$ ATP synthase dimers and C-ring conformation. EMBO Rep. 18, 1077–1089.

Bustos, P.L., Perrone, A.E., Milduberger, N., Postan, M., Bua, J., 2015. Oxidative stress damage in the protozoan parasite Trypanosoma cruzi is inhibited by cyclosporin A. Parasitology 142, 1024–1032.

Bustos, P.L., Volta, B.J., Perrone, A.E., Milduberger, N., Bua, J., 2017. A homolog of cyclophilin D is expressed in Trypanosoma cruzi and is involved in the oxidative stress-damage response. Cell Death Dis. 3, 16092.

Cali, T., Brini, M., Carafoli, E., 2017. Regulation of cell calcium and role of plasma membrane calcium ATPases. Int. Rev. Cell Mol. Biol. 332, 259–296.

Carafoli, E., Lehninger, A.L., 1971. A survey of the interaction of calcium ions with mitochondria from different tissues and species. Biochem. J. 122, 681–690.

Carafoli, E., Balcavage, W.X., Lehninger, A.L., Mattoon, J.R., 1970. Ca^{2+} metabolism in yeast cells and mitochondria. Biochim. Biophys. Acta 205, 18–26.

Carafoli, E., Tiozzo, R., Lugli, G., Crovetti, F., Kratzing, C., 1974. The release of calcium from heart mitochondria by sodium. J. Mol. Cell. Cardiol. 6, 361–371.

Chiurillo, M.A., Lander, N., Bertolini, M.S., Storey, M., Vercesi, A.E., Docampo, R., 2017. Different roles of mitochondrial calcium uniporter complex subunits in growth and infectivity of *Trypanosoma cruzi*. MBio 8, e00574–e00617.

Chiurillo, M.A., Lander, N., Bertolini, M.S., Vercesi, A.E., Docampo, R., 2019. Functional analysis and importance for host cell infection of the Ca^{2+}-conducting subunits of the mitochondrial calcium uniporter of *Trypanosoma cruzi*. Mol. Biol. Cell 30, 1676–1690.

Chiurillo, M.A., Lander, E.S., Vercesi, A.E., Docampo, R., 2020. IP3 Receptor-Mediated Ca^{2+} release from acidocalcisomes regulates mitochondrial bioenergetics and prevents autophagy in *Trypanosoma cruzi*. Cell Calcium 92, 102284.

Clapham, D.E., 2007. Calcium signaling. Cell 131, 1047–1058.

Clarkson Jr., A.B., Bienen, E.J., Pollakis, G., Grady, R.W., 1989. Respiration of bloodstream forms of the parasite *Trypanosoma brucei brucei* is dependent on a plant-like alternative oxidase. J. Biol. Chem. 264, 17770–17776.

Csordas, G., Golenar, T., Seifert, E.L., Kamer, K.J., Sancak, Y., Perocchi, F., Moffat, C., Weaver, D., de la Fuente Perez, S., Bogorad, R., Koteliansky, V., Adijanto, J., Mootha, V.K., Hajnoczky, G., 2013. MICU1 controls both the threshold and cooperative activation of the mitochondrial Ca^{2+} uniporter. Cell Metab. 17, 976–987.

De Stefani, D., Raffaello, A., Teardo, E., Szabo, I., Rizzuto, R., 2011. A forty-kilodalton protein of the inner membrane is the mitochondrial calcium uniporter. Nature 476, 336–340.

Deluca, H.F., Engstrom, G.W., 1961. Calcium uptake by rat kidney mitochondria. Proc. Natl. Acad. Sci. U. S. A. 47, 1744–1750.

Denton, R.M., 2009. Regulation of mitochondrial dehydrogenases by calcium ions. Biochim. Biophys. Acta 1787, 1309–1316.

Detke, S., Elsabrouty, R., 2008. Identification of a mitochondrial ATP synthase-adenine nucleotide translocator complex in Leishmania. Acta Trop. 105, 16–20.

Docampo, R., Lukes, J., 2012. Trypanosomes and the solution to a 50-year mitochondrial calcium mystery. Trends Parasitol. 28, 31–37.

Docampo, R., Vercesi, A.E., 1989a. Ca^{2+} transport by coupled *Trypanosoma cruzi* mitochondria in situ. J. Biol. Chem. 264, 108–111.

Docampo, R., Vercesi, A.E., 1989b. Characteristics of Ca^{2+} transport by *Trypanosoma cruzi* mitochondria in situ. Arch. Biochem. Biophys. 272, 122–129.

Docampo, R., Lopes, J.N., Cruz, F.S., Souza, W., 1977. *Trypanosoma cruzi*: ultrastructural and metabolic alterations of epimastigotes by beta-lapachone. Exp. Parasitol. 42, 142–149.

Docampo, R., Vercesi, A.E., Huang, G., 2014. Mitochondrial calcium transport in trypanosomes. Mol. Biochem. Parasitol. 196, 108–116.

Dumoulin, P.C., Burleigh, B.A., 2018. Stress-induced proliferation and cell cycle plasticity of intracellular *Trypanosoma cruzi* amastigotes. MBio 9, e00673–18.

Fan, C., Fan, M., Orlando, B.J., Fastman, N.M., Zhang, J., Xu, Y., Chambers, M.G., Xu, X., Perry, K., Liao, M., Feng, L., 2018. X-ray and cryo-EM structures of the mitochondrial calcium uniporter. Nature 559, 575–579.

Fernandes, M.P., Leite, A.C., Araujo, F.F., Saad, S.T., Baratti, M.O., Correia, M.T., Coelho, L.C., Gadelha, F.R., Vercesi, A.E., 2014. The *Cratylia mollis* seed lectin induces membrane permeability transition in isolated rat liver mitochondria and a cyclosporine A-insensitive permeability transition in *Trypanosoma cruzi* mitochondria. J. Eukaryot. Microbiol. 61, 381–388.

Gannavaram, S., Vedvyas, C., Debrabant, A., 2008. Conservation of the pro-apoptotic nuclease activity of endonuclease G in unicellular trypanosomatid parasites. J. Cell Sci. 121, 99–109.

Giorgio, V., Burchell, V., Schiavone, M., Bassot, C., Minervini, G., Petronilli, V., Argenton, F., Forte, M., Tosatto, S., Lippe, G., Bernardi, P., 2017. Ca^{2+} binding to

F-ATP synthase beta subunit triggers the mitochondrial permeability transition. EMBO Rep. 18, 1065–1076.

Gnipova, A., Subrtova, K., Panicucci, B., Horvath, A., Lukes, J., Zikova, A., 2015. The ADP/ATP carrier and its relationship to oxidative phosphorylation in ancestral protist *Trypanosoma brucei*. Eukaryot. Cell 14, 297–310.

Greenawalt, J.W., Carafoli, E., 1966. Electron microscope studies on the active accumulation of Sr^{++} by rat-liver mitochondria. J. Cell Biol. 29, 37–61.

Gu, J., Zhang, L., Zong, S., Guo, R., Liu, T., Yi, J., Wang, P., Zhuo, W., Yang, M., 2019. Cryo-EM structure of the mammalian ATP synthase tetramer bound with inhibitory protein IF1. Science 364, 1068–1075.

Hashimi, H., McDonald, L., Stribrna, E., Lukes, J., 2013. Trypanosome Letm1 protein is essential for mitochondrial potassium homeostasis. J. Biol. Chem. 288, 26914–26925.

Huang, G., Docampo, R., 2018. The mitochondrial Ca^{2+} uniporter complex (MCUC) of *Trypanosoma brucei* is a hetero-oligomer that contains novel subunits essential for Ca^{2+} uptake. MBio 9, e1700–e1718.

Huang, G., Docampo, R., 2020. The mitochondrial calcium uniporter interacts with subunit c of the ATP synthase of trypanosomes and humans. MBio 11, e00268–e00320.

Huang, G., Bartlett, P.J., Thomas, A.P., Moreno, S.N., Docampo, R., 2013a. Acidocalcisomes of *Trypanosoma brucei* have an inositol 1,4,5-trisphosphate receptor that is required for growth and infectivity. Proc. Natl. Acad. Sci. U. S. A. 110, 1887–1892.

Huang, G., Vercesi, A.E., Docampo, R., 2013b. Essential regulation of cell bioenergetics in *Trypanosoma brucei* by the mitochondrial calcium uniporter. Nat. Commun. 4, 2865.

Hubbard, M.J., McHugh, N.J., 1996. Mitochondrial ATP synthase F$_1$-beta-subunit is a calcium-binding protein. FEBS Lett. 391, 323–329.

Irigoin, F., Inada, N.M., Fernandes, M.P., Piacenza, L., Gadelha, F.R., Vercesi, A.E., Radi, R., 2009. Mitochondrial calcium overload triggers complement-dependent superoxide-mediated programmed cell death in Trypanosoma cruzi. Biochem. J. 418, 595–604.

Jiang, D., Zhao, L., Clapham, D.E., 2009. Genome-wide RNAi screen identifies Letm1 as a mitochondrial Ca^{2+}/H$^+$ antiporter. Science 326, 144–147.

Kaczanowski, S., Sajid, M., Reece, S.E., 2011. Evolution of apoptosis-like programmed cell death in unicellular protozoan parasites. Parasit. Vectors 4, 44.

Kamer, K.J., Mootha, V.K., 2015. The molecular era of the mitochondrial calcium uniporter. Nat. Rev. Mol. Cell Biol. 16, 545–553.

Kamer, K.J., Grabarek, Z., Mootha, V.K., 2017. High-affinity cooperative Ca^{2+} binding by MICU1-MICU2 serves as an on-off switch for the uniporter. EMBO Rep. 18, 1397–1411.

Kirichok, Y., Krapivinsky, G., Clapham, D.E., 2004. The mitochondrial calcium uniporter is a highly selective ion channel. Nature 427, 360–364.

Ko, Y.H., Delannoy, M., Hullihen, J., Chiu, W., Pedersen, P.L., 2003. Mitochondrial ATP synthasome. Cristae-enriched membranes and a multiwell detergent screening assay yield dispersed single complexes containing the ATP synthase and carriers for Pi and ADP/ATP. J. Biol. Chem. 278, 12305–12309.

Lamour, N., Riviere, L., Coustou, V., Coombs, G.H., Barrett, M.P., Bringaud, F., 2005. Proline metabolism in procyclic *Trypanosoma brucei* is down-regulated in the presence of glucose. J. Biol. Chem. 280, 11902–11910.

Lander, N., Li, Z.H., Niyogi, S., Docampo, R., 2015. CRISPR/Cas9-Induced disruption of paraflagellar rod protein 1 and 2 genes in *Trypanosoma cruzi* reveals their role in flagellar attachment. MBio 6, e01012.

Lander, N., Chiurillo, M.A., Storey, M., Vercesi, A.E., Docampo, R., 2016. CRISPR/Cas9-mediated endogenous C-terminal tagging of *Trypanosoma cruzi* genes reveals the acidocalcisome localization of the inositol 1,4,5-trisphosphate receptor. J. Biol. Chem. 291, 25505–25515.

Lander, N., Chiurillo, M.A., Bertolini, M.S., Storey, M., Vercesi, A.E., Docampo, R., 2018. Calcium-sensitive pyruvate dehydrogenase phosphatase is required for energy metabolism, growth, differentiation, and infectivity of Trypanosoma cruzi. J. Biol. Chem. 293, 17402–17417.

Lehninger, A.L., Rossi, C.S., Greenawalt, J.W., 1963. Respiration-dependent accumulation of inorganic phosphate and Ca ions by rat liver mitochondria. Biochem. Biophys. Res. Commun. 10, 444–448.

Leroux, A.E., Maugeri, D.A., Cazzulo, J.J., Nowicki, C., 2011. Functional characterization of NADP-dependent isocitrate dehydrogenase isozymes from Trypanosoma cruzi. Mol. Biochem. Parasitol. 177, 61–64.

Linstead, D.J., Klein, R.A., Cross, G.A., 1977. Threonine catabolism in Trypanosoma brucei. J. Gen. Microbiol. 101, 243–251.

Liu, J.C., Liu, J., Holmstrom, K.M., Menazza, S., Parks, R.J., Fergusson, M.M., Yu, Z.X., Springer, D.A., Halsey, C., Liu, C., Murphy, E., Finkel, T., 2016. MICU1 serves as a molecular gatekeeper to prevent in vivo mitochondrial calcium overload. Cell Rep. 16, 1561–1573.

Mallilankaraman, K., Cardenas, C., Doonan, P.J., Chandramoorthy, H.C., Irrinki, K.M., Golenar, T., Csordas, G., Madireddi, P., Yang, J., Muller, M., Miller, R., Kolesar, J.E., Molgo, J., Kaufman, B., Hajnoczky, G., Foskett, J.K., Madesh, M., 2012a. MCUR1 is an essential component of mitochondrial Ca^{2+} uptake that regulates cellular metabolism. Nat. Cell Biol. 14, 1336–1343.

Mallilankaraman, K., Doonan, P., Cardenas, C., Chandramoorthy, H.C., Muller, M., Miller, R., Hoffman, N.E., Gandhirajan, R.K., Molgo, J., Birnbaum, M.J., Rothberg, B.S., Mak, D.O., Foskett, J.K., Madesh, M., 2012b. MICU1 is an essential gatekeeper for MCU-mediated mitochondrial Ca^{2+} uptake that regulates cell survival. Cell 151, 630–644.

Matesanz-Isabel, J., Arias-del-Val, J., Alvarez-Illera, P., Fonteriz, R.I., Montero, M., Alvarez, J., 2016. Functional roles of MICU1 and MICU2 in mitochondrial Ca^{2+} uptake. Biochim. Biophys. Acta 1858, 1110–1117.

Mazet, M., Morand, P., Biran, M., Bouyssou, G., Courtois, P., Daulouede, S., Millerioux, Y., Franconi, J.M., Vincendeau, P., Moreau, P., Bringaud, F., 2013. Revisiting the central metabolism of the bloodstream forms of Trypanosoma brucei: production of acetate in the mitochondrion is essential for parasite viability. PLoS Negl. Trop. Dis. 7, e2587.

McCormack, J.G., Denton, R.M., 1986. Ca^{2+} as a second messenger within mitochondria. Trends Biochem. Sci. 11, 258–262.

Millerioux, Y., Morand, P., Biran, M., Mazet, M., Moreau, P., Wargnies, M., Ebikeme, C., Deramchia, K., Gales, L., Portais, J.C., Boshart, M., Franconi, J.M., Bringaud, F., 2012. ATP synthesis-coupled and -uncoupled acetate production from acetyl-CoA by mitochondrial acetate:succinate CoA-transferase and acetyl-CoA thioesterase in Trypanosoma. J. Biol. Chem. 287, 17186–17197.

Miranda, K., Benchimol, M., Docampo, R., de Souza, W., 2000. The fine structure of acidocalcisomes in Trypanosoma cruzi. Parasitol. Res. 86, 373–384.

Mochizuki, K., Inaoka, D.K., Mazet, M., Shiba, T., Fukuda, K., Kurasawa, H., Millerioux, Y., Boshart, M., Balogun, E.O., Harada, S., Hirayama, K., Bringaud, F., Kita, K., 2020. The ASCT/SCS cycle fuels mitochondrial ATP and acetate production in Trypanosoma brucei. Biochim. Biophys. Acta Bioenerg. 1861, 148283.

Moore, C.L., 1971. Specific inhibition of mitochondrial Ca++ transport by ruthenium red. Biochem. Biophys. Res. Commun. 42, 298–305.

Moreno, S.N., Docampo, R., Vercesi, A.E., 1992. Calcium homeostasis in procyclic and bloodstream forms of *Trypanosoma brucei*. Lack of inositol 1,4,5-trisphosphate-sensitive Ca^{2+} release. J. Biol. Chem. 267, 6020–6026.

Nguyen, N.X., Armache, J.-P., Lee, C., Yang, Y., Zeng, W., Mootha, V.K., Cheng, Y., Bai, X.-c., Jiang, Y., 2018. Cryo-EM structure of a fungal mitochondrial calcium uniporter. Nature 559, 570–574.

Nowikovsky, K., Froschauer, E.M., Zsurka, G., Samaj, J., Reipert, S., Kolisek, M., Wiesenberger, G., Schweyen, R.J., 2004. The LETM1/YOL027 gene family encodes a factor of the mitochondrial K$^+$ homeostasis with a potential role in the Wolf-Hirschhorn syndrome. J. Biol. Chem. 279, 30307–30315.

Oxenoid, K., Dong, Y., Cao, C., Cui, T., Sancak, Y., Markhard, A.L., Grabarek, Z., Kong, L., Liu, Z., Ouyang, B., Cong, Y., Mootha, V.K., Chou, J.J., 2016. Architecture of the mitochondrial calcium uniporter. Nature 533, 269–273.

Paillard, M., Csordas, G., Szanda, G., Golenar, T., Debattisti, V., Bartok, A., Wang, N., Moffat, C., Seifert, E.L., Spat, A., Hajnoczky, G., 2017. Tissue-specific mitochondrial decoding of cytoplasmic Ca^{2+} signals is controlled by the stoichiometry of MICU1/2 and MCU. Cell Rep. 18, 2291–2300.

Palty, R., Silverman, W.F., Hershfinkel, M., Caporale, T., Sensi, S.L., Parnis, J., Nolte, C., Fishman, D., Shoshan-Barmatz, V., Herrmann, S., Khananshvili, D., Sekler, I., 2010. NCLX is an essential component of mitochondrial Na$^+$/Ca^{2+} exchange. Proc. Natl. Acad. Sci. U. S. A. 107, 436–441.

Patron, M., Checchetto, V., Raffaello, A., Teardo, E., Vecellio Reane, D., Mantoan, M., Granatiero, V., Szabo, I., De Stefani, D., Rizzuto, R., 2014. MICU1 and MICU2 finely tune the mitochondrial Ca^{2+} uniporter by exerting opposite effects on MCU activity. Mol. Cell 53, 726–737.

Payne, R., Hoff, H., Roskowski, A., Foskett, J.K., 2017. MICU2 Restricts Spatial Crosstalk between InsP$_3$R and MCU channels by regulating threshold and gain of MICU1-mediated inhibition and activation of MCU. Cell Rep. 21, 3141–3154.

Perocchi, F., Gohil, V.M., Girgis, H.S., Bao, X.R., McCombs, J.E., Palmer, A.E., Mootha, V.K., 2010. MICU1 encodes a mitochondrial EF hand protein required for Ca^{2+} uptake. Nature 467, 291–296.

Pittis, A.A., Goh, V., Cebrian-Serrano, A., Wettmarshausen, J., Perocchi, F., Gabaldon, T., 2020. Discovery of EMRE in fungi resolves the true evolutionary history of the mito-chondrial calcium uniporter. Nat. Commun. 11, 4031.

Plovanich, M., Bogorad, R.L., Sancak, Y., Kamer, K.J., Strittmatter, L., Li, A.A., Girgis, H.S., Kuchimanchi, S., De Groot, J., Speciner, L., Taneja, N., Oshea, J., Koteliansky, V., Mootha, V.K., 2013. MICU2, a paralog of MICU1, resides within the mitochondrial uniporter complex to regulate calcium handling. PLoS One 8, e55785.

Raffaello, A., De Stefani, D., Sabbadin, D., Teardo, E., Merli, G., Picard, A., Checchetto, V., Moro, S., Szabo, I., Rizzuto, R., 2013. The mitochondrial calcium uniporter is a multimer that can include a dominant-negative pore-forming subunit. EMBO J. 32, 2362–2376.

Ramakrishnan, S., Asady, B., Docampo, R., 2018. Acidocalcisome-mitochondrion mem-brane contact sites in *Trypanosoma brucei*. Pathogens 7, 33.

Ridgley, E.L., Xiong, Z.H., Ruben, L., 1999. Reactive oxygen species activate a Ca^{2+}-dependent cell death pathway in the unicellular organism *Trypanosoma brucei brucei*. Biochem. J. 340, 33–40.

Rizzuto, R., De Stefani, D., Raffaello, A., Mammucari, C., 2012. Mitochondria as sensors and regulators of calcium signalling. Nat. Rev. Mol. Cell Biol. 13, 566–578.

Sancak, Y., Markhard, A.L., Kitami, T., Kovacs-Bogdan, E., Kamer, K.J., Udeshi, N.D., Carr, S.A., Chaudhuri, D., Clapham, D.E., Li, A.A., Calvo, S.E., Goldberger, O., Mootha, V.K., 2013. EMRE is an essential component of the mitochondrial calcium uniporter complex. Science 342, 1379–1382.

Satrustegui, J., Pardo, B., Del Arco, A., 2007. Mitochondrial transporters as novel targets for intracellular calcium signaling. Physiol. Rev. 87, 29–67.

Schenkman, S., Robbins, E.S., Nussenzweig, V., 1991. Attachment of Trypanosoma cruzi to mammalian cells requires parasite energy, and invasion can be independent of the target cell cytoskeleton. Infect. Immun. 59, 645–654.

Seidman, D., Johnson, D., Gerbasi, V., Golden, D., Orlando, R., Hajduk, S., 2012. Mitochondrial membrane complex that contains proteins necessary for tRNA import in Trypanosoma brucei. J. Biol. Chem. 287, 8892–8903.

Shlomai, J., 2004. The structure and replication of kinetoplast DNA. Curr. Mol. Med. 4, 623–647.

Smirlis, D., Soteriadou, K., 2011. Trypanosomatid apoptosis: 'Apoptosis' without the canonical regulators. Virulence 2, 253–256.

Smirlis, D., Duszenko, M., Ruiz, A.J., Scoulica, E., Bastien, P., Fasel, N., Soteriadou, K., 2010. Targeting essential pathways in trypanosomatids gives insights into protozoan mechanisms of cell death. Parasit. Vectors 3, 107.

Stephens, J.L., Lee, S.H., Paul, K.S., Englund, P.T., 2007. Mitochondrial fatty acid synthesis in Trypanosoma brucei. J. Biol. Chem. 282, 4427–4436.

Surve, S., Heestand, M., Panicucci, B., Schnaufer, A., Parsons, M., 2012. Enigmatic presence of mitochondrial complex I in Trypanosoma brucei bloodstream forms. Eukaryot. Cell 11, 183–193.

Teague, W.M., Pettit, F.H., Wu, T.L., Silberman, S.R., Reed, L.J., 1982. Purification and properties of pyruvate dehydrogenase phosphatase from bovine heart and kidney. Biochemistry 21, 5585–5592.

Territo, P.R., Mootha, V.K., French, S.A., Balaban, R.S., 2000. Ca^{2+} activation of heart mitochondrial oxidative phosphorylation: role of the F_0/F_1-ATPase. Am. J. Physiol. Cell Physiol. 278, C423–C435.

Territo, P.R., French, S.A., Dunleavy, M.C., Evans, F.J., Balaban, R.S., 2001. Calcium activation of heart mitochondrial oxidative phosphorylation: rapid kinetics of mVO2, NADH, AND light scattering. J. Biol. Chem. 276, 2586–2599.

Tsai, M.F., Jiang, D., Zhao, L., Clapham, D., Miller, C., 2014. Functional reconstitution of the mitochondrial Ca2+/H+ antiporter Letm1. J. Gen. Physiol. 143, 67–73.

Turkan, A., Hiromasa, Y., Roche, T.E., 2004. Formation of a complex of the catalytic subunit of pyruvate dehydrogenase phosphatase isoform 1 (PDP1c) and the L2 domain forms a Ca^{2+} binding site and captures PDP1c as a monomer. Biochemistry 43, 15073–15085.

Urbani, A., Giorgio, V., Carrer, A., Franchin, C., Arrigoni, G., Jiko, C., Abe, K., Maeda, S., Shinzawa-Itoh, K., Bogers, J.F.M., McMillan, D.G.G., Gerle, C., Szabo, I., Bernardi, P., 2019. Purified F-ATP synthase forms a Ca^{2+}-dependent high-conductance channel matching the mitochondrial permeability transition pore. Nat. Commun. 10, 4341.

Van Hellemond, J.J., Opperdoes, F.R., Tielens, A.G., 1998. Trypanosomatidae produce acetate via a mitochondrial acetate:succinate CoA transferase. Proc. Natl. Acad. Sci. U. S. A. 95, 3036–3041.

van Hellemond, J.J., Opperdoes, F.R., Tielens, A.G., 2005. The extraordinary mitochondrion and unusual citric acid cycle in Trypanosoma brucei. Biochem. Soc. Trans. 33, 967–971.

Vanderheyden, N., Wong, J., Docampo, R., 2000. A pyruvate-proton symport and an H^+-ATPase regulate the intracellular pH of Trypanosoma brucei at different stages of its life cycle. Biochem. J. 346, 53–62.

Vasington, F.D., Murphy, J.V., 1962. Ca ion uptake by rat kidney mitochondria and its dependence on respiration and phosphorylation. J. Biol. Chem. 237, 2670–2677.

Vercesi, A.E., Docampo, R., 1992. Ca^{2+} transport by digitonin-permeabilized *Leishmania donovani*. Effects of Ca^{2+}, pentamidine and WR-6026 on mitochondrial membrane potential in situ. Biochem. J. 284, 463–467.

Vercesi, A.E., Macedo, D.V., Lima, S.A., Gadelha, F.R., Docampo, R., 1990. Ca^{2+} transport in digitonin-permeabilized trypanosomatids. Mol. Biochem. Parasitol. 42, 119–124.

Vercesi, A.E., Docampo, R., Moreno, S.N., 1992. Energization-dependent Ca^{2+} accumulation in *Trypanosoma brucei* bloodstream and procyclic trypomastigotes mitochondria. Mol. Biochem. Parasitol. 56, 251–257.

Vercesi, A.E., Moreno, S.N., Bernardes, C.F., Meinicke, A.R., Fernandes, E.C., Docampo, R., 1993. Thapsigargin causes Ca^{2+} release and collapse of the membrane potential of *Trypanosoma brucei* mitochondria in situ and of isolated rat liver mitochondria. J. Biol. Chem. 268, 8564–8568.

Verner, Z., Basu, S., Benz, C., Dixit, S., Dobakova, E., Faktorova, D., Hashimi, H., Horakova, E., Huang, Z., Paris, Z., Pena-Diaz, P., Ridlon, L., Tyc, J., Wildridge, D., Zikova, A., Lukes, J., 2015. Malleable mitochondrion of *Trypanosoma brucei*. Int. Rev. Cell Mol. Biol. 315, 73–151.

Xiong, Z.H., Ruben, L., 1998. *Trypanosoma brucei*: the dynamics of calcium movement between the cytosol, nucleus, and mitochondrion of intact cells. Exp. Parasitol. 88, 231–239.

Xiong, Z.H., Ridgley, E.L., Enis, D., Olness, F., Ruben, L., 1997. Selective transfer of calcium from an acidic compartment to the mitochondrion of Trypanosoma brucei. Measurements with targeted aequorins. J. Biol. Chem. 272, 31022–31028.

Ying, W.L., Emerson, J., Clarke, M.J., Sanadi, D.R., 1991. Inhibition of mitochondrial calcium ion transport by an oxo-bridged dinuclear ruthenium ammine complex. Biochemistry 30, 4949–4952.

Yoo, J., Wu, M., Yin, Y., Herzik Jr., M.A., Lander, G.C., Lee, S.Y., 2018. Cryo-EM structure of a mitochondrial calcium uniporter. Science 361, 506–511.

Zakharov, S.D., Ewy, R.G., Dilley, R.A., 1993. Subunit III of the chloroplast ATP-synthase can form a Ca^{2+}-binding site on the lumenal side of the thylakoid membrane. FEBS Lett. 336, 95–99.

Zhuo, Y., Cordeiro, C.D., Hekmatyar, S.K., Docampo, R., Prestegard, J.H., 2017. Dynamic nuclear polarization facilitates monitoring of pyruvate metabolism in *Trypanosoma brucei*. J. Biol. Chem. 292, 18161–18168.

CPI Antony Rowe
Eastbourne, UK
September 16, 2022